A GUIDE TO COMMERCIAL-SCALE ETHANOL PRODUCTION AND FINANCING

A Product of the
Solar Energy Information Data Bank

Solar Energy Research Institute
Operated for the U.S. Department of Energy by the
Midwest Research Institute

Under Contract No. EG-77-C-01-4042

Books for Business
New York-Hong Kong

A Guide to Commercial-Scale Ethanol Production
and Financing

Edited by
Solar Energy Research Institute

ISBN: 0-89499-200-7

Reprinted from the 1981 edition

Books for Business
New York - Hong Kong
http://www.BusinessBooksInternational.com

TABLE OF CONTENTS

TABLES

FIGURES

ACKNOWLEDGMENTS

This effort was initiated by Mr. Bert Greenglass, Acting Director, Office of Alcohol Fuels, and was funded as part of the Solar Energy Research Institute's contract for the management and support of the National Alcohol Fuels Information Center. Mr. William Holmberg, Acting Director, Office of Market Development, Office of Alcohol Fuels, served as the Department of Energy director for this project. We acknowledge the financial information support provided by Mr. Ted D. Tarr of the Office of Alcohol Fuels.

The Solar Energy Research Institute's Solar Energy Information Data Bank (SEIDB) staff was requested to prepare this guide to the commercial-scale production and financing of fermentation ethanol. The effort at SEIDB was directed by Mr. Paul Notari and managed by Mr. Richard Piekarski of the Information Dissemination Branch.

A competitive contract was awarded to TRW Inc. to prepare this guidebook on commercial-scale fermentation ethanol production and financing. A team was formed to act as advisors to the TRW staff and to SERI in preparation of this text. The team consisted of the following individuals:

Program Manager	**Task Manager**
Mr. V. Daniel Hunt	Mr. Warren Standley
TRW Energy Systems Group	TRW Energy Systems Group

Participants/Consultants

Mr. Milton David	Mr. Robert Mabee, Attorney
Development Planning and Research Associates	Quaintance and Swanson
Mr. Samuel F. Eakin	Mr. Strud Nash
Energy Research Group	E. F. Hutton
Mr. David Freedman	Mr. Paul Notari
Center for the Biology of Natural Systems	Solar Energy Research Institute
Dr. Cathryn Goddard, et. al.	Mr. Richard Piekarski
A.T. Kearney	Solar Energy Research Institute
Dr. M. Edward Goretsky	Dr. Jean Simons
TRW Energy Systems Group	TRW Energy Systems Group
Dr. Jean-Francois Henry	Dr. William Stark
TRW Energy Systems Group	PEDCo International
Ms. Ann Heywood	Mr. Ted D. Tarr
TRW Energy Systems Group	Department of Energy
Office of Alcohol Fuels	
Mr. David Jenkins	Mr. Michael Thomas
Battelle Memorial Institute	Arthur Young & Company
Dr. Raphael Katzen	Mr. Arch Wood
Raphael Katzen Associates	TRW Energy Systems Group

Editorial and Production Support

Our appreciation is extended to

- Production editors Ms. Christine Fuller and Ms. Ann Seely for their careful attention to detail and in-depth editing of this book

- Graphic designers, Mr. John Barber, Mr. Vito W. Oporto, Mr. James McComas, and staff

- Word processing, Ms. Carolyn Starr and staff

- Typesetting by Anderson Advertising Art, Lakewood, CO

Photographs

The agricultural photographs are reprinted through the courtesy of Grant Heilman Photography.

Reviewers

We acknowledge the following individuals for their helpful reviews of the draft of *"A Guide to Commercial-Scale Ethanol Production and Financing"*. These individuals do not necessarily approve, disapprove, or endorse the report for which SEIDB assumes responsibility.

Mr. Jerry Allsup
Bartlesville Energy Technology Center

Mr. Jim Childress
National Alcohol Fuel Commission

Dr. Howard Coleman
Office of Alcohol Fuels,
Department of Energy

Mr. Milton David
Development Planning and
Research Associates

Mr. T. D. Striggles
PEDCo International Inc.

Mr. Samuel F. Eakin
Energy Research Group

Mr. Don Fink
U.S. Department of Agriculture

Mr. William H. Foster
U.S. Treasury
Bureau of Alcohol, Tobacco, and Firearms

Mr. William Holmberg
Office of Alcohol Fuels,
Department of Energy

Mr. Maurice Jones
Hydrocarbon Research Inc.

Dr. Raphael Katzen
Raphael Katzen Associates
International, Inc.

Mr. Edward A. Kirchner
Davy McKee Corporation

Dr. Michael R. Ladisch
Purdue University

Mr. Robert J. Lipshutz
Haas, Holland, Lipshutz,
Levison & Gilbert

Mr. Robert Mabee, Attorney
Quaintance and Swanson

Mr. Bert Mason
Solar Energy Research Institute

Mr. Joe Gibbs
Amoco Oil Company

Dr. Cathryn Goddard
A. T. Kearney, Inc.

Mr. C. Paul Green
Navarro Jr. College

Mr. Bert Greenglass
Office of Alcohol Fuels,
Department of Energy

Dr. Harry P. Gregor
Columbia University

Mr. Wiley Harrell
Anheuser Busch, Inc.

Mr. Donald Hertzmark
Solar Energy Research Institute

Mr. Howard Hinten
Midwest Solvents Co. Inc.

Mr. Dwight Miller
U.S. Department of Agriculture
Science and Education Administration

Mr. Strud Nash
E.F. Hutton & Company

Mr. Ted D. Tarr
Office of Alcohol Fuels,
Department of Energy

Mr. Michael R. Thomas
Arthur Young & Co.

Dr. Ruxton Villet
Solar Energy Research Institute

Dr. Harlan Watson
Subcommittee on Energy
Nuclear Proliferation and
Federal Services

Mr. Neil Woodley
Solar Energy Research Institute

This document greatly benefitted from the many previous efforts in alcohol fuels, including working groups at universities and colleges, private marketing efforts, private research and development projects, and individual efforts to collect and organize information pertinent to alcohol fuels.

The chapter on Plant Design Concepts is based on Raphael Katzen Associates' report entitled "Grain Motor Fuel Alcohol-Technical and Economic Assessment Study", prepared for the Department of Energy under Department of Energy Contract No. EJ-78-C-01-6639.

The glossary contains additional input from Michael Pete, Department of Energy, Office of Alcohol Fuels.

CHAPTER I
Introduction

In recent years, fermentation ethanol produced from agricultural feedstocks, in particular grains such as corn, has been demonstrated to be our most promising near-term option for producing synthetic fuels. For all practical purposes, gasohol, a blend of 10 percent anhydrous (water-free) fermentation ethanol and 90 percent unleaded gasoline delivers comparable engine performance as 100 percent unleaded gasoline, with the added benefit of superior anti-knock properties. The production of fermentation ethanol from agricultural crops or wastes is strongly supported by the farming community and the initial response of the general public to the introduction of gasohol as a alternative automotive fuel has been favorable. The Federal government has recognized the potential of fermentation ethanol as a fuel additive or substitute and in November 1978 reduced the excise tax on gasohol by four cents per gallon. Primarily as a result of this encouragement, fermentation ethanol capacity increased from a few million gallons to about 80 million gallons annually by the end of 1979. Of the 80 million gallons of ethanol produced, almost 50 million gallons were blended to obtain gasohol.

Goals for increased production of fermentation ethanol for blending with gasoline have been established by the President and legislation (including P.L. 96-126, Department of the Interior Appropriations for Fiscal Year 1980; P.L. 95-618, The Energy Tax Act; P.L. 96-294, The Energy Security Act of 1980; and P.L. 96-223, The Windfall Profits Tax Act of 1980) has been adopted which encourages the production of fermentation ethanol for fuel. Figure I-1 shows the projected contribution of alcohol fuels to the automotive gasoline market and indicates the magnitude of the President's objectives for ethanol fuel production. Meeting these goals will increase the potential fermentation ethanol fuel market about 10 times over the end of 1979 production in the very near term (1981 to 1982) and about 30 times by 1985 to 1986.

The overall social and marketing climate suggests that fermentation ethanol production is a promising business venture, ranging from small-scale units often farm-based, producing up to a few thousand gallons to commercial-scale units producing more than 15 million gallons of ethanol fuel annually. Small-scale fermentation

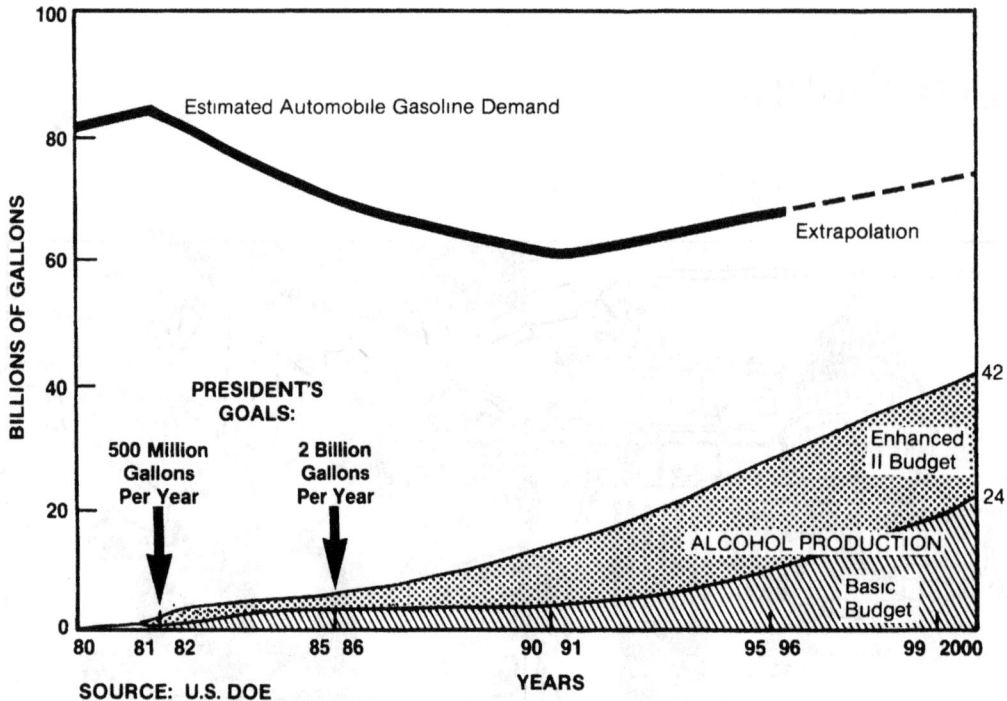

Figure I-1. U.S. DOE Goals for Alcohol Fuels Production from Biomass

ethanol production units and their implementation were discussed in the recently issued SERI report, "Fuel from Farms: A Guide to Small-Scale Ethanol Production", addressed to farmers or small investors. This document, "A Guide to Commercial-Scale Ethanol Production and Financing", is addressed to potential entrepreneurs, designers, contractors, and financiers of commercial-scale (more than 15 million gallons per year) fermentation ethanol fuel production facilities.

OBJECTIVE

The purpose of this document is to provide sufficient information to allow prospective entrepreneurs to make sound judgments about investing in commercial-scale, fuel-grade alcohol production and to develop an understanding of the varied aspects of entering the alcohol fuel production field. This document is organized to lead the reader from the initial stages of interest to the preparation of a feasibility study. The feasibility study will serve a dual purpose: to aid investors in developing a realistic business analysis for the venture and to provide the necessary supporting evidence in raising the capital to launch the venture.

PERSPECTIVES

United States' dependence on foreign oil is illustrated in Figure I-2. The United States can no longer afford such

a degree of dependence on oil reserves outside its control to meet its energy needs. Such dependence is hazardous to the U.S. economy and security. Therefore, the past several years have seen an increased emphasis on domestic alternatives, and pressure to develop synthetic fuels such as fermentation ethanol from biomass resources.

Statistics show a drop in crude oil imports over the past 4 years, perhaps implying that the United States public is responding to the President's calls to conserve energy.

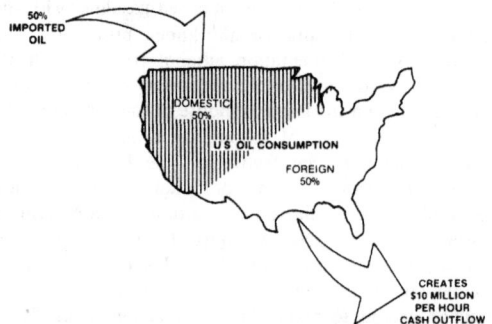

Figure I-2. Impact of Foreign Oil Consumption

COMMERCIAL SCALE ETHANOL PRODUCTION AND FINANCING

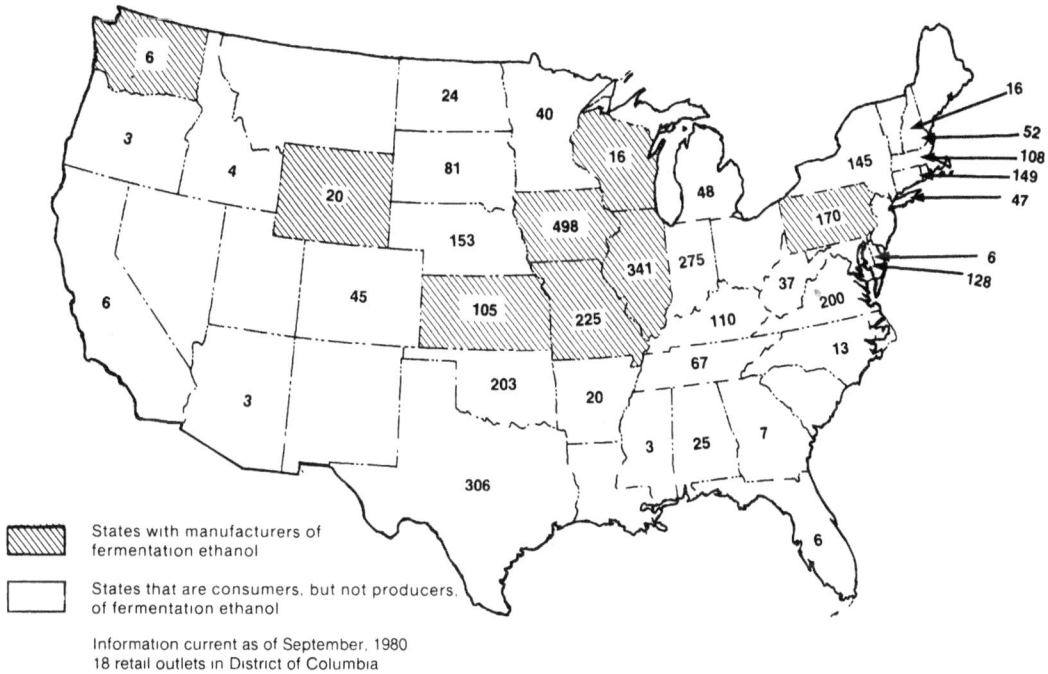

States with manufacturers of fermentation ethanol

States that are consumers, but not producers, of fermentation ethanol

Information current as of September, 1980
18 retail outlets in District of Columbia

Figure I-3. Number of Retail Gasohol Outlets in each State

However, increased development of United States reserves (particularly in Alaska), increases in the price of gasoline, and public response to gasohol play a role in bringing about this decrease.

While little gasohol was marketed in the mid-1970's, by the end of 1979 about 50 million gallons of fermentation ethanol, blended with unleaded gasoline, were sold across the United States. This marketing effort resulted in the establishment of a network of gasohol retail outlets across the United States, as shown in Figure I-3.

In the current political and economic climates, fermentation ethanol fuels have become increasingly attractive. The Department of Energy (DOE) has supported increased production of fermentation ethanol in small-scale facilities. DOE is also expanding its role by encouraging fermentation ethanol production in commercial-scale facilities. Small-scale is currently defined as under 15 million gallons of fermentation ethanol per year; this category includes the farmer producing 10 to 25 gallons per day strictly for his own use. Commercial-scale plants may range in size from 15 million to 100 million gallons annually.

In recent years, the increase in gasoline price and the desire for energy independence have made alcohol attractive as a transportation fuel, even though this is not a new idea. Henry Ford was an early supporter of

using "home-grown" fuels and his Model-T could be adjusted to burn pure alcohol or gasoline. During World War II many countries relied on alcohol as a fuel source. Since 1975, Brazil has had a huge program in operation to cause a national change from the use of

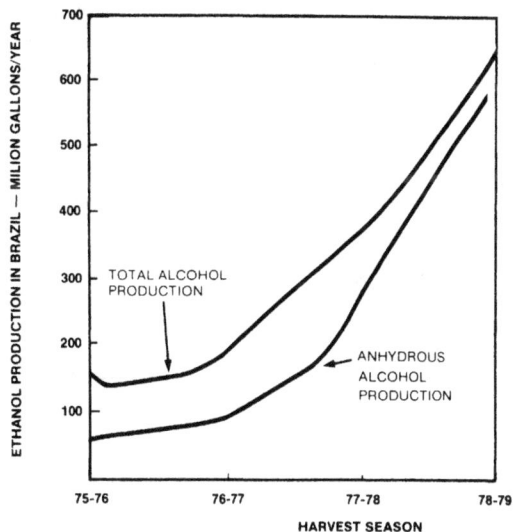

Figure I-4. Ethanol Production in Brazil as Reported in the Brazilian National Alcohol Program

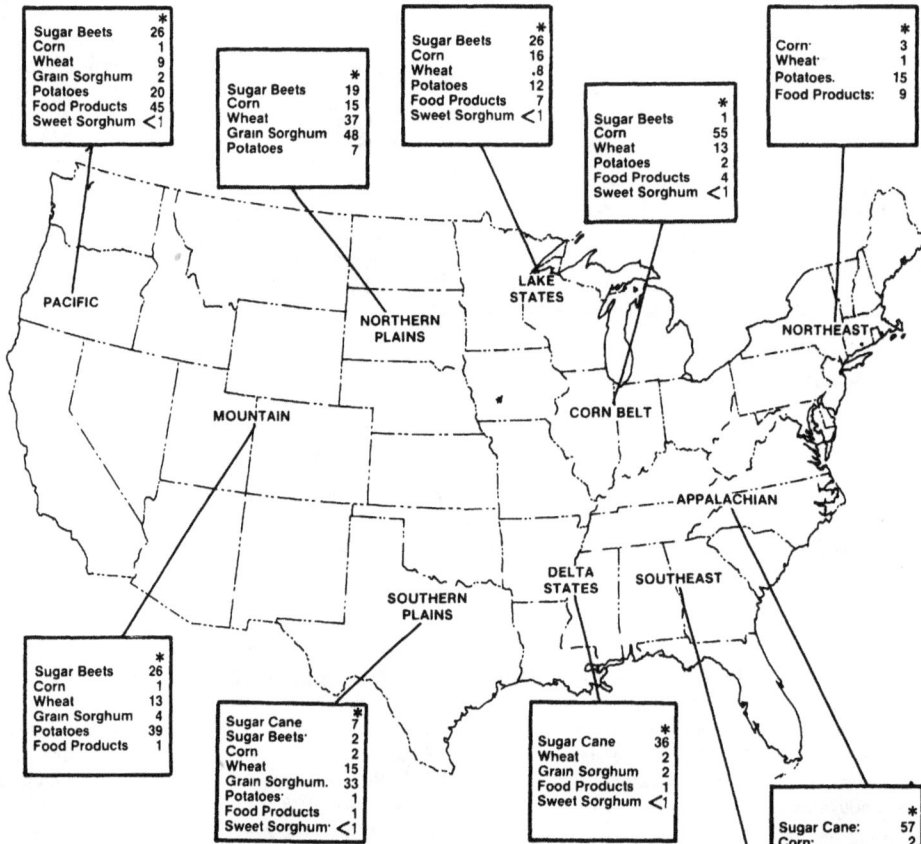

Pacific box:
Sugar Beets 26 *
Corn 1
Wheat 9
Grain Sorghum 2
Potatoes 20
Food Products 45
Sweet Sorghum <1

Northern Plains box:
Sugar Beets 19 *
Corn 15
Wheat 37
Grain Sorghum 48
Potatoes 7

Lake States box:
Sugar Beets 26 *
Corn 16
Wheat .8
Potatoes 12
Food Products 7
Sweet Sorghum <1

Corn Belt box:
Sugar Beets 1 *
Corn 55
Wheat 13
Potatoes 2
Food Products 4
Sweet Sorghum <1

Northeast box:
Corn 3 *
Wheat 1
Potatoes 15
Food Products: 9

Mountain box:
Sugar Beets 26 *
Corn 1
Wheat 13
Grain Sorghum 4
Potatoes 39
Food Products 1

Southern Plains box:
Sugar Cane 7 *
Sugar Beets 2
Corn 2
Wheat 15
Grain Sorghum 33
Potatoes 1
Food Products 1
Sweet Sorghum <1

Delta States box:
Sugar Cane 36 *
Wheat 2
Grain Sorghum 2
Food Products 1
Sweet Sorghum <1

Southeast box:
Sugar Cane: 57 *
Corn: 2
Potatoes: 2
Food Products: 29
Sweet Sorghum: <1

Appalachian box:
Corn: 5 *
Wheat 2
Grain Sorghum: 1
Potatoes 2
Food Products: 3
Sweet Sorghum: <1

Map region labels: PACIFIC, NORTHERN PLAINS, LAKE STATES, NORTHEAST, MOUNTAIN, CORN BELT, APPALACHIAN, DELTA STATES, SOUTHEAST, SOUTHERN PLAINS

*Numbers represent percent of total United States production for specific crops

FEEDSTOCKS	PRODUCTION (1978)
Corn	7,081 9 Million Bushels
Wheat (1977 data)	2,025 8 Million Bushels
Grain Sorghum	748 4 Million Bushels
Food Products (Vegetables and Fruits Excluding Potatoes)	29 7 Million Bushels
Sugar Beets	25 9 Million Bushels
Sugar Cane	16 0 Million Bushels
Potatoes (1977 data)	17 6 Million Bushels
Sweet Sorghum	Minimal at Present

Figure I-5. Potential Ethanol Feedstocks and their Availability by Regions

petroleum to the use of alcohol fuels, Brazil is now using 20 percent alcohol blends on a regular basis and large fleets of government- or industry-owned cars run on 100 percent alcohol. Figure I-4 shows the trends in ethanol production in Brazil. It is apparent that since the announcement of the Brazilian National Alcohol Program in November 1975, the total production of ethanol, as well as anhydrous ethanol for blending, has increased dramatically. This experience shows that a dedicated effort to reduce oil imports by developing national renewable resources can produce significant results in a short period of time.

If the need arises, the United States could institute the same kind of intensive ethanol program as Brazil instituted. A significant reduction in gasoline use could be achieved by changing to gasohol. The 10 percent alcohol/90 percent unleaded gasoline blend requires no engine modifications and has the a potential to reduce oil imports.

The raw materials for commercial-scale ethanol production are readily available in various forms. The major feedstocks are: corn, wheat, grain sorghum, barley, potatoes, sugar beets, sweet sorghum, sugar cane, and fodder beets. Since it is possible to use a wide variety of feedstocks for ethanol production, the plant location is closely linked to the locally available type of feedstock.

For example, a plant in Nebraska would use corn as feedstock, while one in Texas might use grain sorghum, and a plant in Idaho could consider potatoes or sugar beets. Figure I-5 shows the potential ethanol feedstocks and their availability by regions in the United States.

To meet the President's goal of 500 million gallons of ethanol for 1981, there is a need for increased production of fermentation ethanol. Figure I-6 shows the number of plants of various sizes required to attain annual production goals ranging from 200 to 2700 million gallons. It is apparent from the figures used to develop the chart, that a significant fraction of ethanol fuel is expected to be produced in commercial-sized plants (the medium- and large-sized plants mentioned in the chart).

ISSUES

The major issues relating to alcohol fuels and the position of the Office of Alcohol Fuels are outlined in Table I-1. There was a feeling of uncertainty about the Government's policies toward alcohol fuels before the President expressed himself firmly in favor of increased fermentation ethanol production and proposed a permanent exemption from the 4¢/gallon Federal gasoline excise tax for gasohol. A number of States are providing additional incentives in the form of waivers or reduction

PLANT SIZE	CY 1980		CY 1982		CY 1985	
	TOTAL NO. PLANTS	PRODUCTION CAPACITY, MILLION GALS/YEAR	TOTAL NO. PLANTS	PRODUCTION CAPACITY, MILLION GALS/YEAR	TOTAL NO. PLANTS	PRODUCTION CAPACITY, MILLION GALS/YEAR
10,000 – 15 Million Gallons/Year	96	175	200	375	300	600
> 15 – 50 Million Gallons/Year	6	150	13	450	23	900
> 50 Million Gallons/Year	1	100	2	200	5	500
Cumulative Total Installed Capacity (Projected in Million Gallons Per Year)		425		1,025		2,000
National Goals Million Gallons/Year		300*		1,000**		2,000*

* President's Goals Announced 11 January 1980
** P.L. 96-294.

Figure I-6. Number of Plants Required to Meet Goal for Alcohol Production

Table I-1. Major Issues Relating to Alcohol Fuels

ISSUE	OFFICE OF ALCOHOL FUELS VIEW
Food Versus Fuel	States that there are no problems until ethanol production exceeds 2 billion gal/year. Beyond that, increasing supplies of food and fuel are possible with scientific and technical advances already occurring. America can do both. Later cellulosic waste will replace part of grain and sugar-containing feedstocks.
Net Energy Balance	States that if non-petroleum fuels are used to the extent reasonable in farming operations and in the conversion of carbohydrates into ethanol, each gallon of alcohol produced that goes into making gasohol will save 1.53 gallons of oil.
Critical Need to Reduce America's Use of Foreign Fuels	Highlights the importance of alcohol-formulated fuels to reduce dependence on imported oil, revitalize rural America, reinforce the family farm, and strengthen the nation's national defense posture
Current and Projected Technological Advances	Recognizes current developments and assumes scientific and technical advances before 1985 and beyond as the nation moves out of the petroleum era into an energy mix where renewable resources play an important role.
Potential for Production Cost Reduction	Cites evidence indicating that the cost of corn in modern fuel-grade alcohol plants will be well under 50 percent of the manufacturing cost of ethanol. Also, other non-grain feedstocks, such as food processing waste, sweet sorghum, fodder beets, hybrid sweet potatoes, Jerusalem artichokes, honey locust, and cassava will be used to produce fuel and feed. Technological advances, through process research, will reduce production costs, also save energy
Value and Uses of CO_2	Indicates that CO_2 is a valuable coproduct of the fermentation process. Can be used in oil recovery, for storage of food and feed, refrigeration, in irrigation water, and in air and water injection systems to accelerate the growth of terrestial and aquatic biomass
Realistic Production Targets	Believes that ethanol production will meet the Administration's goal of 500 million gallons during 1981 and 2 billion gallons during 1985, as well as the congressional goal of 920 million gallons in 1982. This would save the nation 3.5 billion gallons, or 230,000 barrels of oil a day

SOURCE U S DOE, Office of Alcohol Fuels

in State taxes. A list of the states and the incentives they provide is shown in Appendix D. The Energy Security Act of 1980 (P.L. 96-294) provides significant financial support for the National Alcohol Fuels Program. The Act establishes national goals for alcohol production, creates the Office of Alcohol Fuels within DOE at the Assistant Secretary level, and provides significant financial support in the form of loan guarantees and loans.

The future of ethanol questions arise as to the short-and long-term future of ethanol fuels. In the short term, there is an established need to supplement our imported petroleum supplies. The DOE Alcohol Fuels Program addresses the need to increase production of alcohol fuels, and the goal to reduce the price of ethanol in the long term. These goals are to be attained by a substantial Federal program utilizing loan guarantees, and cooperative agreements to increase production capability to meet the President's goals. The funding of realistic near-term research and development will provide innovative feedstock and new production techniques to help reduce the long-term price of ethanol. A

breakdown of ethanol production costs is shown in Figure 1-7.

The emphasis for the near term is on the increased production of fermentation ethanol, followed by production of ethanol from lignocellulose and, after the 1990's, a significant increase in methanol from biomass, as shown in Figure 1-8.

ORGANIZATION OF THE BOOK

Alcohol fuels production can be an attractive, money-making venture provided it is approached in a business-like fashion and full advantage is taken of the various incentives presently available.

The economics and, therefore, the profitability of alcohol fuels production are sensitive to a number of factors ranging from site-specific conditions to national and international policies. Raising the necessary starting capital through conventional financing channels may be difficult because of the unknown risks associated with alcohol production.

COMMERCIAL SCALE ETHANOL PRODUCTION AND FINANCING

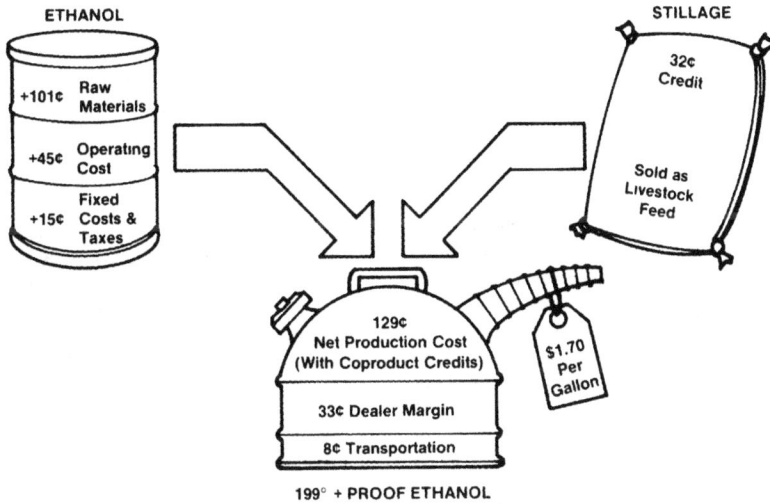

ETHANOL

+101¢ Raw Materials

+45¢ Operating Cost

+15¢ Fixed Costs & Taxes

STILLAGE

32¢ Credit

Sold as Livestock Feed

129¢ Net Production Cost (With Coproduct Credits)

33¢ Dealer Margin

8¢ Transportation

$1.70 Per Gallon

199° + PROOF ETHANOL

SOURCE U S DOE 1st Annual Report to Congress (Revised)

Figure I-7. Breakout of Production Costs for Ethanol

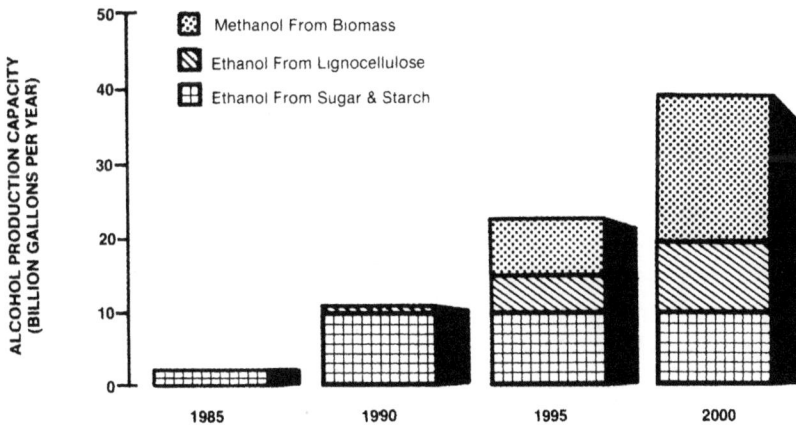

Methanol From Biomass

Ethanol From Lignocellulose

Ethanol From Sugar & Starch

ALCOHOL PRODUCTION CAPACITY (BILLION GALLONS PER YEAR)

1985   1990   1995   2000

Figure I-8. Long-Term Contribution of Ethanol, Lignocellulose, and Methanol from Biomass

This document is designed to lead the potential investor through all the steps necessary to develop a business plan and prepare a feasibility analysis for a site-specific project. Therefore, emphasis is placed on marketing, financing, management, and incentives rather than primarily technical matters. In this context, it is hoped that the book will provide the investor with the required background for involvement in the fermentation ethanol fuel business.

The introduction provides an overview of the perspectives and issues in the alcohol fuels industry. Chapter II, seeks to surface factors which affect the decisionmaking process and which the investor must face before entering the alcohol fuels industry. The chapter attempts to lead the investor step-by-step through the series of decisions and choices to be made before reaching the final decision to enter the business. This final decision to obtain financing for the plant is made after a detailed feasibility study has been performed.

Chapters III to IX discuss in detail some of the important elements to be addressed during the decisionmaking process. Chapter III describes the types of feedstocks available and relates them to areas within the United States. Trends and fluctuations in the price of the major grain feedstocks (which are used predominantly as animal feeds) are also discussed in terms of their potential use and value compared to other feeds. Chapter IV discusses the market potential of ethanol and its coproducts, and examines how the locations of the ethanol markets in relation to those of the feedstock supplies may influence selection of a plant site.

Various aspects of plant design are discussed in Chapter V. A 50 million gallon per year plant is analyzed to provide the general technical background and costing data required in analyzing plants of various sizes and designs. Safety aspects and environmental concerns are treated in Chapters VI and VII. The regulations are reviewed and their impact on plant design and operation is discussed. The basic elements of a business plan are described in Chapter VIII, which leads to an approach for development of the feasibility study in Chapter IX. Other information on financial assistance, regulations, current legislation, and reference material is given in the Appendices.

It should be stressed that the initial step of performing a realistic estimate of the feasibility of the project is essential if the project is to become a successful business reality. Irrespective of the possible outcome of the analysis, sufficient funds must be allocated at the outset to perform a realistic and meaningful feasibility analysis.

In view of the focus for this document, emphasis is placed on describing and analyzing the steps involved in implementing a commercial-scale project.

CHAPTER II
Decision to Produce

A NEW VENTURE

Fermentation ethanol fuel production utilizing biomass feedstocks appears to be an attractive business venture for a wide range of potential investors. The production of ethanol fuel from agricultural crops or wastes is strongly supported by the farming community and the public's initial response to the introduction of gasohol as a substitute automotive fuel has been favorable.

Ethanol production raises numerous issues which may complicate the task of the investor when deciding whether to enter the ethanol market. The production of ethanol through fermentation of agricultural feedstocks requires interaction with a variety of producer and consumer groups and local, State, and Federal agencies. The feedstock for production originates from the agricultural community and some of the coproducts of ethanol production may impact on the same farming community. The production and marketing of ethanol fuel are subjected to the same basic regulations as the production of alcoholic beverages; simultaneously, the marketing and distribution of ethanol fuel may require

the cooperation of gasoline producers and distributors. The economics of ethanol fuel production benefit from State or Federal incentives, and will also be influenced by uncontrollable factors such as climate or world events which impact on the price of agricultural feedstocks.

The initial idea of entering the ethanol fuel business may stem from a variety of motivations: desire to make a profit; desire to diversify investments; make use of idle or surplus facilities or resources; fill the needs of a perceived market; and others. Whatever the motivation, the decision to produce ethanol must be based on a careful and thorough analysis of all aspects of the proposed business. In view of the factors impacting on the project, it will be necessary for the prospective investor to seek and secure the help of individuals or teams familiar with the various aspects of ethanol fuel production. This groundwork will include the collection of data, analysis of feedstock and production options, and evaluation of financial opportunities which are summarized in the form of the comprehensive feasibility analysis. The conclusion drawn in the feasibility analysis should form the basis upon which the decision to produce will be made. The feasibility study is also needed to interest other investors or sources of financing to take part in the project. It must therefore be recognized that a certain amount of front-end money will be required to reach a substantiated decision to enter the fermentation ethanol business.

The objective of the present chapter is to describe the major steps required in deciding to produce ethanol, discuss the data to be gathered and questions to be answered at each of these steps, and indicate the kind of effort needed to complete the feasibility analysis on which the final decision will be based.

OVERVIEW OF ETHANOL FUEL PRODUCTION

The Operating Plant and Its Environment

Fermentation ethanol fuel production involves multiple interactions among a variety of "players." Each of these aspects must be addressed during the decision-making process in order to avoid overlooking a problem which could be detrimental to the economic viability of the project.

Figure II-1 presents an overview of an ethanol business, and identifies some of the major elements and entities involved. Four major areas of concern are identified.

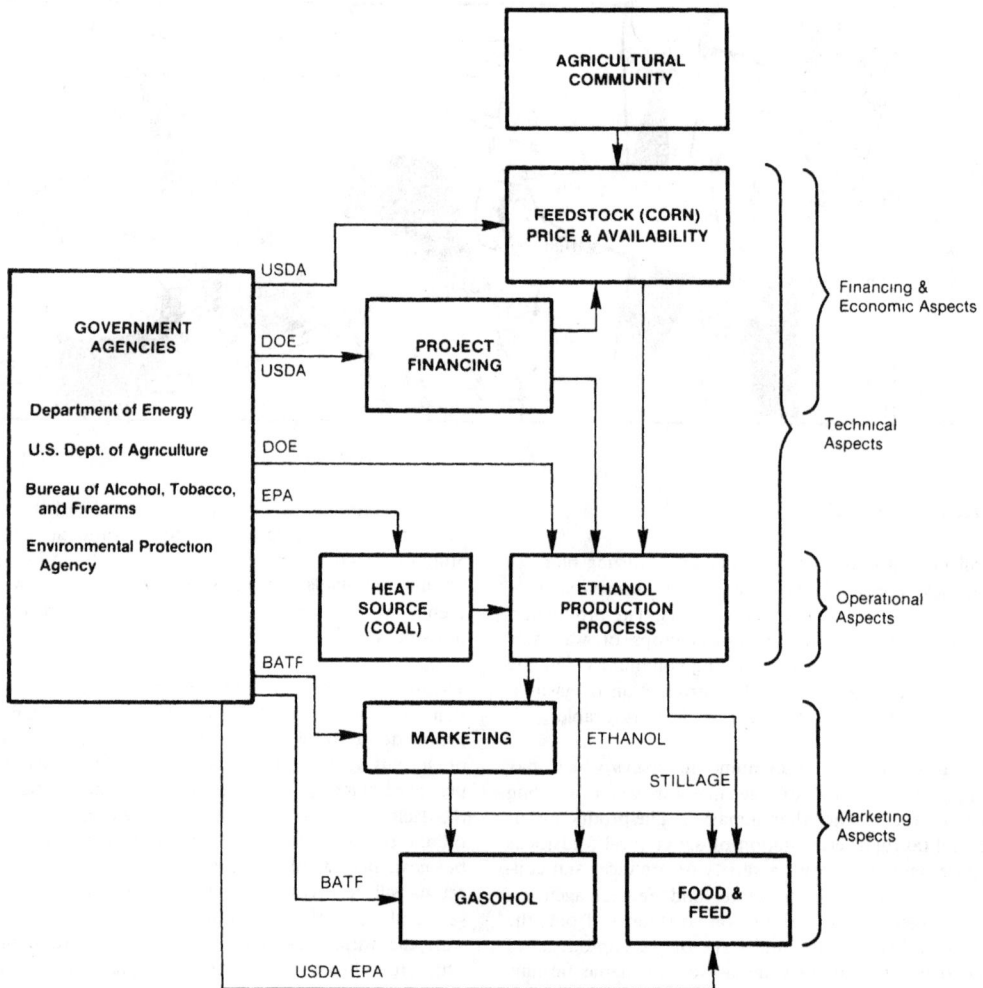

Figure II-1. Basic Organization Interfaces for Production of Gasohol

Technical Aspects. Ethanol production through fermentation of grain is a well-established technology practiced by the alcoholic beverage industries in the United States. The production of fermentation ethanol fuel is an extension of this technology. The economics involved, however, are quite different. Alcoholic beverage industries are geared toward satisfying a luxury market; therefore, their profitability is not as critically dependent on process efficiency or price of supplies as is that of ethanol fuel plants which must compete with a still relatively cheap product—gasoline. When evaluating the feasibility of a fermentation ethanol fuel plant, it will be essential to carefully assess the method of procurement of the feedstock (such as corn) over a long period, the availability of resources such as fuel for the process heat source and water, the sale of the stillage, the efficiency of the ethanol production process, and other technical aspects which can impact on the overall economics of the plant.

As indicated in Figure II-1, various Government agencies influence the process of fermentation ethanol production. The U.S. Department of Agriculture (USDA) may impact on feedstock availability and prices, the Bureau of Alcohol, Tobacco, and Firearms (BATF) requires permits for production, the Environmental Protection Agency (EPA) impacts on stillage disposal, the Department of Energy (DOE) may encourage the use of renewable process fuels, etc.

Marketing Aspects. Successful marketing of both fermentation ethanol fuel as gasohol and the stillage coproducts of ethanol is essential to ensure economic viability of the project. A careful identification and evaluation of the long-term market will be required to assess the economic potential of the production plant. Contractual agreements for the sale of ethanol and stillage, prior to plant operation, will develop a sound basis for the venture.

Various government agencies may have an impact on the marketing process. Exemption from the excise tax on fuel blends containing ethanol is controlled by the Federal and/or State governments, government policies concerning the export of food and feed products, and income tax credits for fuel blenders may modify the marketing plans adopted for the products of the plant.

Financing and Economic Aspects. Obtaining financing from private sources for the proposed project will, to a large extent, depend on the quality and thoroughness of the feasibility analysis.

In the course of performing the feasibility analysis, Federal loan guarantee or loan programs should be reviewed to determine the eligibility of the fermentation ethanol project being considered. The purpose of the loan guarantee programs is to provide funds to sectors of the economy that otherwise could not obtain credit or could only obtain it under unfavorable conditions through existing commercial financial organizations. Federal agencies having the authority to provide funds to prospective fermentation ethanol fuel producers include DOE, the Small Business Administration (SBA), the Farmers Home Administration (FmHA) of USDA, the Department of Housing and Urban Development (HUD)'s Urban Development Action Grants, and the Economic Development Administration (EDA) of the Department of Commerce. Securing loan guarantees or loans for the project must be considered a priority item for the investor. Indeed, the terms of these loans are usually attractive and can improve the economics of the project, and loan guarantees will facilitate the task of raising funds from private sources. Securing loans or loan guarantees, however, requires the preparation of documented applications which may delay the decision-making.

Operational Aspects. The investor must address a number of aspects related to the operation of a fermentation ethanol plant. The availability of trained or trainable labor must be considered. Can the regional labor force provide the technical personnel for the plant? Will the creation of a new industry upset the local or regional organizations and therefore risk antagonizing the community?

The above factors, as well as the possible impact of agencies such as the Occupational Safety & Health Administration (OSHA) must be carefully evaluated.

Approach to Implementing an Ethanol Plant

The preceding discussion has shown that many factors may affect the profitable operation of an ethanol plant. Before finalizing the decision to start an ethanol plant, the four key aspects of ethanol production will have to be discussed and evaluated to choose the options or compromises which will optimize financial return to the investors.

Figure II-2 illustrates the major elements in both of the phases involved in bringing a plant to production. The scope and purpose of these elements are described below.

Decisionmaking Phase. The overall purpose is to reach a final decision whether to become involved in a fermentation ethanol project. This phase involves three major steps.

- **Prefeasibility Study.** Having identified an interest in the production of ethanol, an investor and his partners must be convinced that the project is viable, at least in general terms. Some help from consultants or local/regional sources such

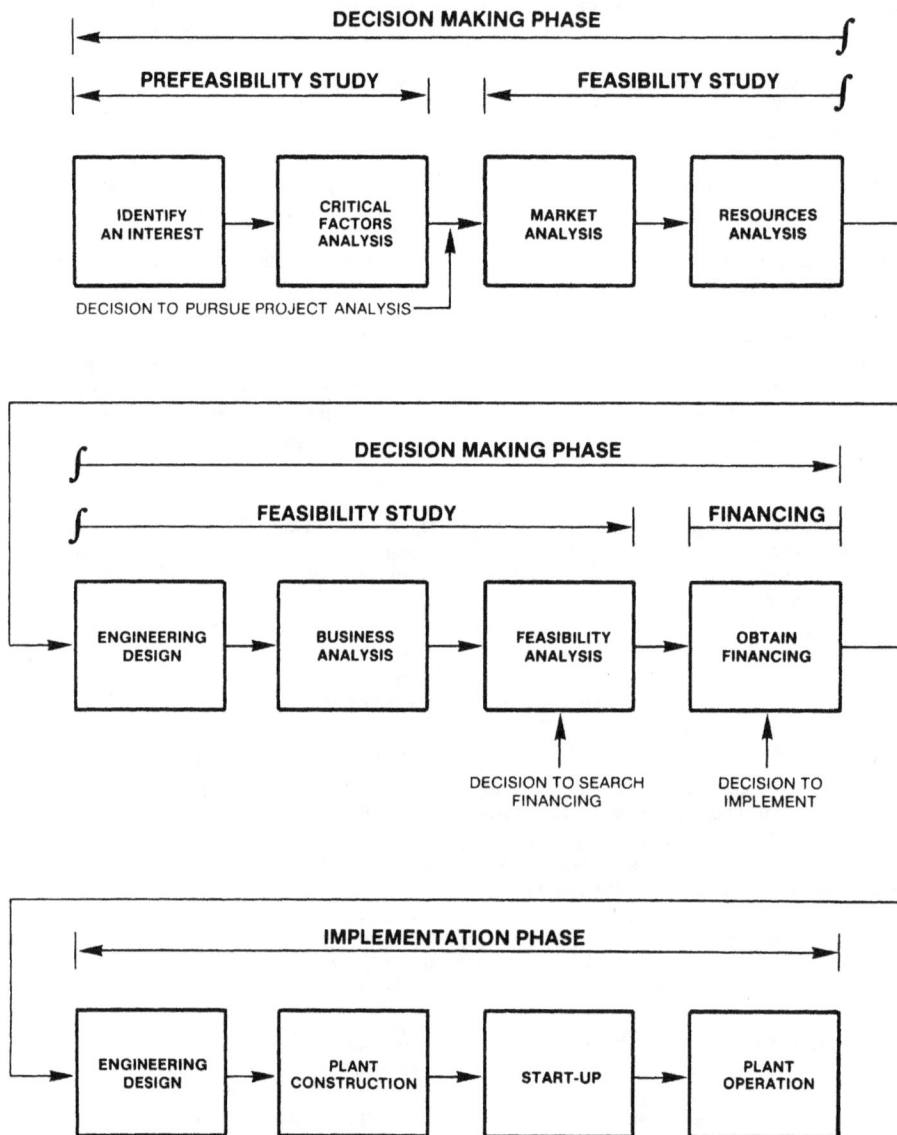

DECISION MAKING PHASE

PREFEASIBILITY STUDY   **FEASIBILITY STUDY**

IDENTIFY AN INTEREST → CRITICAL FACTORS ANALYSIS → MARKET ANALYSIS → RESOURCES ANALYSIS

DECISION TO PURSUE PROJECT ANALYSIS

DECISION MAKING PHASE

FEASIBILITY STUDY   **FINANCING**

ENGINEERING DESIGN → BUSINESS ANALYSIS → FEASIBILITY ANALYSIS → OBTAIN FINANCING

DECISION TO SEARCH FINANCING   DECISION TO IMPLEMENT

IMPLEMENTATION PHASE

ENGINEERING DESIGN → PLANT CONSTRUCTION → START-UP → PLANT OPERATION

Figure II-2. Steps Involved in Bringing a Plant to Production

as extension agents, agricultural school staff, or others may be needed in discussing some specific aspects of the project. The product of this step is a prefeasibility study which should establish the rationale for pursuing the project. Some of the key areas to be addressed include: existence of a market for both ethanol and its stillage coproduct, distribution mechanism for the products, selection and availability of a feedstock, approximate size of the proposed plant, availability of non-petroleum fuels for process steam and process water, general design of the fermentation plant, approximate cost of the plant, business plan, potential methods of financing, and economic viability of the project. At the end of this first step, the potential investor in the project should be able to make the decision whether to pursue the project.

- **Feasibility Study.** The objective of this step is to perform detailed technical and financial analyses of the viability of the project. The feasibility study will address the four major areas of marketing, technical feasibility, financing and economics, and operation; and will develop a realistic plan. The feasibility study will serve three purposes: 1) confirm and refine the estimates of the prefeasibility study; 2) develop a business plan and business planning schedule; and 3) serve as supporting evidence when negotiating the financing of the project.

Performing a detailed feasibility study will require contributions from many areas: engineering firms, market specialists, suppliers, environmentalists, law firms, accountants, etc. An adequate budget must be earmarked for the purpose of generating a credible, well-documented study.

At the end of this step, the potential investor will be in a position to decide whether the project has sufficient commercial potential for the financial community to be interested.

- **Financing.** The last and most important step is to secure the necessary financing for the project. In these negotiations, the investor will need the support of law firms, accountants, and other specialists familiar with the alcohol fuels industry.

When financing has been secured, the final decision to implement the project can be made.

Implementation Phase. This phase involves the same steps followed for any commercial/industrial project: engineering design; construction; start-up; and operation.

The role of the investor(s) is somewhat reduced in this phase, because at this point a management team should become responsible for the project implementation on behalf of the investors. The primary role of the investor is to assist the team in reaching the final decision of implementing the project. To do so, he must be aware of the questions to be asked, issues to be raised, problems to be solved, and players involved in the overall implementation of such a project. Some of these questions, problems, or issues related to the process of deciding to produce ethanol are more complex than for more conventional ventures. This results from the particular significance of fermentation ethanol in the present energy and international context and because of its intimate relationship with various markets and economic sectors.

THE DECISIONMAKING PROCESS

The decisionmaking process involves addressing and answering specific questions to reach a decision concerning implementation of the project.

The questions to be addressed are similar to those relating to any business venture. Should one get involved? Is it really worth it? What is the market? Is the project technically feasible? Is the project economically viable? Is financing available? The approach to answering these questions resembles that used for other similar businesses: market and engineering studies; business analysis; etc. The purpose of this section is to identify some specific questions and options raised by fermentation ethanol production and to indicate sources of data that could help the investors reach a decision.

The Decisionmaking Flow Diagram

Figure II-3 presents a flow diagram of the decisionmaking process. The figure is an expansion of the first phase of the implementation diagram shown earlier (Figure II-2).

The top row of the matrix identifies questions to be addressed or decisions to be reached. Each column of the matrix suggests an approach to answering the corresponding questions, the type of answer or output desired, the sources of information and possible options, and an estimate of the front-end cost required to obtain an answer or reach a decision. These costs are approximate and may include such items as consultant fees, data gathering, trips, etc.

Elements of the Decisionmaking Process

The various steps in the decisionmaking process are discussed below.

Initial Expression of Interest by the Investor: should he get involved at all in the ethanol fuel business?

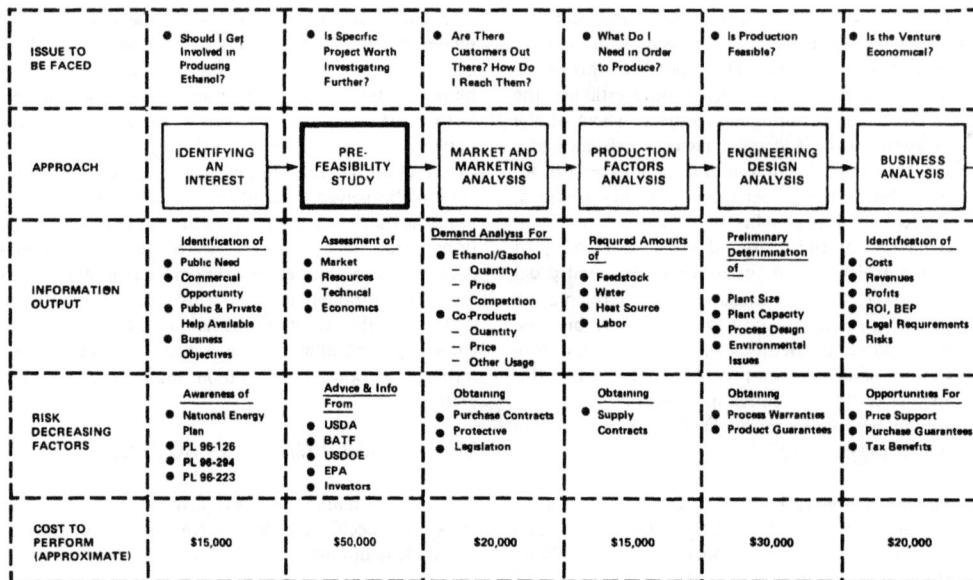

ISSUE TO BE FACED	• Should I Get Involved in Producing Ethanol?	• Is Specific Project Worth Investigating Further?	• Are There Customers Out There? How Do I Reach Them?	• What Do I Need in Order to Produce?	• Is Production Feasible?	• Is the Venture Economical?
APPROACH	IDENTIFYING AN INTEREST	PRE-FEASIBILITY STUDY	MARKET AND MARKETING ANALYSIS	PRODUCTION FACTORS ANALYSIS	ENGINEERING DESIGN ANALYSIS	BUSINESS ANALYSIS
INFORMATION OUTPUT	Identification of • Public Need • Commercial Opportunity • Public & Private Help Available • Business Objectives	Assessment of • Market • Resources • Technical • Economics	Demand Analysis For • Ethanol/Gasohol — Quantity — Price — Competition • Co-Products — Quantity — Price — Other Usage	Required Amounts of • Feedstock • Water • Heat Source • Labor	Preliminary Determination of • Plant Size • Plant Capacity • Process Design • Environmental Issues	Identification of • Costs • Revenues • Profits • ROI, BEP • Legal Requirements • Risks
RISK DECREASING FACTORS	Awareness of • National Energy Plan • PL 96-126 • PL 96-294 • PL 96-223	Advice & Info From • USDA • BATF • USDOE • EPA • Investors	Obtaining • Purchase Contracts • Protective • Legislation	Obtaining • Supply Contracts	Obtaining • Process Warranties • Product Guarantees	Opportunities For • Price Support • Purchase Guarantees • Tax Benefits
COST TO PERFORM (APPROXIMATE)	$15,000	$50,000	$20,000	$15,000	$30,000	$20,000

Figure II-3. Decisionmaking Process for Entering the Ethanol Business

The investor will need to identify the existence of interest in ethanol production at the local, regional, or national level. He will need to evaluate the availability of Federal incentives to support his project. His expression of interest could result from his contacts with grain producers, agribusinesses, or farmers seeking new markets for their products, alternative methods of disposal of stillage or waste products, or distributors trying to secure a long-term reliable source of substitute fuel. The investor will need to evaluate the true level of interest and the commercial opportunity it generates and then must define his objectives. These could range from producing ethanol fuel as a separate business entity, to combining it with current business interests. The alcohol fuel could be marketed through cooperatives for special markets or regional distribution.

In clarifying his objective, the investor will not only have to rely on inquiries but also become familiar with State and Federal programs to determine if his perceived interest matches national and regional policies. Data on the National Alcohol Fuels Program may be found in Appendix C of this document. The cost of this first step is approximately $15,000, which will include travel costs, quick inquiries and surveys, and general data and information gathering.

Attractiveness of the Proposed Project: is the identified project worth investigating further?

Having established that significant interest exists for an ethanol fuel project, the investor must determine if the basic elements for a successful business are present. A prefeasibility study will provide this answer.

The major aspects to be considered in a prefeasibility study include:

* Existence of an identified market for the ethanol fuel
* Availability of a market for the stillage and coproducts
* Availability of sufficient feedstocks such as corn and alternate feedstocks to reduce the project risks
* Availability of a non-petroleum heat source such as coal
* Feasibility of the technical process and plant economics

The initial contact with potential investors or financing institutions should be made during this prefeasibility study to determine if the proposed project has a realistic chance of being financed. The product of this study should be a document or a data base sufficient for the investor to determine whether the project appears attractive enough to justify the expenditure of further front-end money. At this time, the investor should also have some knowledge of how his project will be received by the financing community and whether some other investors or the government may be willing to supply some of the seed money for the forthcoming feasibility study. Formation of a corporation may be justified at this time.

COMMERCIAL SCALE ETHANOL PRODUCTION AND FINANCING

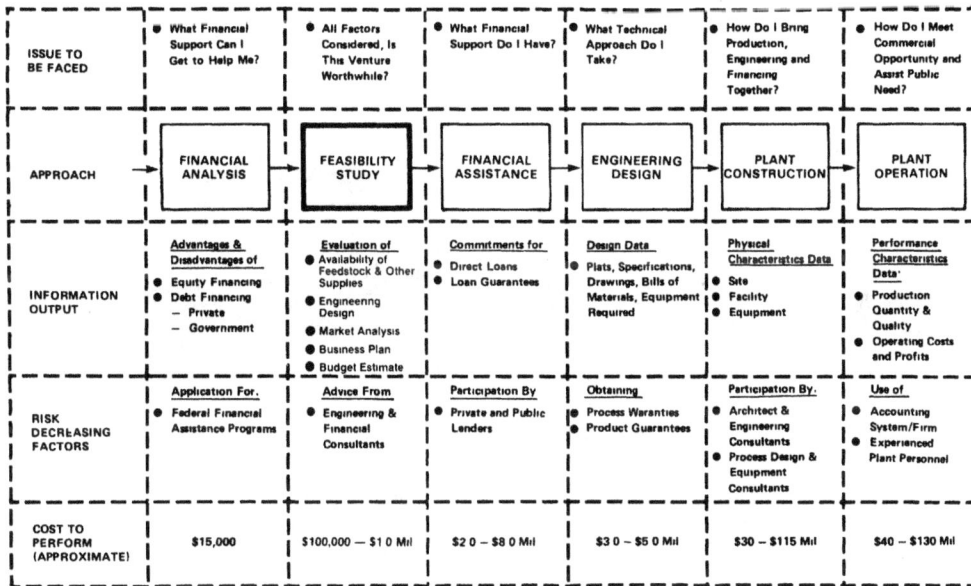

Figure II-3. Decisionmaking Process for Entering the Ethanol Business (Continued)

	What Financial Support Can I Get to Help Me?	All Factors Considered, Is This Venture Worthwhile?	What Financial Support Do I Have?	What Technical Approach Do I Take?	How Do I Bring Production, Engineering and Financing Together?	How Do I Meet Commercial Opportunity and Assist Public Need?
ISSUE TO BE FACED						
APPROACH	FINANCIAL ANALYSIS	FEASIBILITY STUDY	FINANCIAL ASSISTANCE	ENGINEERING DESIGN	PLANT CONSTRUCTION	PLANT OPERATION
INFORMATION OUTPUT	Advantages & Disadvantages of • Equity Financing • Debt Financing – Private – Government	Evaluation of • Availability of Feedstock & Other Supplies • Engineering Design • Market Analysis • Business Plan • Budget Estimate	Commitments for • Direct Loans • Loan Guarantees	Design Data • Plats, Specifications, Drawings, Bills of Materials, Equipment Required	Physical Characteristics Data • Site • Facility • Equipment	Performance Characteristics Data • Production Quantity & Quality • Operating Costs and Profits
RISK DECREASING FACTORS	Application For. • Federal Financial Assistance Programs	Advice From • Engineering & Financial Consultants	Participation By • Private and Public Lenders	Obtaining • Process Warranties • Product Guarantees	Participation By. • Architect & Engineering Consultants • Process Design & Equipment Consultants	Use of • Accounting System/Firm • Experienced Plant Personnel
COST TO PERFORM (APPROXIMATE)	$15,000	$100,000 — $1 0 Mil	$2 0 – $8 0 Mil	$3 0 – $5 0 Mil	$30 – $115 Mil	$40 – $130 Mil

Sources of information to contact in performing this prefeasibility study may include the DOE, USDA, SBA, and EDA through their regional offices or extension services. State agencies, engineering firms, BATF, and local agricultural economists could be contacted to broaden the information base. Some specialized help will be required, the cost of which could reach about $50,000. The timeframe needed for this study may be as short as one or two months but in most cases will probably extend over a four- to six-month period. The following chapters of this book provide the basic data needed to pursue the project, as well as references to information sources.

Having determined that the project is worth pursuing, the investor must analyze in further detail the various aspects of the proposed business. These detailed analyses will result in a feasibility study which will be the basis for the financing and implementation of the project.

Identification of Markets: are there markets for the ethanol and coproducts of the plant and how can the markets be reached?

A market analysis is required to determine such elements as the size of the proposed plant, its location, and the marketing structure required. The size of the potential market, market penetration, market price, location of the market, competition, and other relevant factors will be estimated for both ethanol fuel and its coproducts. When feasible, letters of intent, letters of interest,

or tentative contractual agreements will be obtained from prospective customers. This market analysis will require expert help from firms specialized in marketing and distribution of both fuel and feed. Background data on these market aspects can be found in Chapters III and IV of this book. The cost of the market analysis has been estimated at about $20,000.

Production Factors: what are the resources needed to produce?

Having sized the market, the investor can then size the capacity of his proposed plant. It will be necessary to estimate the resources required for plant operation. These resources include feedstocks, a non-petroleum process heat source, water, land, and labor. A market for stillage-derived coproducts or a facility for stillage disposal is also required. Materials balances should be performed to estimate the needed gross quantities of production resources.

Where possible, tentative supply contracts will be drawn for feedstock and stillage, zoning or preliminary zoning permits will be searched, and local environmental issues will be reviewed. The result of this analysis will be the identification of a tentative site or sites and a certain degree of confidence that all required production elements are available.

Data concerning some of these aspects can be found in Chapters III, IV, V, and VII, and Appendix C of this

The Written Commitment of a Supplier to Provide Feedstock
Will Help Establish the Viability of the Plant

book. With technical and financial assistance, it is estimated that this step may cost about $15,000. If permits are to be obtained, the time frame to finish this task may extend from several months to a year in the worst case.

Technical Feasibility: is production feasible?

Although conventional ethanol production from sugar-starch feedstock is well-established technology, site-specific conditions or constraints such as quality of feedstock, source of process heat, water quality, etc., require modifications to the conventional plant design. A preliminary process and engineering design analysis, which will include comparison of available options, must be performed by a reputable, well-recognized engineering firm. This analysis will include an optimized process design, materials requirements and energy flow diagrams, plant capacity, expected on-stream factor, and other aspects required in a complete engineering package. Process warranties and product guarantees should also be stated. Availability of electric power and water should also be discussed.

Background data relative to the plant design aspects can be found in Chapter V, but it is stressed that this analysis must be performed by a well-established firm to lend credibility to the study and facilitate financing. Such a study could cost about $50,000 and may extend over several months, depending on the number of options to be analyzed.

Economic Viability: is the project economically viable?

This is probably the most crucial element of the feasibility study in terms of securing adequate financing. A complete business plan must be proposed, including period of erection of the plant, planned date of operation, legal status (corporation, cooperative, etc.), capital and working capital requirements, and operating costs. This data then will be used to determine estimated return on investment (or price of ethanol required to obtain a proposed return) using discounted cash flow or other method suitable to the type of business envisioned. The financial method used to estimate the economics of the plant must be flexible enough that sensitivity analyses can be performed to evaluate the potential impact of factors such as fluctuations in feedstock and stillage prices, tax credits, and inflation rate on the economics and return on investment from the plant. It is important that the perceived risks be quantified as much as possible to ensure the development of a viable business plan.

Performing this business analysis will require a data base on feedstocks, chemical and fuel supplies, and market prices of products, as well as appropriate financial data such as inflation rates for fuels, products and labor, prevalent rate of interest, and others. Major accounting and engineering firms have developed computer software for the business analysis discussed here. Legal and financial advice also will be needed in performance of this task. The business analysis could cost $20,000 and extend over several weeks once the necessary data has been collected.

Sources of Financing: what is available?

Possible sources of financing should be reviewed, including private, governmental, or mixed funding of the project. Loans or loan guarantees should be considered. The impact of each of these options must be reviewed in terms of the overall economic viability of the project. Appendix A provides an in-depth discussion of Federal financial incentives for the investor.

The product of this analysis will be the identification of a preferred method of financing and some alternative methods, if warranted.

COMMERCIAL SCALE ETHANOL PRODUCTION AND FINANCING

Such an analysis will require legal, financial, and analytical inputs from specialized firms or consultants and could cost $20,000 or more.

Overall Feasibility of the Project: is the venture technically feasible and economically viable?

Having performed the detailed market, resources, technical, and economic analyses, an overall evaluation of the project must now be performed. This feasibility study will help in deciding whether or not to seek financing, and in describing the total project and its prospects to future investors.

The feasibility study must address all aspects of the proposed business, attempt to answer all foreseeable questions, and establish the credibility of the originator(s) of the project and its supporting team (legal, technical, etc.). Letters of commitment or intent from suppliers, markets, permit and zoning agencies, and others should be an important part of this package. The proposed management team and its credentials should be described.

Performing such a feasibility study will require contributions from various teams and individuals and could be quite expensive. An outline for a typical feasibility study is described in Chapter IX.

Financing: Prior to completion of the feasibility study, a team must be assembled to prepare a detailed economic and financial analysis of the project and to negotiate and arrange the financing. The qualifications of this financing team are contingent upon the Company's internal management team and the complexity of the financing to be arranged, which generally includes specialists in the following areas: economic and financial analysis; negotiation and placement of debt; placement of stock, partnership interest, or bonds; and grants.

TIMEFRAME FOR THE DECISIONMAKING PROCESS

Developing a credible feasibility study and negotiating the required financing, i.e., reaching the decision to implement the plant, may be a time-consuming process.

The prefeasibility study must be performed before engaging in the feasibility study.

While performing the feasibility study, several tasks may be conducted in parallel: market evaluation, feedstock supplies, application for permits, negotiation of tentative sales and purchase contracts, etc. This approach could accelerate the process but runs the risk of expending more funds than otherwise necessary if at any point one of the elements of the feasibility package is missing or one of the analyses suggests abandoning the project. For instance, refusal of the necessary permits at a late date or rejection of the project as environmentally unsound could result in the loss of significant amounts of money if all tasks required under the feasibility study are conducted in parallel. The investor will need to decide whether reaching the market at an earlier date is worth risking a large fraction of the front-end investment.

CHAPTER III
Feedstock Choices

Selection of the feedstock is one of the most critical elements in the process of assessing the feasibility of an ethanol production facility. Closely linked with this decision are: an evaluation of the quantities of the selected feedstock that can be secured annually, or on a long-term basis, to support a plant, and an estimation of the cost of the resource. This first step is essential as it will provide an order of magnitude for the size of the project considered, and for the market required to absorb the ethanol fuel and its marketable coproducts. The selection of a feedstock will affect the design and characteristics of the projected plant and the economics of the overall project.

Many feedstocks can be used to produce fermentation ethanol, as shown in Table III-1. Despite the fact that corn presently offers the largest potential for U.S. grain production, other crops or residuals could be considered. (See Figure III-1.) Particular attention should be given to site- or region-specific conditions that make particular crops or their residues especially attractive as feedstock material. Disposal of food processing resi-

dues, damaged crop, or crop residues through ethanol production could result in a credit or attractive feedstock prices for the ethanol producer. Such favorable situations may in some cases make small commercial plants more economically attractive than larger ones using more conventional feedstocks such as grain or sugar cane.

The prospective investor should therefore approach the problem of feedstock selection with an open mind. Also, he should attempt to integrate his operation in the context of the region considered, in order to take advantage of local or regional conditions that could be beneficial. The investor should also consider the existence of other producers in the same area and their feedstock requirements.

Selection of a particular feedstock will have a direct bearing on the design of the plant. Therefore, this chapter discusses some feedstock characteristics, knowledge of which is necessary to the decisionmaking process of the prospective investor.

TYPES OF FEEDSTOCKS

The biological production of ethanol is accomplished through the fermentation of six-carbon sugar units (principally glucose) in the presence of yeast. All agricultural crops and crop residues contain six-carbon sugars or compounds of these sugars, and can be used to produce ethanol, provided the six-carbon sugars they contain are accessible for fermentation. Agricultural crops and residues can be subdivided into three broad classes: sugar crops; starch crops; and cellulosic material, as shown in Table III-1. In sugar crops, the six-carbon sugars or fermentable sugars occur individually or in bonded pairs. Minimal mechanical treatment will release the fermentable sugars. In starch crops, the six-carbon sugars are linked in long, branched chains (called starch). These chains must be broken down into individual or pairs of six-carbon sugars before yeast can use the sugars to produce ethanol. These crops, therefore, will require additional treatment (mechanical, chemical, or biological) before fermentation can occur. In cellulosic materials, the six-carbon sugars are linked in extremely long chains involving strong chemical

Corn Offers the Largest Potential for Ethanol Production in the United States

Table III-1. Summary of Feedstock Characteristics

Type of Feedstock	Processing Needed Prior to Fermentation	Principal Advantage	Principal Disadvantage
Sugar Crops (e.g., sugar beets, sweet sorghum, sugar cane, fodder beets, jerusalem artichoke)	Milling to extract sugar	• Preparation is minimal • High yields of ethanol per acre. • Crop coproducts have value as fuel, livestock feed, or soil amendment.	• Storage may result in loss of sugar. • Cultivation practices vary widely, especially "nonconventional" crops.
Starch Crops Grains: corn, wheat, grain sorghum, barley Tubers: culled potatoes, potatoes	Milling, liquefaction, and saccharification.	• Storage techniques well developed. • Cultivation practices are widespread with grains. • Livestock coproduct is relatively high in protein.	• Preparation involves additional equipment, labor, and energy costs. • DDG from aflatoxin-contaminated grain is not suitable as animal feed.
Cellulosic Crop Residues: corn stover, wheat straw Forages: alfalfa, Sudan grass, forage, sorghum	Milling and hydrolysis of the cellulosic linkages.	• Use involves no integration with the livestock feed market. • Availability is widespread.	• No commercially cost-effective process exists for hydrolysis of the cellulosic linkages.

bonding. Releasing the six-carbon sugars for fermentation requires extensive pre-treatment. The optimum method for recovering the fermentable sugars from cellulosic materials has not been commercially demonstrated and various research programs are being pursued to improve the process.

Specifics regarding crop selection and its impact on the project are given below.

Sugar Crops

Processing and Storage Requirements. As previously mentioned, fermentable sugars are easily released from sugar crops. Preparation of the crop for fermentation involves milling or crushing and sugar extraction. Variations of the sugar recovery process are used for different crops, i.e., slicing of sugar beets followed by sugar extraction, pressing of citrus crops, etc. The sugar recovery process involves relatively low capital, labor, and energy costs.

The ease of recovery of fermentable sugars from sugar crops is counter-balanced by a significant disadvantage. The high moisture content of these easily accessible sugars make them very susceptible to infestation by micro-organisms, resulting in crop spoilage during storage. Crop spoilage in turn will result in reduced ethanol production. Sugar loss during storage can be reduced or eliminated by treatment of the extracted sugar solution. Two processes for treating the extracted sugars may be used—pasturization of the solution or evaporation of part of the water to obtain a concentrated sugar solution. Both processes, however, are costly in terms of equipment and energy.

Potential Sugar Crops. The two major sugar crops that have been cultivated at a significant commercial level of production are sugar cane and sugar beets. Other alternative sugar crops that are or could be cultivated in the United States include sweet sorghum, Jerusalem artichokes, fodder beets, and fruits.

Sugar Cane - Sugar cane is considered an attractive feedstock because of its high yield of sugar per acre and a correspondingly high yield of residue known as bagasse, which can be used as fuel to generate process heat. The major drawback of this feedstock is the limited availability of land suitable for economical production of the crop. The potential for expansion of the production of sugar cane to support a large ethanol industry appears limited due to specific geographic conditions necessary to its cultivation.

Sweet Sorghum - The name sweet sorghum refers specifically to varieties of sorghum bicolor. Sweet sorghum is grown on a small scale for syrup and silage. Other sorghums are grown for grain. Sweet sorghum is

a potentially attractive feedstock because of its high yield of ethanol per acre and its adaptability to a wide range of climates and soils. It could make a significant contribution to the feedstock resource for future ethanol production. However, while this feedstock is easier to store than sugar cane, it deteriorates rapidly in storage. The juice from sweet sorghum and milo once extracted could be concentrated and then preserved for later fermentation. Systems for doing this have only been tested at the laboratory level at this time. However, given the initial success of these systems and the development of an appropriate processing infrastructure, sweet sorghum and milo could readily become the principal biomass sources for ethanol production. The full potential of the crop as a feedstock has not yet been realized because genetic improvements and improvements in crop management have not been implemented.

Fodder Beets - Fodder beets are a high-yielding forage crop obtained by crossing two other beet species, sugar beets and mangolds. Fodder beets have higher sugar yields per acre, better storage characteristics, and are less demanding than sugar beets. When fully developed, fodder beets could contribute significantly to the feedstock resource base for ethanol production.

Sugar Beets - Sugar beets tolerate a wide range of climatic and soil conditions and therefore offer a possibility of expanded production to support ethanol production facilities. Widespread expansion of sugar beet cultivation, however, is limited to some extent by the necessity to rotate this with non-root crops on a three-year basis. Sugar beets are considered an attractive feedstock because of their high yield of sugar per acre and correspondingly high yield of beet pulp and beet tops coproducts.

Fermentable Sugars Can Be Recovered from Sugar Beets

Jerusalem Artichokes - The Jerusalem artichoke has shown potential as an alternative sugar crop. It is well adapted to northern climates and a variety of soils, and is not demanding of soil fertility. With expanded production, this crop could make a significant contribution to the feedstock resource for future ethanol production.

Fruit Crops - Fruit crops are not likely to be used as direct feedstock for ethanol fuel production because of their high market value for direct human consumption. The coproducts of processing fruit crops (citrus molasses, for example) could be used as feedstocks because fermentation is an economical method for reducing the potential environmental impact of disposal of untreated wastes containing fermentable sugars. Although prime fruit crops are too valuable to utilize, distressed fruit crops are an excellent feedstock for ethanol production.

Starch Crop

The starch crop used as feedstock for ethanol production includes grains (corn, wheat, barley, grain sorghum, etc.) and tubers (potatoes and sweet potatoes).

Processing and Storage Requirements. Yeasts cannot directly use starch, such as long-branched chains of six-carbon sugars, to produce ethanol. Before fermentation, the starch chains must be broken down into individual or pairs of six-carbon sugar units. This step involves the reaction of the starch-containing material with water (hydrolysis) in the presence of enzymes to produce a simple sugar solution. In the case of grain feedstocks, the process involves milling of the grains to a fine meal to expose the starch, slurrying the meal with water to form a mash, and hydrolysis. Hydrolysis involves the liquefaction of the mash into a solution of high molecular-weight sugars (dextrins) followed by their conversion to fermentable sugars. Both steps are conducted in the presence of enzymes (protein catalysts) under controlled temperature. The preparation process prior to fermentation may include several variations of the general procedure just described. The grains may be prepared under dry or wet milling conditions resulting in the production of various coproducts (germs, oil, hulls, etc.) These process variations are further discussed below, under grains.

The conversion of potatoes or, more generally, tubers to ethanol is similar to that of grains, with minor modifications to the process. The potatoes are washed to remove dirt and soil microbes, sliced, and cooked before the hydrolysis step.

A distinct advantage of starch crops is the relative ease with which these crops can be stored with minimal loss of the fermentable portion. Ease of storage is related to

The Starch in Grains Can Be Converted to Fermentable Sugar in a Simple Two-Step Process

the fact that a conversion step is needed before fermentation. Many micro-organisms including yeasts can utilize individual or small groups of sugar units but not long chains. Some micro-organisms present in the environment produce the enzymes needed to break up the chains, but, unless certain conditions such as moisture, temperature, and acidity are just right, the rate of conversion during storage is very low. When crops and other feeds are dried to about 12-percent moisture (the percentage below which most micro-organisms are not active) the deterioration of starch and other valuable components such as proteins and fats, is minimal. Grains are routinely dried before storage and therefore little risk of loss is expected for these feedstocks. Potatoes can usually be stored about six months before significant losses occur.

Potential Starch Crops

Grains - Potential grain feedstocks include rye, wheat, milo, rice, barley, and corn. Corn presently constitutes the largest potential feedstock supply. The following

Milo Is A Popular Grain Crop In The Southwest

discussion nevertheless is applicable to other grains with relatively minor variations of the process described for corn.

Corn kernels are composed of three major constituents: the germ (2 to 3 percent of the kernel) which is rich in oil and proteins; the hull and bran layers (13 to 17 percent of the kernel) which are rich in protein, cellulose, and minerals; and the endosperm (the remainder of the kernel) which is mostly starch fixed in a matrix of protein. Various food or feed products may be extracted from the kernels through different processing methods. The simplest way of processing corn kernels prior to fermentation to alcohol is shown in Figure III-1. Whole kernels are ground and the resulting meal is treated through slurrying and hydrolyzing to convert the starch contained in the grain to fermentable sugars. The non-fermentable part of the grain or spent grain contains most of the non-starch nutritive elements originally present in the kernels. This coproduct of alcohol production may be retrieved after distillation of the alcohol, as shown in Figure III-1, or at various other steps in the alcohol production chain, such as after hydrolysis or fermentation. More intricate approaches to converting corn into food or feed are the dry- and wet-milling processes.

In some modern dry milling processes, the germ is first removed and treated to produce oil and germ meal used for food or feed (see Figure III-2). After grinding of the degerminated kernel, the hull is separated and used for feed or food additives. The remaining endosperm fraction of the kernel is then milled and used for making a variety of food products, such as corn meal, grits, corn flakes, etc.

A part of the endosperm fraction of the kernel may be used as feedstock for an ethanol production unit. The residue of ethanol production, i.e., the stillage itself, is a valuable coproduct which can be used as animal feed or feed supplement. Figure III-2 therefore indicates that ethanol production can be integrated into a corn dry-milling process having food or feed production as its main objective. The advantage of dry milling over the simple processing described in Figure III-1 is that a variety of high-valued coproducts (oil, food additives, feed, or feed supplements) are produced in addition to ethanol. This advantage, however, is counter-balanced by both the higher capital required for an integrated dry-milling/ethanol production unit and the need for sophisticated marketing to dispose of the coproducts.

In the wet-milling process, the starch contained in the endosperm is further separated from the protein matrix (gluten) in which it is embedded. Figure III-3 illustrates a wet-milling process in which an alcohol production line using part of the starch has been incorporated. Several plants of this type are either in operation or in

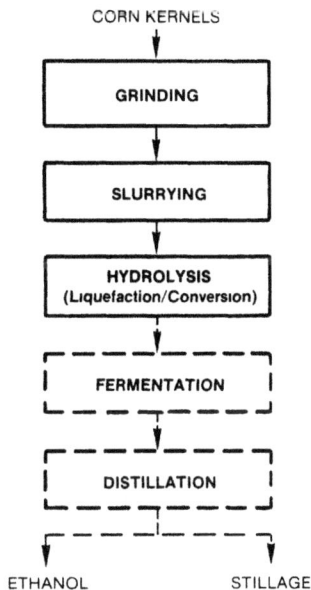

Figure III-1. Simple Processing of Corn Prior to Fermentation

the planning stages. Multiple high-value coproducts are generated during ethanol production via both the wet- and dry-milling processes. The largest traditional wet-milling plants process between 150,000 and 209,000 bushels of corn per day. This would correspond to a production capacity of 380,000 to 620,000 gallons per day for a plant of simplified design.

In a combined wet-milling/fermentation ethanol process, however, emphasis is placed on food or feed production, and the ethanol output is only 70 to 80 percent that of a plant of simplified design with the same corn input devoted primarily to ethanol production. The capital cost of the combined wet-milling/fermentation ethanol plant will be higher than that for a simple ethanol plant, and a careful analysis of the potential benefits of generating valuable feed or food coproducts will be required before selecting the corn wet-milling processing route.

Tubers. Potential tuber feedstocks include potatoes or potato wastes from food processing plants, sweet potatoes, and other starchy tubers. The grinding, slurrying, and hydrolysis operations prior to fermentation are similar to those for grains, but modified to account for the size and moisture content of the tuber feedstock.

Cellulosic Feedstocks

Enormous amounts of cellulosic materials are potentially available for ethanol production. Although no practical process for converting cellulosic materials to

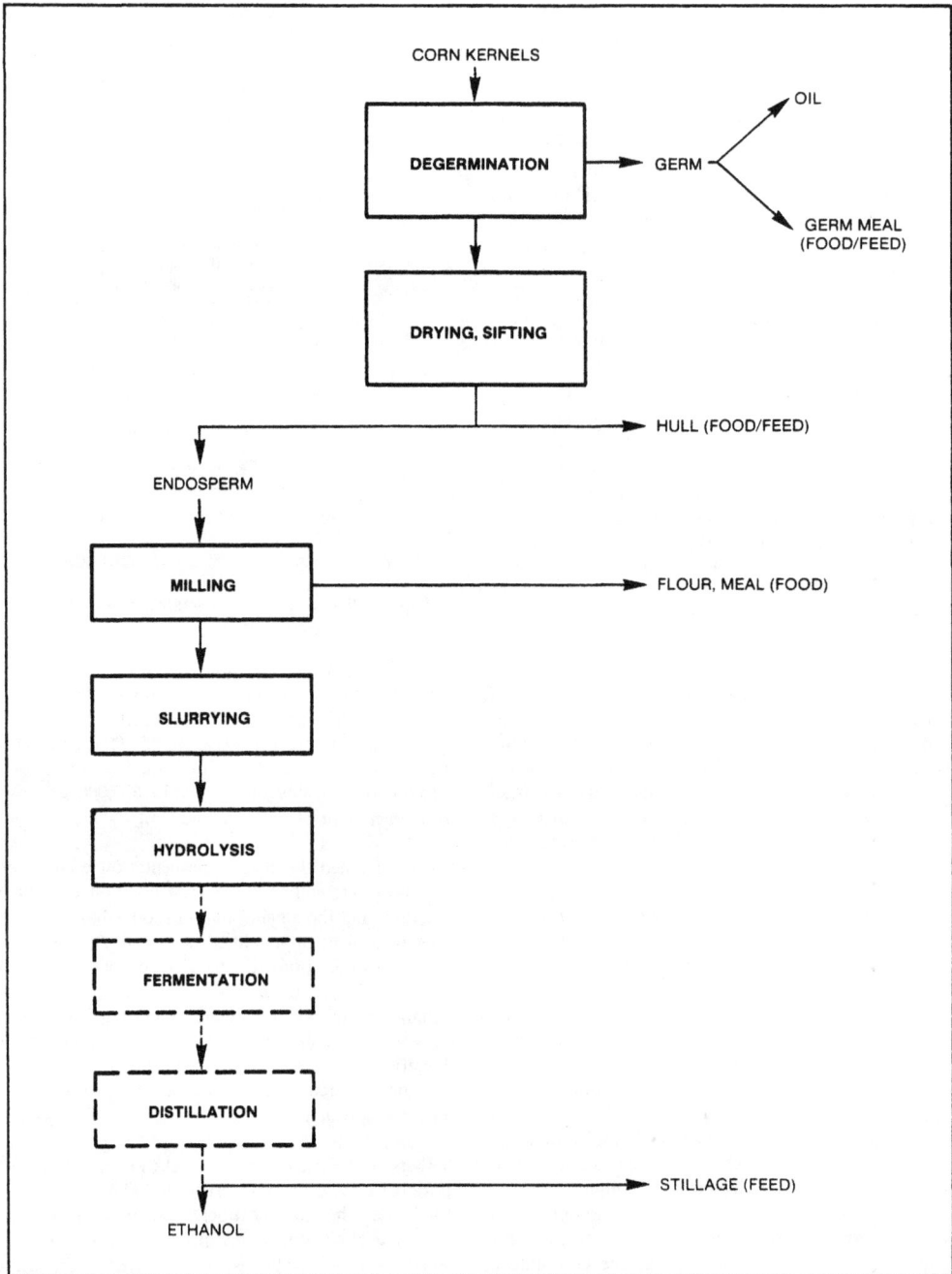

Figure III-2. Dry-milling Process With Ethanol Production

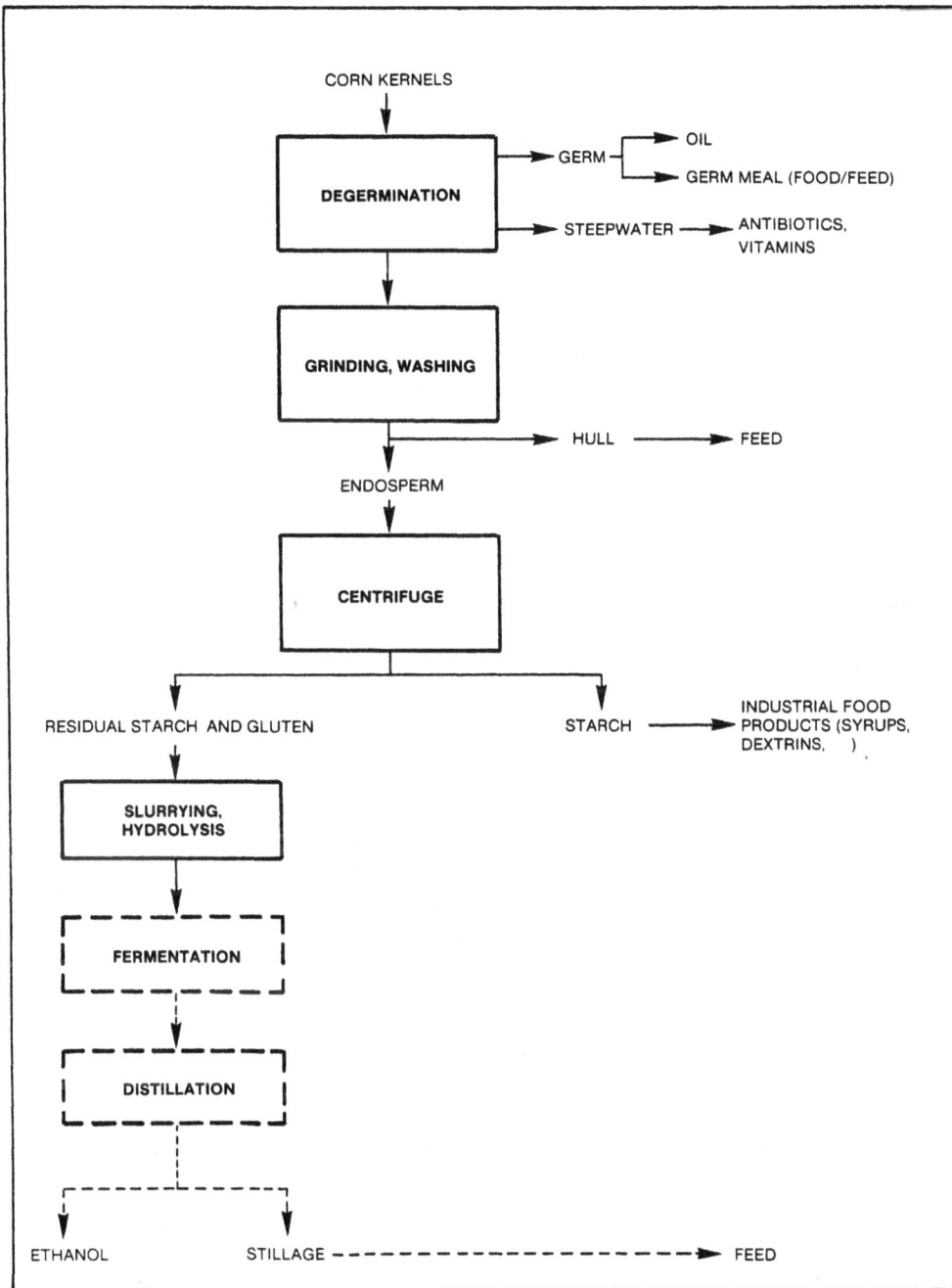

Figure III-3. Wet-Milling Process With Ethanol Production

ethanol on an industrial scale has been demonstrated, research is underway. Cellulosic materials require extensive preprocessing to release the fermentable sugars. Several processes which may include mechanical, chemical, and biological treatments are presently being investigated. This technology, therefore, must be considered a future method of ethanol production. It is impossible to project its real potential or economic feasibility at present. It should be noted, however, that available data suggest that the preprocessing of cellulose prior to fermentation will be different from that practiced for starch feedstocks. Modification of a plant from starch to a cellulosic crop would require the addition of an appropriate front-end processing area to the original plant and very little if any modifications to the existing starch processing front-end. Moreover, that plant then would have the capability of processing both starch/sugar and cellulosic feedstocks, which could be an advantage.

Stover is a Potential Feedstock for Cellulosic Conversion Technologies Presently Under Development

Potential feedstocks include such materials as corn stover, straw, sugar cane bagasse, mill residues, forest residues, and industrial and urban wastes.

YIELDS OF PRODUCTS

Ethanol

The potential conversion rate of feedstocks to ethanol is an important element when selecting a feedstock. The absolute conversion rate expressed in gallons of ethanol per unit quantity of feedstock, such as gallons per bushel of grain, will establish a relationship between the amount of feedstock required and the size of the plant contemplated or, conversely, will determine the size of the plant which can be supported by the resources available in a certain region. Also, when several feedstocks

could be available, all other factors being equal, the feedstock having the highest conversion rate, and requiring the least handling and logistics for procurement of bulky raw material, will be preferred. The relative conversion rate expressed in gallons of ethanol per acre of land devoted to a given crop is also an important aspect of crop selection. This yield per unit land area provides an estimate of the geographic crop area required to support a plant of given capacity, and, therefore, of the potential impact of an ethanol plant on other users of the crop of interest in the region. A comparison of land requirements for feedstocks, when a choice is available, will indicate which feedstock is most appropriate to the farming patterns of the region and which feedstock has the highest potential for long-term availability.

Table III-2 summarizes conversion yield data (quoted as average values) for various feedstocks. Fluctuations around these values will occur due to such factors as the degree of preprocessing of the feedstock prior to fermentation and the quality of the feedstock. As an example, conversion yields as high as 2.7 gallons of ethanol per bushel for corn and other grains, and yields ranging from 75.0 to 60.3 gallons of ethanol per ton for sorghum and citrus molasses, respectively, have been quoted. The conversion yields quoted in the table assume that the main objective of the conversion plant is ethanol production. As indicated in an earlier section, the yields of ethanol fuel per unit feedstock will be lower when ethanol production is integrated with food or feed production processes. (This lower yield of ethanol per unit feedstock does not result from lower sugar to ethanol conversion efficiency, but rather from a different utilization of the components of the feedstock. In these food/feed/ethanol integrated processes, a fraction of the starch—the source of fermentable sugars—is recovered for food and feed uses and, therefore, is not available for ethanol production.)

The data on Table III-2 also show that the wastes of feedstocks or feedstocks of lesser quality (depressed grains, for instance) may have yields lower than those mentioned in the table. For instance, potato wastes from french fry processing plants produce about half of the fermentation ethanol obtained from whole potatoes for equal feedstock inputs. These lower yields per unit input and the associated costs of handling larger quantities of feedstock must be compared to the lower feedstock costs for waste materials. In some cases, the utilization of wastes may result in a credit because ethanol production may provide a method for disposal of the wastes from food-processing plants. The ethanol production rates per acre are based on averaged U.S. yields of the various crops per acre. Significant regional differences in yields have been recorded. As an example, in 1977, while the United States average yield of corn for grain was 90.8 bushels per acre, yields of 29 bushels

Table III-2. Probable Commercial Yields of 200-Proof Ethanol From Various Feedstocks

FEEDSTOCK	GALLONS PER BUSHEL*	GALLONS PER TON	GALLONS PER ACRE**
Corn	2 5	89	228
Grain Sorghum	2 4	86	135
Wheat	2 4	80	74
Rye	2 2	79	54
Oats	1 0	64	57
Barley	2 1	88	92
Rice	1 8	80	175
Potatoes	0 69	23	299
Potato Wastes	—	13	—
Sweet Potatoes	0 94	34	190
Yams	0 75	27	NA
Jerusalem Artichokes	0 60	20	NA
Sugar Beets	—	22	412
Sugar Cane	—	15	555
Sweet Sorghum	NA	NA	NA
Apples	0 35	14	NA
Peaches	0 28	12	NA
Molasses	—	68	NA

Average Yields
**Based on Average 1977 Crop Yields

SOURCE U S Department of Agriculture, "Small-Scale Fuel Alcohol Production," prepared with the assistance of Development Planning and Research Associates, Inc , Manhattan, KS, 66502, March 1980, Washington, D C 20250

per acre were reported for Alabama and yields of 116 and 105 bushels per acre were reported for Colorado and Ohio, respectively.

On the basis of the data from Table III-2, a 50-million-gallon-per-year plant would require about 21.3 million bushels of corn per year, the crop harvested from about 233,000 acres of farm land having a productivity of 91 bushels per acre. Assuming that the conversion plant is located in the center of a corn production area and that the crop land is about 50 percent of all land, the ethanol plant would consume all the grain produced within a circle of about 30.4 miles in diameter. For the range of productivities mentioned above (i.e., 30 to 130 bushels per acre-year) the diameter of the land area required to sustain the ethanol plant could range from about 13 to 27 miles. These simplified considerations give an order of magnitude for the crop area which will be affected by the installation of an ethanol plant, and should be kept in mind when siting the plant to avoid direct competition with other users of the feedstock.

The table also shows that sugar crops, such as sugar cane grown at present in the United States, have a higher ethanol productivity per acre than the starch crops presently produced. This yield advantage of sugar crops, however, is counter-balanced by the fact that sugar crops spoil more quickly in storage.

Of the starchy feedstocks listed in the table, corn and potatoes have the highest potential for fermentation ethanol production per unit land area.

Coproducts

The fermentation process resulting in the production of ethanol also yields several coproducts, including carbon dioxide, fusel oil, yeast, and stillage. Other coproducts such as food or feed components also may be generated when ethanol production is integrated in a food processing chain such as the dry or wet milling of grains described earlier. The coproducts of ethanol production may have a beneficial effect on the overall economics of ethanol production if they can be recovered economically in significant quantities, and if a commercial market exists to absorb these coproducts. The amount, quality, and therefore, market value of the coproducts vary widely and depend on the feedstock and the processing steps used in producing ethanol.

Carbon Dioxide. Carbon dioxide is used in carbonated beverages, in fire extinguishers, in the manufacture of dry ice, and as a food preservative. The recovery of carbon dioxide is likely to be practical only for large-scale plants and only when a local market is readily available. In most cases, carbon dioxide recovery will not be economically justifiable and, except for special conditions, no credit can be expected from carbon dioxide.

Fusel Oil. Fusel oil is a poisonous liquid mixture of alcohols consisting mostly of normal amyl and iso-amyl alcohol. Where corn is the feedstock for ethanol production, less than one percent of the total amount of alcohol produced is fusel oil. Fusel oil can be used as a denaturing agent for the ethanol produced.

Yeast. Yeast present in the fermentation medium may be recycled or recovered for commercial uses. In either case, its value as a coproduct should be determined. If yeast is not recovered, it will contribute to the high protein content of the stillage.

Stillage. The stillage from fermentation contains fibrous carbohydrate material, high-protein yeast, proteins from the original feedstock, and non-fermentable solids and solubles including various minerals and other nutrients. The actual nutritional value of the stillage will vary among feedstocks and with the ethanol production process used. Whole thin stillage usually contains about 90 percent water. Although whole stillage is currently used as animal feed or feed supplement, its usefulness is limited by the high water content which prevents animals from consuming large quantities of it. Moreover, whole stillage is very susceptible to microbial degradation and must be delivered and consumed within a short time after removal from the distillery. In summer, it is recommended that the stillage be consumed within 24 hours after recovery from the still. This can result in complex and expensive logistic problems for commercial-scale operations generating large quantities

of stillage daily. Stillage may be dried to avoid or reduce the storage, marketing, and distribution problems.

Four types of coproducts are presently recovered from the stillage of distilleries using grain as feedstock: Distillers Dried Solubles (DDS), obtained by evaporating and drying the thin stillage to recover the minerals and nutrients in solution; Condensed Distillers Solubles, obtained by concentrating the thin stillage to a semi-solid form; Distillers Dried Grains (DDG), obtained by separating the coarse grains from the whole stillage and drying the solid fraction recovered; and Distillers Dried Grains with Solubles (DDGS), which result from the blending of DDG and DDS prior to drying. The paths of utilization of stillage and its coproducts are illustrated in Figure III-4. DDG or DDGS is generally marketed at 10 percent moisture and has a protein content of the order of 25 to 28 percent dry basis, i.e., lower than soybean meal (about 45 percent) but significantly higher than corn (about 9 percent). The dried coproducts recovered from stillage are easier to store, transport, and market than fresh stillage.

Table III-3 summarizes the production rates of stillage and stillage-derived coproducts for some of the major feedstocks presently available for ethanol production. A 50-million-gallon-per-year ethanol plant using corn feedstock will generate about 500 million gallons of fresh stillage annually or about 1.3 million gallons of stillage daily. The same plant will produce about 163,000 tons of dried distillers products annually or

Table III-3. Production Rates of Stillage and Stillage-Derived Products for Ethanol Production Feedstocks

FEEDSTOCK	UNIT INPUT	WEIGHT INPUT[1] (LB)	VOLUME OF STILLAGE[2] (GAL.)	WEIGHT OF STILLAGE[2] (LB)	WEIGHT OF DISTILLERS COPRODUCTS[3] (LB)
Corn	0.38 bu	21.5	10 4	92.7	6.5
Wheat	0.38 bu	23.1	10.4	94.2	8.0
Grain Sorghum	0.38 bu	21.5	10.4	92.7	6.5
Potatoes	0.71 cwt	71.4	7.0	58.4	10.6
Sugar Beets	0.049 ton	98.5	10.3	85 6	13.0
Molasses	2.5 gal	29.3	4.8	39.0	—

[1] These yields are slightly different from average yields quoted in Table III-2 and reflect expected variations in ethanol production for various feedstocks.

[2] Volume and weight of stillage will depend on amount of water used in fermentation mash and amount of water recycled to process after distillation.

[3] 10% moisture for grain feedstocks; 75% moisture for sugar feedstocks.

SOURCE

M L David, G S Hammaker, R J Buzenberg, and J P Wagner, "Gasohol Economic Feasibility Study", Report prepared for Energy Research and Development Center, University of Nebraska, Development Planning and Research, Inc , p 261, July 1978, Manhattan, KS, 66502 U S Department of Agriculture, "Small Scale Fuel Alcohol Production", prepared with the assistance of Development Planning and Research Associated, Manhattan, KS 66502, March 1980, Washington, D C 20250

COMMERCIAL SCALE ETHANOL PRODUCTION AND FINANCING

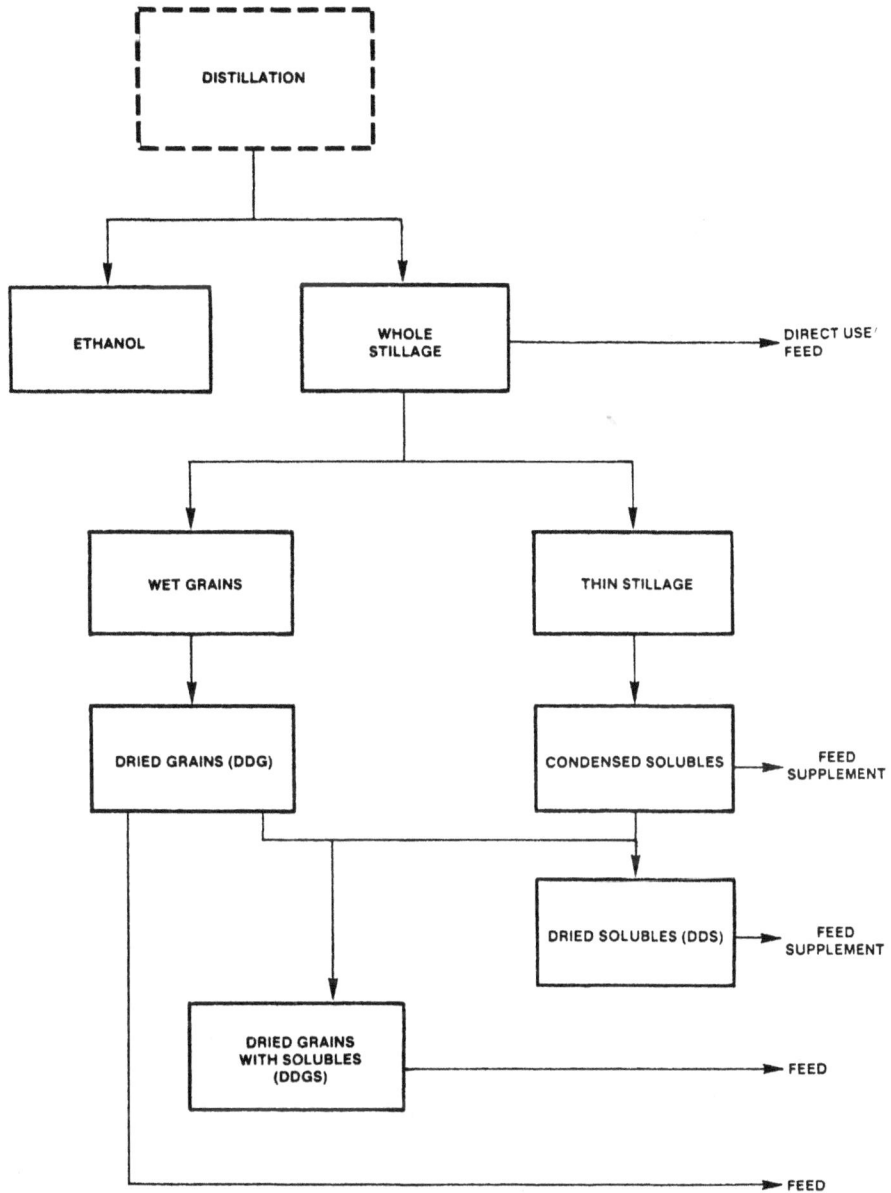

Figure III-4. Paths of Utilization of Residual Stillage from Grain Feedstock Distillery

about 450 tons of products daily. This comparison suggests that treatment of the stillage to dried distillers products will simplify the logistics of marketing the coproducts of a commercial-size plant.

The output of coproducts shown in Table III-3 refers to a simple fermentation ethanol plant design, i.e., a plant designed primarily for ethanol production. The outputs from an ethanol production plant integrated with a grain processing plant (dry- or wet-milling process) will be quite different. As an example, the treatment of one bushel (56 lbs) of corn through a wet-milling process will result in the production of about 9.2 pounds of feed at about 21 percent protein; 2.7 pounds of gluten meal at about 60 percent protein, 3.5 pounds of germ at about 50 percent oil; and 31.5 pounds of starch. This starch can be partially used for food or totally converted to ethanol. In the latter case, about 2.5 to 2.7 gallons of ethanol would be produced per bushel of corn treated. As most of the food or feed components of the grain has been removed through the wet milling process, only about 3.2 pounds of residual distillers coproducts would be recovered from the stillage. The above-mentioned outputs of a wet-milling process are only indicative and will vary according to the design of the system.

The major market for stillage or stillage-derived coproducts is animal feed or animal feed supplement. The quantity of stillage or stillage-derived products, which can usefully be consumed by animals, is a function of a number of factors, such as: the type of animal, diet needs within its lifecycle, ability to digest and absorb the products, and other factors. Considerable research is in progress to optimize the use of distillers residues in formulating animal diets. There is, therefore, some uncertainty as to the exact amount of stillage or its derivatives which can be tolerated by various types of animals.

Some typical quantities of coproducts which could be fed to various types of animals in combination with other feeds required to supply a balanced ration are

shown in Table III-4. The animal population required to absorb the distiller's coproducts generated by ethanol plants of various sizes is reflected in the data of Tables III-3 and III-4. Some typical values are shown in Table III-5.

The table shows that the animal population required to absorb the distillers coproducts from commercial-size plants is large. The problem of marketing these coproducts will have to be carefully evaluated before choosing an ethanol plant site.

The available markets for stillage will probably be expanded several fold in the next few years by technological developments which have already been proven at the laboratory level and are going into the pilot testing at this time. Membrane processes show promise for the low cost denaturing of stillage together with the capability of reducing its salt content. Denaturing costs by membrane processes will be of the order of a few dollars per thousand gallons as contrasted with well over twenty dollars when accomplished by evaporators. These membrane denaturing plants can also be quite small as contrasted to evaporators which are efficient only in relatively large plants. Membrane processes remove salts, and it has been shown that the salt content is one of the elements which limits the amount of distillers coproducts which can be fed to animals. There is evidence that stillage without excess salt can be fed to animals as readily as any animal feed material.

AVAILABILITY OF FEEDSTOCKS

Overview of Feedstock Availability

Near-Term Availability. Table III-6 summarizes estimates of the quantities of feedstocks potentially available at present or in the near term in the United States. These estimates assume that none of the feedstocks is diverted from its present food or feed uses (i.e., the amounts shown in the table are essentially surplus production) and that present patterns of agriculture are

Table III-4. Animal Consumption of Distiller's Coproducts

ANIMALS	STILLAGE 10% SOLIDS (GAL/DAY/ANIMAL)	STILLAGE 20% SOLIDS (GAL/DAY/ANIMAL)	DRIED DISTILLERS GRAINS WITH SOLUBLES (LB/DAY/ANIMAL)
Calf (550 lb)	6.3	3.1	5.8
Steer (770 lb)	9 2	4.6	8.5
Cow (1,300 lb)	7.2	3.6	6.6
Pig (60 lb)	1.2	0.6	1.1
Pullet (3.7 weeks)	—	—	0.13
Pullet (7.5 weeks)	—	—	0.22

SOURCE U S Department of Agriculture, "Small-Scale Fuel Alcohol Production," prepared with the assistance of Development Planning and Research Associates, Inc , Manhattan, KS, 66502, March 1980, Washington, D C 20250

COMMERCIAL SCALE ETHANOL PRODUCTION AND FINANCING

Table III-5. Animal Population (in Thousands) Required to Utilize the Distillers Coproducts from Various Sized Ethanol Plants

PLANT SIZE IN MM GAL/YR
COPRODUCT

ANIMAL	5		15		25		50		100	
	STILL-AGE[1]	DDGS	STILL-AGE	DDGS	STILL-AGE	DDGS	STILL-AGE	DDGS	STILL-AGE	DDGS
Calf (550 lb)	25	17	75	51	125	85	250	170	500	340
Steer (770 lb)	17	12	51	86	85	60	170	120	340	240
Cow (1,300 lb)	22	15	66	45	110	75	220	150	440	300
Pig (60 lb)	131	90	393	270	655	450	1,310	900	2,620	1,800
Pullets (3 7 Weeks)	NA	758	NA	2,274	NA	3,790	NA	7,580	NA	15,160
Pullets (7 5 Weeks)	NA	448	NA	1,344	NA	2.240	NA	4,480	NA	8,960

Assumes corn feedstock and 330 days production per year
[1] Ten percent solids

SOURCE This data was taken from Table III-3 and III-4

not changed. The table shows that a goal of over one billion gallons of ethanol annually is a realistic objective in the near future. The table also shows that corn, grain sorghum, and other resources which include grains other than corn constitute the backbone of our resources. Citrus wastes, although significant in quantity, are available only on a seasonal basis, which limits their economic attractiveness.

Long-Term Availability. If the goal proposed by the President of 1.8 billion gallons of ethanol per year is to be reached, other sources of feedstock must become available. Traditionally, U.S. agriculture has shown a great degree of flexibility in adjusting to new market demand and technologies. Increasing demand for corn feedstock for ethanol production accompanied by increased availability of distiller's grains feed supplement would result in a gradual shift from soybean and grain sorghum production to corn or other grains. Similarly, the existence of a steady and significant market for ethanol feedstocks may induce farmers to develop marginally used land areas.

Recent estimates by the USDA[1] indicate that about 78 million acres presently not used for crops have a high potential for cropland development. The development of these areas into cropland could translate to a 33 percent increase of corn production or a 43 percent increase of soybean production in the Northern Plains and Southern Plains, respectively, and a 350 percent in-

crease of soybean production in the Southeast. Despite increasing national and foreign demand for farm products, the potential exists to significantly increase the ethanol feedstock resources. It should, however, be mentioned that some development problems such as erosion, periodic flooding, and rocky soil may have to be resolved when bringing some of the potential areas into cropland use.

The above discussion indicates that on a national level, sufficient resources exist to justify an ethanol program in the near term and prospects for maintaining or even expanding an ethanol program in the long term appear favorable.

Regional Availability of Feedstocks. The availability of feedstocks varies from region to region, as climatic and soil characteristics result in specialized crop production. Also, while some crops such as corn are grown in many states, some regions or states will register a surplus of production and others, although producers, will import corn. Surplus regions or the proximity of surplus regions where competition for the feedstock with other users is reduced and prices are less sensitive to changes in demand should obviously be preferred when siting a plant.

---

[1] L K Lee, "A Perspective on Cropland Availability", U S Department of Agriculture-Economics, Statistics, and Cooperative Services, Agricultural Economic Report No 406, Washington, DC, 1978

Figure I-5 shows the distribution of potential ethanol feedstocks production by USDA farm production regions. The productions indicated in notes to the table are the totals. The quantity of feedstock available for ethanol production is only a fraction of the total production, i.e., surplus, distressed crops, or wastes from processing. The data shows that corn, wheat, and grain sorghum are the major potential feedstocks and that half or more of these crops are produced in the Northern Plains, Lake States, and Corn Belt regions. Sugar beet production is spread over many regions but the amounts potentially available are small. Sugar cane production is geographically very limited and its feedstock potential is very small. Potatoes (surplus, distressed crops, or wastes from processing) and wastes from food products processing could supply the feedstock needed in a variety of regions although the total potential resource is limited. Food processing wastes also have the drawback that they are often seasonal feedstocks. As a result, plants relying on these feedstocks could be idle for part of the year if no alternative feedstock is available to fill the idle periods.

Sweet sorghum, a potentially very attractive feedstock, has been grown in several regions as indicated in Figure I-5. This crop offers an alternative for other feedstocks in those regions.

Factors Affecting Feedstock Availability

Climate and Productivity. Increased soil management and particularly increased usage of nitrogen fertilizers have resulted in significant increases in average productivity of grains, especially corn. This increased soil management, however, has also resulted in increasing the amplitude of fluctuations in grain yields due to climate variations because positive response to fertilizer and other management techniques depends on favorable weather conditions.

Figure III-5 shows historical trends of corn yields per acre during the 1963 to 1978 period. During the 1970's, yields varied between 72 and 100 bushels per acre, i.e., a variation of about 11 to 26 percent below and above the average productivity of 81 bushels per acre. Such fluctuations in productivity and production will have an impact on both food and fermentation ethanol feedstock prices and availability.

Market Demand. Markets for grains include the domestic and foreign food and feed markets and ethanol production. These markets are defined and predictable to a certain extent. However, unforeseen variations resulting from foreign crop deficits, cancellation of foreign deliveries as a result of political decisions, emergency situations, or climate may change the yearly supply and demand balance for grains in the

Table III-6. Quantities of Feedstocks for Ethanol Production Potentially Available in the Near-Term and Their Ethanol Equivalent

Feedstocks	Quantity (Million Dry Tons/Year)	Ethanol Equivalent (Million Gallons/Year)
Corn	1.8	180
Grain Sorghum	0.3	30
Citrus Waste	1.9	210
Whey	0.9	90
Others	1.7	150
TOTAL		**660**

Source: U.S Department of Energy, Assistant Secretary for Policy Evaluation, "Report of the Alcohol Fuels Policy Review", DOE/PE-0012, Washington, DC, June 1979.

United States. It should be noted that once a significant ethanol industry is established, it will provide a predictable market for feedstocks which can be included in the projections of future United States markets.

Government Policies. As a result of the shift toward capital- and energy-intensive agricultural practices, U.S. agriculture has shown periods of overproduction. The Federal price support and stabilization policy designed to offset this overproduction has included long-term land retirement programs to remove cropland from intensive cultivation and renewable yearly set-aside programs. Little or no grain cropland is removed from production under either the long-term diversion or yearly set-aside program at present.

Historic Trends of Feedstock Supplies

Table III-7 shows trends of corn production, utilization, and carry-over stocks from 1965 to 1978. During the 1970's, an average of about 65 million acres was harvested and the data of Table III-7 reflects the variations in productivity reported in Figure III-5. During the 1972 to 1974 period, a sharp decline in productivity and total production was recorded. During that period, foreign demand increased significantly. As a result, stocks, i.e., one of the elements through which fluctuations in supplies and prices can be damped, declined sharply. Concurrently, as discussed below, prices rose by over 60 percent between 1972 to 1974. The supply and price of other feedstocks could undergo similar fluctuations.

These fluctuations in feedstock supplies are a fact of life which must be faced by ethanol producers. Food proc-

Table III-7. Historical Trends of Corn Production, Utilization, Stocks, and Costs

Year	Production (Million Bushels)	Utilization (Million Bushels)	Stocks[1] (Million Bushels)	Price[2] ($/Bushel)
1960	3,907	3,678	1,787	---
1965	4,103	4,409	1,147	1.32
1970	4,152	4,494	1,005	1.47
1971	5,646	5,188	667	1.23
1972	5,580	6,000	1,126	1.91
1973	5,671	5,896	708	2.95
1974	4,701	4,826	484	3.12
1975	5,829	5,793	361	2.75
1976	6,266	5,784	399	2.30
1977	6,357	6,110	884	---
1978	7,082	---	---	---

[1]Carryover Stocks, Oct. 1
[2]Price on the Chicago Market

essing plants such as for coffee face some of the same uncertainties in supplies, fluctuations due to variations in productivity, and changes in policies by producers of the raw material. Potential ethanol producers must recognize this and explore ways of reducing the impact of these unavoidable fluctuations on the operation (and profitability) of their plant.

COST OF FEEDSTOCK

Historic Perspective of Feedstock Costs

Table III-8 shows wholesale prices for corn, soybean meal, and distiller's grains on the Chicago market for the 1970 to 1979 period for January and July of each year. Also included is the approximate price for stillage, derived from the historical relationship between stillage price and those for soybean meal and corn. The data of the table show that significant fluctuations in prices of corn and the coproducts of distillation occur on a yearly basis. These have been related earlier to fluctuations in production as well as changes in demand (national and foreign). Seasonal variations also occur as shown by comparing the data for January and July; in many cases, summer prices are higher than winter.

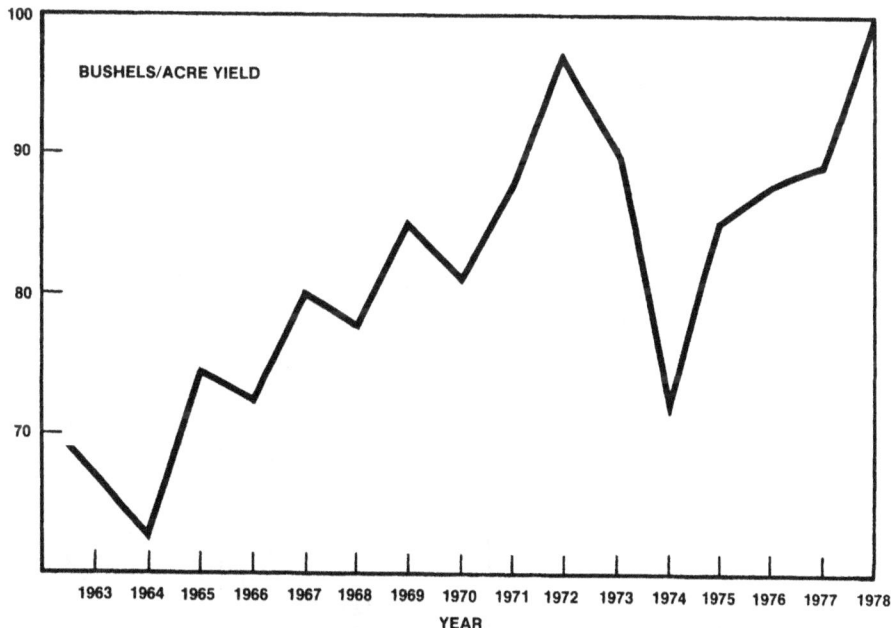

SOURCE U S Department of Agriculture, "Statement of Bob Bergland, Secretary of Agriculture before the Committee on Science and Technology, Subcommittee on Energy Development and Applications, Honorable Richard Ottinger, Chairman, House of Representatives, USDA 1032-79, May 4, 1979

Figure III-5. U.S. Corn Crop Yields, 1963 to 1978

Another important fact apparent from the data is the correlation between DDG or stillage prices and those of corn and soybean meal: at constant soybean meal price, DDG or stillage prices increase with that of corn and vice versa. At constant corn price, the price of DDG or stillage increases and falls with that of soybean meal. This is an important relationship since under certain circumstances, an increase in corn feedstock price resulting from the expected fluctuations in price for this commodity could be partly compensated by the increased value of the coproduct from ethanol production.

As an example, the data of Table III-8 for January 1973 and 1974 show that the per-gallon cost of raw material for ethanol production would have jumped from $0.62 in 1973 to $1.09 in 1974, assuming a conversion rate of 2.5 gallons per bushel. The value of DDG, i.e., a credit toward feedstock price, simultaneously would have increased per gallon of ethanol from $0.33 in 1973 to $0.45 in 1974 on the basis of a production of 6.5 lbs of DDG per gallon of ethanol. The net price of the raw material (feedstock price minus credit) would therefore have been $0.29 and $0.64 per gallon in 1973 and 1974, respectively. This is a large increase in feedstock price, i.e., 121 percent, but nevertheless smaller than the increase which would have been incurred had the price of DDG been kept at its 1973 level, i.e., 162 percent.

These related feedstock and coproduct prices also could be advantageous to the producer. Between 1975 and 1976 (January prices, Table III-9), the price of corn decreased by about 22 percent while that of DDG remained essentially constant. As a result, the net price

per gallon for the feedstock (feedstock minus credit) decreased from $0.98 to $0.69 over the same period, i.e., a decrease of about 30 percent.

These two examples indicate that the impact of fluctuations in feedstock prices on alcohol production costs, while potentially very significant, must be evaluated in the context of the overall market situation for the grains and coproducts. The discussion also suggests that the profitability of the ethanol plant should be examined on a plant-life basis, i.e., over a period long enough that the fluctuations in feedstock and coproduct prices compensate each other. This somewhat unpredictable economic performance may be difficult to make acceptable to potential investors. To a certain extent, the investor can anticipate the fluctuations in feedstock prices around a generally predictable trend. (As an example, see Figure III-6 for corn average selling prices.)

Methods of Procurement of the Feedstock

As indicated in the previous section, it is necessary for the ethanol producer to explore methods of procurement of feedstocks which will ensure a continuous flow of raw material despite expected fluctuations in production of the crop.

Feedstock purchase on the commodity market is the most direct method of raw material procurement. This option, however, puts the ethanol producer in direct competition with other users of the commodity and therefore, the commodity price will be extremely sensitive to supply and demand relationships.

Table III-8. Wholesale Prices of Corn and Other Commodities on the Chicago Market

YEAR	JANUARY 1					JULY 1				
	#2 YELLOW CORN	SOYBEAN MEAL	DISTILLERS GRAINS	STILLAGE		#2 YELLOW CORN	SOYBEAN MEAL	DISTILLERS GRAINS	STILLAGE	
	$/BU[1]	$/TON	$/TON	$/TON	$/1000 GAL[2]	$/BU	$/TON	$/TON	$/TON	$/1000 GAL
1979	2 24	80.00	194 60	141.00	45.76	3.04	108.60	218.30	151.10	53.03
1978	2.18	77.90	182.60	130.00	43.64	2.40	85.70	186.10	120.00	45.15
1977	2.48	88.60	209.70	140.00	49.09	2.20	78.60	184.70	140.00	43.64
1976	2.58	92.10	136.00	105.00	40.61	3 05	108.90	232.70	130.00	55.45
1975	3.32	118.70	139.80	107.50	42.73	2.81	100.00	124.30	103.00	40.61
1974	2.72	97.10	174.80	138.00	46.67	3.06	109.30	107.80	92.50	39.41
1973	1.54	55.00	189 10	100.00		2.38	85.00	306.60	138.00	
1972	1.20	42.90	87.50	64.00		1.26	45.00	106.20	71.00	
1971	1.58	56.40	85.90	70.00		1.52	54.30	88.00	67.00	
1970	1.23	44.00	84 50	64.00		1.39	49.60	85.70	61.00	

[1] Assumes 56 lb per bushel.

[2] Approximate price derived from historical relationship with soybean meal and corn prices.

SOURCE USDA "Small-Scale Fuel Alcohol Production," Washington, D C, March 1980

COMMERCIAL SCALE ETHANOL PRODUCTION AND FINANCING

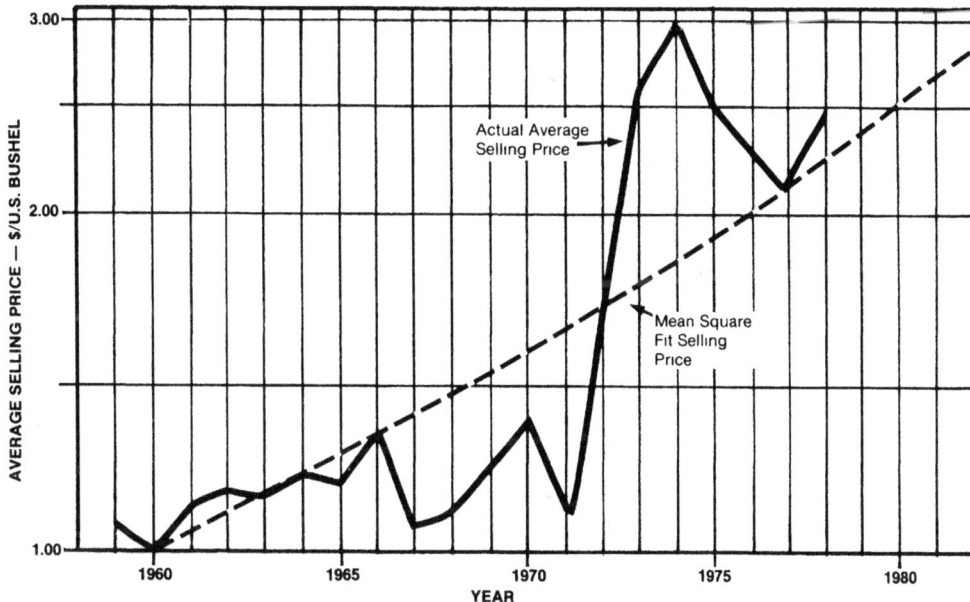

Figure III-6. The Selling Price of Corn

SOURCE Raphael Katzen Associates, "Grain Motor Fuel Alcohol Technical and Economic Assessment Study,"
Department of Energy Contract No EJ-78-C-01-6639, December 1978

Another option is to secure long-term contracts with farmers for the supply of a fraction of the needed raw material, and rely on the commodity market for the remainder. In this case, purchases on the commodity market can be scheduled to benefit from downward price trends. Long-term contracts with farmers may have to include indexing clauses to account for inflationary trends. Because of the high energy inputs required by intensive farming (fuels and fertilizers), specific indexing agreements addressing these energy costs may have to be negotiated. Also, direct supply contracts with producers may raise problems of storage of the delivered raw material.

Other options which address specific or special circumstances may be considered. A cooperative ethanol production facility including farmer members may require delivery from the farmer members of a specified amount of raw material over a certain time as part of the cooperative charter. Agreements could be negotiated between an ethanol producer and food processors whereby a long-term supply of wastes is guaranteed in exchange for the service of disposing of the wastes through ethanol production.

Procurement of feedstocks for a commercial-size plant is a critical function which impacts drastically on the economics of ethanol production. The organizational chart of the proposed plant should therefore identify this function and the needed specialized personnel.

Feedstock Cost Basis

Table III-9 shows average feedstock costs per gallon of alcohol for some of the major potential feedstocks presently considered. A 15-year average (1963 to 1977) of prices paid to farmers for the commodities, converted to 1979 dollars by using the GNP price deflator, was used, as were average yields of ethanol per unit of feedstock (data from Table IV-4).

The table shows that corn, grain sorghum, and rye are the least expensive feedstocks. Sugar cane and sugar beets compare to wheat and barley. Fruit and potato wastes have been assumed to be available at $1/cwt (hundred pounds) or $20/ton. On that basis, the cost of these feedstocks is comparable to those of sugar crop and wheat.

Factors Affecting the Cost of Feedstocks

Productivity or production, supply and demand relations, and to a certain extent, government policies such as the agricultural stabilization program, will affect feedstock availability and therefore feedstock prices. These have been discussed earlier. Two other factors may also affect the cost of feedstocks or the overall cost of ethanol production: government incentives promoting the use of wastes and the cost of collection of the feedstock for large-scale operations.

Table III-9. Historical Trends

FEEDSTOCK	AVERAGE YIELD (GAL./TON)	AVERAGE 1963 TO 1977 PRICE PAID TO FARMERS (1979 $)	$/GAL.
Rye	78.8	2.36/bu	1.07
Grain Sorghum	79.5	2.40/bu	1.08
Corn	84.0	2.69/bu	1.14
Barley	79.2	2.35/bu	1.24
Wheat	85.0	3.46/bu	1.36
Oats	63.6	1.46/bu	1.43
Sugar Beets	22.1	31 58/ton	1.43
Fruit Wastes	13.0	20.00/ton	1.54
Sugar Cane	15.2	23.68/ton	1.56
Potato Wastes	12.5	20.00/ton	1.60
Rice	79.5	229.00/ton	2.88
Potatoes	22.9	99.60/ton	4.35
Sweet Potatoes	34.2	195.60/ton	5.72

SOURCE USDA "Small-Scale Fuel Alcohol Production," Washington, D C , March 1980

Government Incentives. Prior to the Windfall Profits Tax Act, Federal law permitted the financing of solid waste disposal facilities through tax-exempt industrial development bonds. The Act expands the definition of solid waste disposal facilities to include property used primarily to process solid waste to alcohol. To be qualified as an alcohol producer under this provision, a facility must satisfy three requirements:

- The primary product obtained from the facility must be alcohol (there is no minimum proof requirement)

- More than half of the feedstock used in the production of alcohol must be solid waste or a feedstock derived from solid waste

- Substantially all of the solid waste-derived feedstock must be produced at a facility located at or adjacent to the site of the alcohol-producing facility and both facilities must be owned and operated by the same person (ownership is meant for tax purposes)

The Windfall Profits Tax Act specifies that such bonds will not be tax-exempt if they are guaranteed by the Federal government or if any payment of the principal or interest is made with funds from a Federal, State, or local energy program. The Windfall Profits Tax Act also expands the definition of a solid waste disposal facility that can be financed through tax-exempt industrial development bonds to include solid waste disposal facilities that produce steam or electricity.

These provisions of the Windfall Profits Tax Act, while not directly affecting the feedstock cost, nonetheless encourage the development of ethanol-producing plants integrated with solid waste-producing operations. The overall economics of an ethanol plant using wastes from food processing plants could therefore be more attractive than suggested by the feedstock costs shown in Table III-9. The provisions of the Act also can favor a plant using wastes to produce the process steam and/or electricity required by an ethanol plant. The use of bagasse, i.e., cellulosic residue from sugar cane processing, to fuel process steam boilers would benefit from the provisions of the Act, for instance.

Grain Assembly Costs. A commercial 50-million-gallon-per-year plant will require about 20 million bushels of grain annually. The cost of gathering this quantity of feedstock will depend on the density of production of grain per unit land area (bushels-per-acre or tons-per-square mile) and on the fraction of the grain which can be purchased for ethanol production.

The estimated crop production areas required to support a 50 million-gallon-per-year plan (corn feedstock) are shown in Figure III-7 for the States of Illinois, Indiana, and Ohio. The density of corn production per unit land area was estimated for each State by dividing the corn grain production for 1978 (USDA -"Agricultural Statistics 1979" - Washington, DC, 1979) by the area of each state. Production densities ranged from about 21,000 bushels per square mile for Illinois, to about 9,000 bushels per square mile for Ohio. To derive the data in the figure, it is further assumed that ten per-

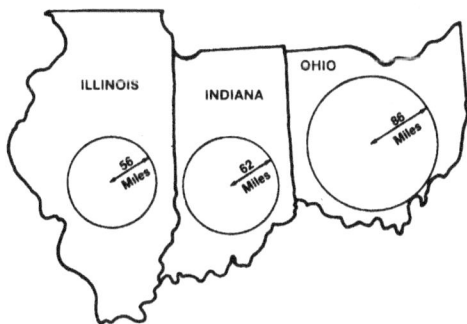

*Assumptions 1978 productivities. uniform distribution of crop-producing land within the state, 10 percent of crop available for ethanol production

Figure III-7. Corn Production Area Required to Support a 50-million-Gallon-Per-Year Plant (Corn Feedstock)*

cent of the corn produced is available for feedstock use. The average hauling distances (i.e., radius of the circle supplying one half of the feedstock required) range from 40 miles for Illinois to 61 miles for Ohio. On the basis of the data shown in Figure IV-3, corn hauling costs would add about 4 and 5 cents per gallon to the cost of ethanol produced in Illinois and Ohio, respectively, for the corn assembly scenario used in the present example. Although these supplementary costs are small in absolute value, they may become significant in terms of overall alcohol production economics.

These transportation costs will become more significant as the capacity of the ethanol plant increases. It therefore will be necessary to carefully evaluate the optimum plant size for which the economy of scale associated with large plants is overridden by the supplementary cost for assembling the feedstock and disposing of the coproducts and effluents.

The cost of assembling the feedstock also will have to be weighed against the cost of marketing the products of the plant: proximity to the markets for the products, i.e., ethanol, and coproducts may be economically more attractive than proximity to the feedstock source. This problem will be discussed further in the marketing section (Chapter IV).

Trends and Fluctuations in Price

The data from Table III-8—corn prices for January 1 and July 1, 1970 to 1979 on the Chicago Market—are plotted on Figure III-8. The full line shows the trend in prices as obtained by regression of the set of data points. The slope of the line suggests a general price increase at a rate of about 8.4 percent per year, i.e., a rate

comparable to the average inflation rate during the period considered.

During the period, wide fluctuations around the general trend line are observed. On the basis of a yield of 2.6 to 2.7 gallons per bushel, a fluctuation of 10 cents per bushel around the price trend line corresponds to a fluctuation of 4 cents in the cost of ethanol. The data on the table show that corn purchased in the winter of 1975 would have been about $1.00 over the expected (trend) price line and therefore that the cost of ethanol would have been boosted by about $0.40 per gallon during that period. Similarly, corn purchased in January 1979 would have cost about $0.80 less than the price expected from the general trend line, resulting in a production cost for ethanol of $0.32 per gallon lower than projected. It is therefore apparent that over a period of time, a certain amount of compensation for the fluctuations in feedstock costs could take place, resulting in relatively constant feedstock costs over that period.

Figure III-9 shows the value (July 1979 dollars) of corn feedstock over the 1970 to 1979 period. The full line shows the present values projected on the basis of the price trend shown in Figure III-8 (full line on the figure). These projected present values can be used to estimate the yearly cash flows relating to the purchase of the feedstock. The average present value for the 10-year period is $3.99 per bushel. The data points on the figure show the present value obtained from these data points is $4.24 per bushel. Over the 10-year period, therefore, the averaged impact of the fluctuations in corn prices around the projected trend amounts to only about $0.25 per bushel, or about $0.10 per gallon of ethanol. As suggested above, the impact of yearly fluctuations in corn prices could be drastically reduced when averaged over a long period, which could be the lifetime of the plant. Therefore, on that basis, the long-term economics of ethanol production may be more predictable than suggested by the scatter of data points on Figure III-8. The occurrence of the periodic fluctuations in feedstock prices, however, requires that sufficient working capital be available during periods of large fluctuations in feedstock price. This will add a financial burden to the plant cash flow stream which must be recognized and planned for.

As was indicated earlier, the most important cost element in determining the economics of an ethanol plant is the net price of the feedstock, i.e., the feedstock price minus the credit received for the coproducts (DDG, for instance). The impact of high corn prices, i.e., prices well above the expected historical trend, may be partially offset or enhanced by the corresponding price for DDG. High DDG prices will tend to reduce the impact of high corn prices, while depressed DDG prices will enhance it. Figure III-10 shows the value (July 1979 dollars) of net corn feedstock prices expressed as dollars per gallon of ethanol for the 1970 to 1979 period. The

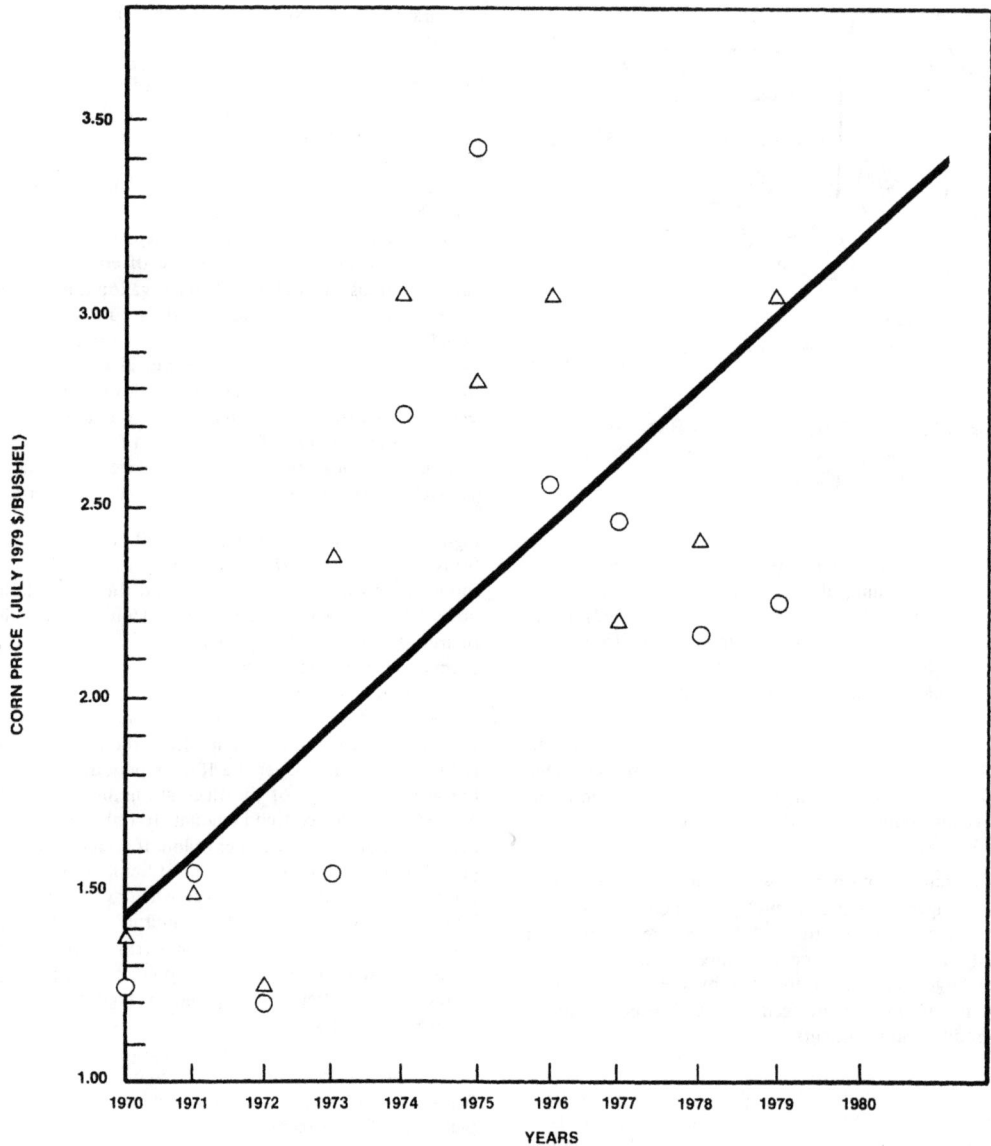

Figure III-8. Trends and Fluctuations in Corn Prices[1]

COMMERCIAL SCALE ETHANOL PRODUCTION AND FINANCING

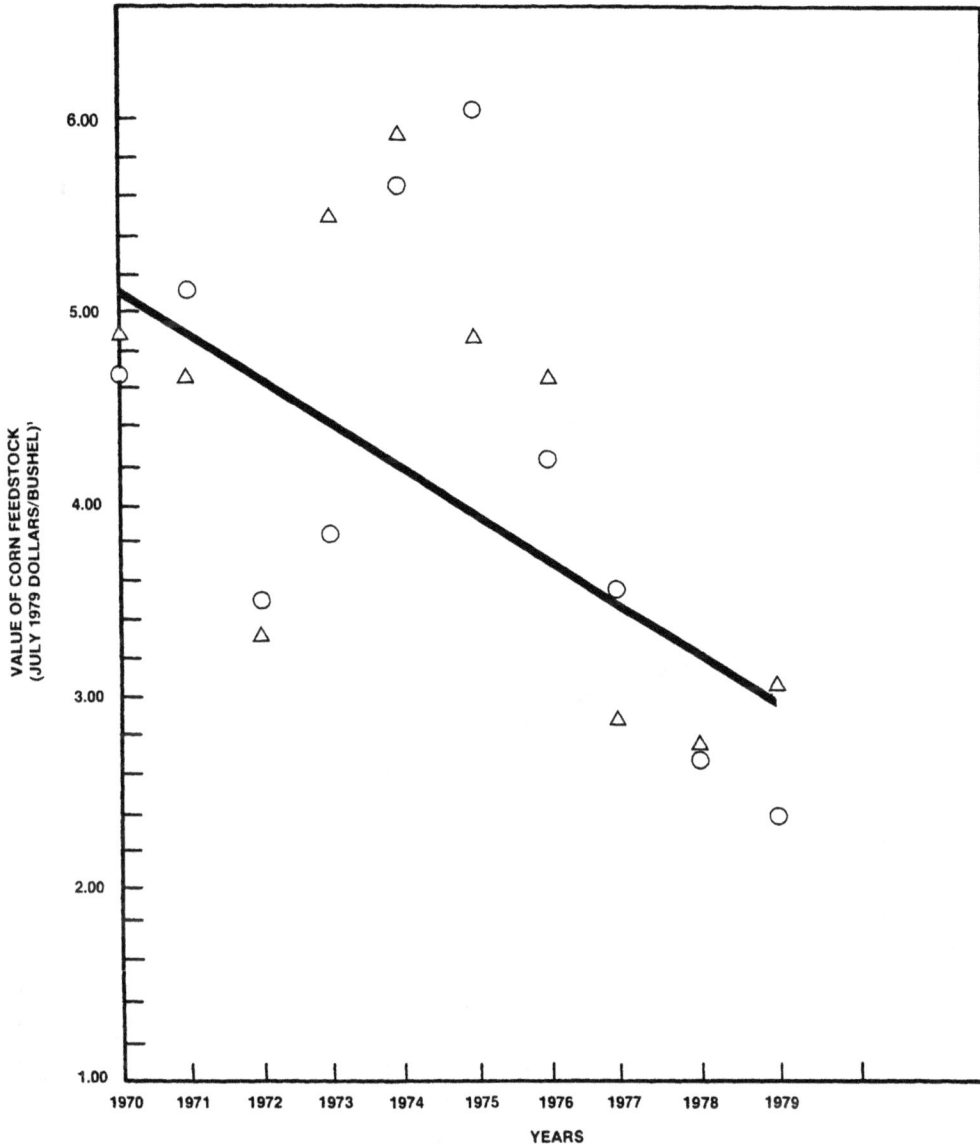

Figure III-9. Value of Corn Feedstock

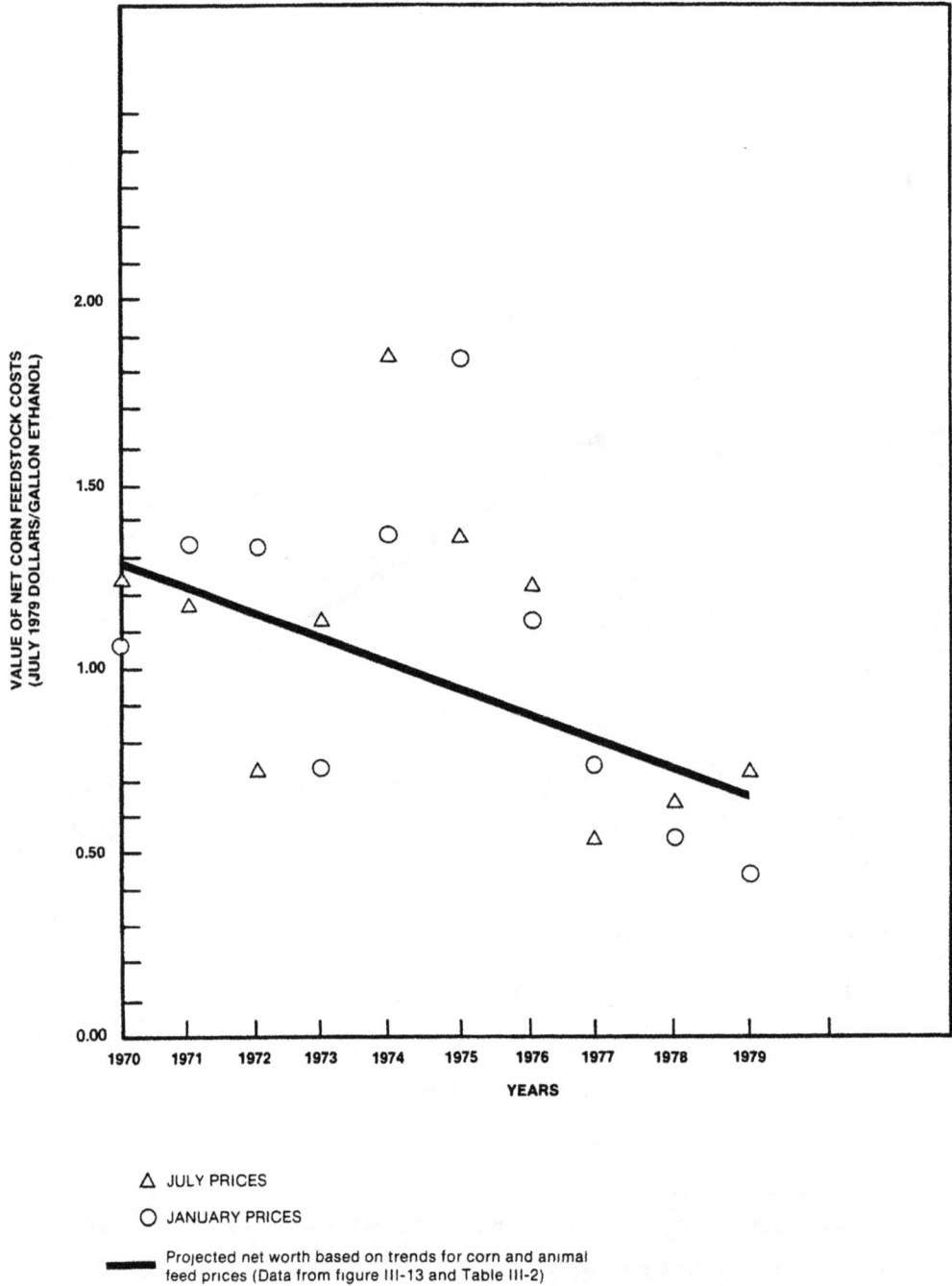

Figure III-10. Value of Net Corn Feedstock Cost

data of Table III-8 were used to estimate the data points (a 15 percent discount rate is assumed). The full line shows the projected trend for the value of the feedstock as determined by regression analysis of the data. The average value of the projected net feedstock cost over the 10-year period is about $0.95 per gallon of ethanol. Therefore, the average impact of the fluctuations in net feedstock cost, for the 10 years considered, amounts to an average increase in net feedstock of about $0.12 per gallon. The value of the working capital needed to purchase a six-month feedstock supply for a typical 50-million-gallon plant at the projected net feedstock cost in 1974 is about $25 million (the value of the net feedstock cost is about $1.00, as shown in the figure). As a result of the wide fluctuations in net feedstock cost

recorded in that year, a working capital having a July 1979 value of about $46.3 million would have been required to purchase the same quantity of feedstock at 1974 summer prices. This example illustrates the magnitude of the financial burden which may have to be incurred to secure the feedstock supply under fluctuating market conditions.

As indicated before, this burden could be alleviated if a sizable fraction of the feedstock supply could be obtained under long-term contracts at predictable prices (for instance, at prices following the projected trend lines shown in Figures III-9 and III-10). Corn prices usually rise sharply when there is a poor crop and the supply of grain for food and exports become tight.

CHAPTER IV
Markets for Ethanol and Coproducts

Establishing an economically viable ethanol plant requires four key elements of availability: a reliable, sufficient feedstock supply; the support resources such as process water, process energy, and labor; an adequate market for ethanol and its coproducts; and transportation for feedstock and ethanol. These elements impact directly the selection of a site for the proposed plant and the overall economics of the project. Compromises in the following areas may need to be made to approach an ideal situation: proximity to the feedstock supply; access to plentiful water resources and renewable process energy sources; and direct access to large enough markets for ethanol and its coproducts to absorb the plant's output.

The problem of availability of the required feedstock resources has been addressed in the preceding chapter. This chapter discusses some of the characteristics of the markets for ethanol and its coproducts and examines some of the issues to be addressed by a potential investor.

ETHANOL FUEL

Ethanol as a Fuel and as a Chemical

Ethanol may be used in various forms for fuel:
- As a blend with gasoline in various proportions
- As hydrated lower-proof ethanol
- As neat anhydrous ethanol
- As fuel supplement in dual-carbureted diesel engines

Ethanol also is used as a chemical in such industries as pharmaceuticals and perfumes. Each of these potential markets for ethanol has some constraints which bear on the decisionmaking process of producing ethanol. Some of these constraints result from the differences in the properties of fuel ethanol compared to petroleum-based fuels now used.

Fuel Properties of Ethanol. Table IV-1 summarizes some of the properties of ethanol and other fuels.

Table IV-1. Summary of Ethanol and Other Fuel Properties

PROPERTY	GASOLINE	ETHANOL	NO. 1 DIESEL
Molecular Weight	126	46	170
Heating Value			
Higher (Btu/lb)	20,260	12,800	19,240
Lower (Btu/lb)	18,900	11,500	18,250
Lower (Btu/gal)	116,485	76,152	133,332
Latent Heat of Vaporization (Btu/lb)	142	361	115
Research Octane	85-94	106	
Motor Octane	77-86	89	10-30
Stoichiometric Air/Fuel Ratio	14.7	9.0	
Flammability Limits (Volume Percent)	1.4 to 7.6	3 3 to 19	

Blending ethanol with other fuels results in modifications to the properties of the original fuel. Adding 10 percent ethanol to gasoline results in a lowered energy content (about 112,000 Btu per gallon for gasohol versus about 116,000 Btu per gallon for gasoline) but a higher octane number. The stoichiometric air-to-fuel ratio for the blend also will be quite different from that of gasoline as the percentage of ethanol in the blend is increased. Therefore, as more ethanol is added to the blend, the air/fuel mixture of a carburetor set for gasoline becomes less favorable and the driveability of the vehicle may be affected. Comparative fuel economy for gasohol- or gasoline-operated vehicles is still a matter of controversy. In some instances, improved mileage has been claimed for gasohol-operated vehicles but in most cases, mileage has not changed. The direct use of ethanol in diesel engines is difficult without major engine modifications. Under present engine designs, ethanol does not meet the fuel specifications of diesel engine manufacturers. Therefore, ethanol cannot presently be substituted for diesel fuel. Ethanol-diesel fuel mixtures have unfavorable self-ignition characteristics due to the low self-ignition tendency of ethanol, and the performance of blend-fed engines is strongly affected. The best approach at present appears to be carbureting ethanol in the diesel engines. This, however, requires engine modifications and separate ethanol and diesel fuel tanks.

If the market for diesel fuel warrants it, there are well established alternative fermentation technologies which can produce diesel fuels from biomass. For example, butanol is readily produced by fermentation and this is an excellent diesel fuel. If the producers wished to switch from ethanol to a diesel fuel, this would require only a change from yeast to another biocatalyst and some modifications in distillation equipment.

Ethanol-Gasoline Blends. Because of the phase separation problems occurring when hydrated ethanol (less than 200° proof ethanol) is mixed with gasoline, ethanol-gasoline blends usually include anhydrous (200° proof) ethanol. Such blends have been used for many years in various countries. A blend of 90 percent unleaded gasoline and 10 percent anhydrous ethanol is marketed at present in many states under the name gasohol.

Despite the controversy concerning the fuel efficiency of gasohol versus gasoline, gasohol has been well received by the public and is generally accepted as a substitute for gasoline. The present political climate and the desire to reduce the nation's dependence on foreign energy supplies may be some of the motivating factors behind the adoption of gasohol. If the present trends are maintained, i.e., if the use of gasohol continues to be encouraged by state and Federal governments and if the public continues to accept gasohol as a substitute for unleaded gasoline, the potential market for ethanol is about 10 billion gallons per year if a 10 percent blend is used.

Hydrated Ethanol. Hydrated ethanol (less than 200° proof) can be burned efficiently in internal combustion engines with minor engine modifications. The carburetor jet size must be enlarged slightly when converting from gasoline to ethanol because the ethanol component contains less useful energy per unit volume than gasoline (see Table IV-1). With most engines, it is also desirable to modify the intake manifold to ensure proper vaporization of the ethanol so that all cylinders will be operated with the same air-to-fuel ratios.

The use of hydrated ethanol fuel would have major advantages for the producer. The last step in the usual process of ethanol refining that is the dehydration of the azeotrope could be eliminated. The most likely hydrated ethanol fuel would be one of about 186° proof, which can easily be obtained at relatively low energy consumption directly by single distillation of the fermenter beer. Major savings both in equipment and energy consumption would result with significant reduction in cost for this product. The use of 186° proof as 93 percent ethanol is likely because this is the aim of the Brazilian effort and both General Motors and Volkswagen are already producing or will produce automobiles for the Brazilian market which are designed to make use of 186° proof fuel. Also, the engine modifications required to make appropriate use of this fuel are modest compared to those required for the use of 160° proof fuel or the like.

The use of hydrated ethanol has certain drawbacks. There will be a reduced mileage range per tank as compared to gasoline. Problems due to the possible accumulation of water are possible but are easily circumvented by appropriate designs and with this fuel there is no danger of freezing at low temperatures. Thus, while the direct hydrated ethanol market may have only a limited attractiveness for a period of time for commercial-sized ethanol plants, a major conversion to the use of this fuel rather than anhydrous ethanol in gasohol could be possible.

Anhydrous Ethanol Fuel. Anhydrous ethanol can be burned directly in spark-ignition engines using essentially the same modifications discussed above for the use of hydrated ethanol. Ethanol may also be used in furnaces, boilers, or gas turbines. In the latter case, efficiencies slightly higher than those obtained with hydrocarbons have been recorded. At present, however, the market for anhydrous ethanol fuel appears limited to blending with gasoline.

Diesel Fuel Supplement. As indicated above, diesel engines can operate on separately carbureted anhydrous ethanol and diesel fuel. Engine modifications and separate tanks for the two fuels are required. Thus, the market for ethanol in this application probably will be limited.

Industrial Applications of Ethanol. The chemical industry consumes large quantities of industrial ethanol as either feedstock or solvent. In the latter case, one of the major consumers is the pharmaceutical industry, requiring extremely pure ethanol. Industrial ethanol also can be the feedstock to produce two important industrial chemicals: acetic acid and ethylene. Acetic acid can be obtained directly from ethanol by fermentation; ethylene may also be derived from ethanol.

At present, most industrial ethanol is produced from petroleum or natural gas-derived ethylene. The cost of industrial ethanol therefore is directly related to those of petroleum and natural gas. As petroleum-derived industrial ethanol costs continue to climb paralleling the cost of petroleum and natural gas, fermentation-derived ethanol will become more attractive as an industrial feedstock or chemical. The potential annual market for industrial ethanol alone is on the order of 200 million gallons.

Impact of Ethanol Fuel on Petroleum Import Requirements. The previous discussion suggests that the most attractive market for commercial-size fermentation ethanol plants is ethanol to be used as a blend with gasoline, i.e., gasohol. Despite the slightly lower thermal value per unit volume of gasohol compared to gasoline, no significant fuel mileage decrease has been recorded when gasoline is replaced by gasohol. Ethanol, therefore, can displace a quantity of gasoline equivalent to the proportion of ethanol in the gasohol blend.

The addition of ethanol to gasoline increases the octane rating of the blend because anhydrous ethanol is a higher octane fuel. In the past, the octane rating of fuels was increased by adding tetraethyl lead. Because of the adverse effects of lead compounds on humans, the conversion to unleaded gasoline was mandated some years ago. The changes in refinery operations required to produce fuel of a given octane without lead additives reduce the quantity of fuel produced from a barrel of crude oil. The octane-boosting process requires additional energy in the refining process, energy lost from every barrel processed. The addition of ethanol to gasoline gives the required octane boost without the supplementary energy expenditure in the refining process. Therefore, every barrel of ethanol produced decreases the crude oil demand not only by the quantity of gasoline directly replaced by the ethanol but also by the crude oil saved as a result of the value of ethanol as an octane enhancer.

Problems of Storage, Handling, Blending, and Distribution. Small amounts of water in the ethanol will cause the separation of ethanol-gasoline blends in two layers, water-alcohol and gasoline. This separation will result in poor engine performance. It is therefore necessary that ethanol be kept anhydrous during transportation prior to blending with gasoline. The points at which blending of ethanol with gasoline could occur are at the refinery during loading of trucks, the pipeline terminal as the trucks are loaded, or the retail station by means of a blending pump. In view of the bulk quantities involved when dealing with the output of a commercial operation, the first two blending options are preferred. Hydrated ethanol, on the other hand, offers very little problems and is readily handled by commercial equipment

Blending shortly before use will have an impact on the economics of ethanol fuel. The costs of transportation of ethanol from the production facility to the point of blending and the storage of ethanol at the site of blending will have to be added to the retail price of ethanol.

The cost of transportation has been estimated to be about $0.008 per gallon for the first 20 miles plus about $0.003 per gallon for each additional 20 miles traveling distance (1980 dollars). The cost of storage facilities is estimated to be about $0.10 per gallon capacity for a 50-million-gallon-per-year facility. With adequate maintenance, the life of storage facilities may exceed 40 years. The proximity of a pipeline terminal or refinery may therefore be a desirable feature when siting an ethanol plant.

Major Ethanol Producers and Distributors

The principal promoters of gasohol to date have been independent oil companies. Texaco, Inc. has taken a lead role and is currently retailing gasohol at most of its outlets. Besides Texaco, prominent distributors of gasohol include Atlantic-Richfield Oil Company and Standard Oil Company (division of Amoco Oil Company). The major companies as a group have shown reserve in entering the market. Nevertheless, some major oil companies are test marketing gasohol. These firms include Cities Service Company, Phillips Petroleum Company, Standard Oil Company of Indiana, and Diamond Shamrock.

In addition, some major oil companies are discussing joint-venture arrangements with large food processors. Three of these joint-venture arrangements include:

- Texaco and CPC International
- Chevron and American Maize
- Ashland Oil and Publicker Industries

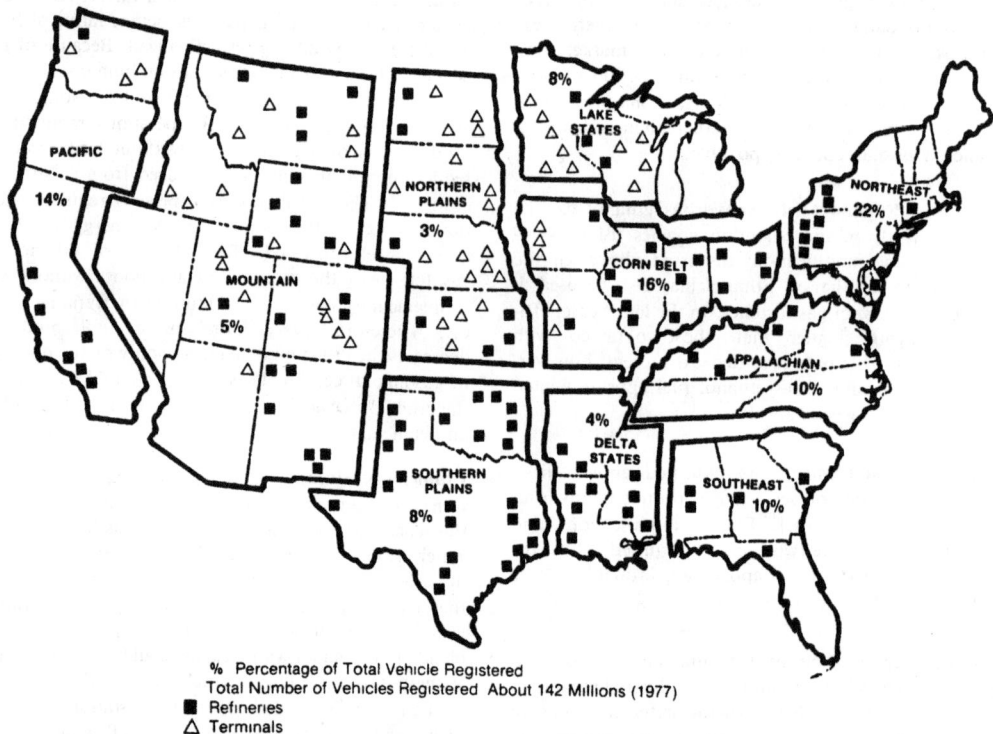

% Percentage of Total Vehicle Registered
Total Number of Vehicles Registered About 142 Millions (1977)
■ Refineries
△ Terminals

SOURCE
U S Department of Commerce, "Statistical Abstract of the United States, 1978", 99th Annual Edition, Washington, D C, 1978
M L David, G S Hammaker, R J Buzenberg, and J P Wagner, "Gasohol Economic Feasibility Study", Report prepared for Energy Research and Development Center, University of Nebraska, Development Planning and Research Associates, Inc, Manhattan Kansas, p 261, July 1978

Figure IV-1. Percentage of Motor Vehicles Registered by USDA Farm Regions and Location of Refineries and Terminals

COMMERCIAL SCALE ETHANOL PRODUCTION AND FINANCING

Some of the major ethanol producers are listed in Table IV-2 and fuel ethanol plants, which have been projected for the near term, are listed in Table IV-3. Neither the total national production capacity including small producers nor the amount of ethanol sold for fuel is well defined at present.

Table IV-2. Major Ethanol Fuel Producers

Producers	Capacity (Million Gallons/Year)
Eastman	25
USI	66
Publicker	60
Union Carbide	120
Archer Daniels Midland Company	100
Georgia Pacific	4.6
Midwest Solvents	15

MARKET FOR ETHANOL

U.S. Market for Gasohol

In 1979, approximately 50 million gallons of ethanol were used to produce gasohol in the United States. The gasohol was sold through the almost 2,000 retail outlets. Most of the gasohol was made with unleaded gasoline so it could compete against premium unleaded fuels. However, some gasohol utilized regular gasoline. The latter mixture primarily served to extend gasoline supplies during shortages.

Consumer reaction to gasohol has been very favorable, especially in the farming community. Generally, motorists rate the product high in terms of engine performance. It is no surprise that gasohol acceptance appears to be increasing, as is the number of companies selling gasohol.

Gasohol test market results show penetration rates of 8 to 30 percent of overall gasoline sales. In areas such as the Midwest where gasohol has been promoted extensively, a near-term penetration rate of 20 percent of gasoline sales appears reasonable. In less developed markets, a conservative estimate of 10 percent penetration may be more realistic in the short term.

The near-term gasohol penetration rate could increase significantly under conditions of gasoline shortage or gasoline price rises relative to ethanol. Many motorists contend that gasohol must overcome its price disadvantage relative to gasoline before its use becomes more widespread.

A factor which once constrained gasohol sales was the difficulty for retailers to get unleaded gasoline allocations for blending. However, gasoline supplies currently appear available. To the extent that supplies remain adequate, this constraint will be moderate. In addition, the DOE is considering changes in gasoline allocation rules to assign automatically unleaded gasoline to blenders of gasohol. The oil companies are expected to oppose these changes, however, because such gasoline assignments would take gasoline away from existing customers.

In the future, the sales potential of gasohol as an unleaded fuel may increase. If for no other reason, the percent of gasoline sales comprised of unleaded fuel is increasing as new vehicles (using only unleaded fuels) replace older ones.

Regional Markets for Ethanol Fuel

One important consideration in the siting and planning of an ethanol plant is the proximity of a market for the fuel produced and of blending sites. Figure IV-1 shows the percentage of the total number of vehicles registered by USDA regions. Also plotted on the figure are approximate locations of refineries and pipeline terminals. A few facts are apparent from the figure:

- The preferred sites for blending of ethanol with gasoline, i.e. refineries and terminals, are widely available within most regions and in particular in the Lake States, Northern Plains, and Corn Belt regions where the largest fraction of the grain feedstocks is produced (refer to Figure III-5).

- Only about 27 percent of the vehicular fleet is registered in the three major grain-producing regions. Marketing of ethanol produced in those regions will therefore have to include shipment of ethanol or gasohol to large vehicular markets such as the Northeast, Appalachian, and Southeast regions (combined fleet: 42 percent of the national total). A compromise between siting of the plant in the vicinity of the feedstock resource and close to the market for ethanol may have to be reached.

Government Policies and the Market for Ethanol. The major impact of government policies on the market for ethanol will be through enhancing the economic attractiveness of ethanol over fossil fuels. The excise tax exemption, as well as a 10 percent additional investment tax credit for facilities that convert alternate feedstocks to liquid fuels are the two major Federal policies encouraging the penetration of ethanol in the fuels market. A number of States have also eliminated the state fuel tax for ethanol blend fuels or gasohol.

Table IV-3. Potential Ethanol Fuel Plants - Near Term
(Received Grants Under P.L. 96-126)

STATE AND COMPANY	FEEDSTOCK	CAPACITY (Million/ Gallons/ Year)	STATE AND COMPANY	FEEDSTOCK	CAPACITY (Million/ Gallons/ Year)
Alabama			Minnesota		
Grasp, Inc	Corn and milo	40	CBA, Inc	Corn	24 5
Arizona			Mississippi		
Arizona Grain, Inc	Corn, grain, barley	12 5	Alcohol Fuels of Miss Inc	Wood chips and dust	1
Arkansas			Missouri		
Arkansas Grain Fuels Inc	Milo	35	Missouri Farmers Assoc	Corn, wood	5
California					
Ultrasystems, Inc	Sugar beets, potatoes, wheat, grain	20	Montana		
			Infinity Oil Co Inc	Wheat, barley	5
Western Concentrates, Inc	Corn	132	Nebraska		
			Nebraska Alcohol Fuels Corporation	Corn	50
Colorado			Nevada		
Grand American, Inc	Grain	10	Geothermal Food Processors, Inc	Corn	5
Georgia					
Cafpro, Inc	Corn	2	New York		
Nuclear Assurance Corp	Wood	25	Andco Environmental Processes	Corn	15
Hawaii			North Carolina		
Hilo Coast Processing	Molasses	23 5	Diversified Fuels, Inc	Corn	30-50
Idaho					
Clearwater Palouse Energy Co-op	Wheat, barley	20	North Dakota		
			Dawn Enterprise, Inc	Wheat, potatoes	50
Illinois			Oklahoma		
Rochell Energy Developers	Corn	1 5	Fulton Energy Corp	Corn, milo	25
Indiana			Oregon		
Agri Answer, Inc	Corn	Unknown	Morrow Ag Energy Corp	Corn, wheat, sugar beet, potatoes	20
New Energy Corp	Corn	50			
Iowa			Pennsylvania		
Agri Grain Power Inc	Corn	50	Lavco, Inc	Corn	20
Kansas			South Carolina		
Planning, Design & Dev Inc	Grain	10	Energy Conversion Corp	Corn	20
Louisiana			South Dakota		
Apex Oil Inc	Various	33	Sodak Resources, Ltd	Corn	20
Independence Energy, Inc	Sugar cane, sorghum, molasses, corn	20	Vermont		
			Alternative Concepts of Energy	Cheese whey	1 5
Maine			Virginia		
D W Small & Sons	Corn, potatoes	25	A Smith Bowman	Corn	20-40
Maryland			Washington		
Americol Ltd	Local feedstocks	10	Omega Fuels	Corn	50
Massachusetts			Wisconsin		
Belcher New England, Inc	Corn, grain, potatoes, cellulose	25	Dvorak Farms	Corn	2
			Wisconsin Agri Energy Corp	Corn	20
Michigan					
U S Ethanol Industries	Corn	40	TOTAL PROJECTED PRODUCTION		950-1030 Gallons/Year

COMMERCIAL SCALE ETHANOL PRODUCTION AND FINANCING

Expected Market Fluctuations. If present price and incentive policies are maintained, tending to make gasohol competitive with gasoline, and if the present reception of gasohol by the public is sustained, no drastic market fluctuations for fuel ethanol are expected. In the long run, the improved average mileage required under Federal law will somewhat tend to decrease the demand for automotive fuel. The impact on the ethanol market will, however, be small.

PRICE OF ETHANOL

Historic Review of Prices. Industrial ethanol is manufactured from ethylene. As prices of petroleum products increase, so do those of ethanol. The price of anhydrous industrial ethanol is about $2.02 per gallon (May 1980, wholesale, f.o.b. plant).

The price of ethanol produced from agricultural feedstocks is dependent on factors such as type and price of the feedstock, plant operating life, financing terms, value of the credit for coproducts, and other factors. Recent estimates suggest market prices ranging from $1.30 to $1.65 per gallon (1980 dollars) for ethanol from corn and wheat feedstocks under various plant financing conditions. These market prices assume that a credit is received for the coproducts. On an energy-content basis ($/Btu), ethanol from grain is still more expensive than gasoline. However, if one accepts equivalence in performance of gasoline and ethanol in a gasohol blend, the price of ethanol is approximately equivalent to that of gasoline. At present, gasohol retail prices (even with the tax exemptions) are equal to or higher than those for unleaded gasoline. As prices of petroleum increase, the price position of ethanol versus gasoline will further improve.

Factors Affecting the Price of Ethanol. As earlier indicated, one of the provisions of the Windfall Profits Tax is to exempt ethanol from renewable feedstocks from the Federal excise tax. This amounts to an exemption of $0.40 per gallon of ethanol or $0.04 per gallon of gasohol. Similarly, many states exempt ethanol or gasohol from state taxes. The amount of the exemption varies. (Refer to Appendix D for exemption data.)

Transportation and distribution costs of ethanol increase the cost of ethanol used in gasohol blends. This is particularly true for markets far from ethanol producing plants. As an example, the price of gasohol in Virginia ($0.829 in May 1979) included 1.7 cents per gallon for shipping the ethanol from Illinois.

Projected Prices and Uncertainty of Price. The market price of gasohol will follow closely that of gasoline, i.e., increase at a rate slightly higher than inflation. A 10-cent rise in price for unleaded gasoline raises that for gasohol by 9 cents. The present and near-term projected production capacities for ethanol suggest that full production will not glut the market and depress the market price of ethanol.

The production cost of ethanol is very sensitive to net feedstock costs (feedstock cost minus credit for coproducts). As was shown in Chapter III, wide fluctuations in net feedstock costs have occurred and must be expected. Over the years, these fluctuations tend to smooth out. Their impact on fermentation ethanol prices has a relatively slight effect on gasohol prices—a 10-cent price rise in ethanol results in only a 1-cent rise in gasohol price.

There is a certain risk involved when entering the ethanol fuel business. The risk is probably related to the uncertainty of the feedstock cost rather than projected changes in demand for fuel ethanol. To minimize the risk, emphasis must be placed on securing long-term reliable feedstock supplies.

COPRODUCTS OF ETHANOL MANUFACTURE

Human Food Coproducts

As discussed in Chapter III, an ethanol production unit can be integrated with a dry or wet milling grain plant. Dry and wet milling operations are very specialized industries requiring unique marketing efforts to dispose of the variety of products generated and adjust to the changes in demand for food/feed products.

The most common case of integration of an ethanol plant with such operations will probably result from the addition of an ethanol unit to an existing mill rather than the creation of an entirely new milling/ethanol complex. The former case is being implemented at the Archer Daniels Midland Company. In this instance, ethanol is really a coproduct of the major products (food and feed) of the plant and as such, is a manifestation of a desire to diversify by taking advantage of an emerging market. The other approach, i.e., creation of a totally new integrated milling/ethanol complex has some drawbacks for the prospective investor. The capital cost is much higher than for an ethanol plant; the problems of marketing are increased because of the variety of products ranging from animal and human food/feed and pharmaceutical-derivatives to fuel. Each of the components of the integrated project, i.e., milling and ethanol production, is a venture in itself including its own risks. Integrating the two components may provide some hedge against these risks but also may result in accumulated problems and risks unattractive to the investor unfamiliar with these industries.

The remainder of this chapter will therefore assume that the main objective of the proposed plant is the production of ethanol fuel and that the major coproduct is animal feed (DDG or its equivalent).

Animal Feed Coproducts

Stillage, the residue of fermentation and distillation in the production of ethanol, contains many nutritive elements. This is particularly true of grain stillage, which has been used as animal feed or feed supplement over the years. Marketing these coproducts is essential to the economics of ethanol production. The present section focuses on coproducts resulting from the production of ethanol from grain. Some of the characteristics of these coproducts have already been discussed in Chapter III but will be briefly summarized as needed here.

Marketing Options for the Coproducts. Fresh stillage is a mixture of various nutrients dissolved or suspended in water. It can be fed directly to animals but is not tolerated in large quantities because of the limited capacity for water intake by cattle and other animals. Fresh stillage also degrades rapidly, particularly in warm climates, and therefore disposing of the stillage output of a commercial plant will result in a complex distribution problem. Fresh stillage can be concentrated or dried. In this form, the product can be stored and shipped, making marketing an easier, more predictable

task. This approach, however, requires a supplementary investment in drying equipment and storage facilities for the coproducts. In the forthcoming sections it is assumed that stillage is marketed as a dried product, DDG or DDGS.

Market for Coproducts. Table IV-4 shows the market for selected animal feeds and the total market in the United States for the years 1963 to 1976. The total market includes oilseed, animal protein, and other mill products. As a point of reference, a 50-million-gallon ethanol plant will produce about 177,000 tons of DDG per year, i.e., about 45 percent of the 1976 market for that commodity. A 600-million-gallon ethanol program—a near-term objective in the United States—will produce almost 2 million tons of DDG (or the equivalent) or about 7 percent of the total 1976 feedmarket and about five times the amount of DDG sold in the United States in 1976. In the 1963 to 1976 period, the domestic market for all animal feeds increased at an average rate of about 1.2 percent annually. The market for soybean meal (with which DDG is more directly competing) expanded at a rate of about 3.5 percent over that period. This expansion is not quite sufficient to absorb the expected expansion in production of DDG as a result of ethanol fuel production. Some substitution between DDG and soybean may take place and as a result, some readjustment of agricultural production patterns will probably take place as the demand for corn feedstock increases.

Table IV-4. Market for Some Commercial Feeds in the United States

					IN THOUSAND TONS			
YEAR	CORN	SOYBEANS	WHEAT MILL FEEDS	GLUTEN FEED MEAL	BREWERS DRIED GRAINS	DISTILLERS DRIED GRAINS	DRIED/MOLASSES BEET PULP	TOTAL COMMERCIAL FEEDS
1964	82,800	9,236	4,716	1,165	295	409	1,289	99,910
1965	94,100	10,274	4,612	1,135	304	426	1,153	119,170
1966	93,200	10,820	4,499	1,193	324	425	1,129	118,998
1967	98,200	10,753	4,490	1,053	336	447	1,130	123,724
1968	100,200	11,525	4,469	963	333	437	1,523	127,226
1969	106,300	13,582	4,633	1,000	361	428	1,675	135,006
1970	100,300	13,467	4,499	1,236	361	382	1,509	128,905
1971	111,400	13,173	4,364	1,067	369	404	1,570	139,788
1972	120,700	11,972	4,397	1,262	361	428	1,566	148,175
1973	117,700	13,854	4,465	1,361	348	456	1,375	146,505
1974	90,300	12,552	4,693	1,502	346	339	1,325	117,872
1975	100,600	15,613	4,933	1,477	321	400	1,860	131,640
1976	100,400	14,056	4,797	1,038	297	374	1,760	129,180
1977[1]	103,800	16,277	4,970	1,223	282	403	1,500	128,455
1978[1]	112,000	17,400	— [2]	—	—	—	—	—

[1] Preliminary.

[2] Not available as yet.

SOURCE: U.S. Department of Agriculture, "Agricultural Statistics, 1979," United States Government Printing Office, Washington, 1979.

COMMERCIAL SCALE ETHANOL PRODUCTION AND FINANCING

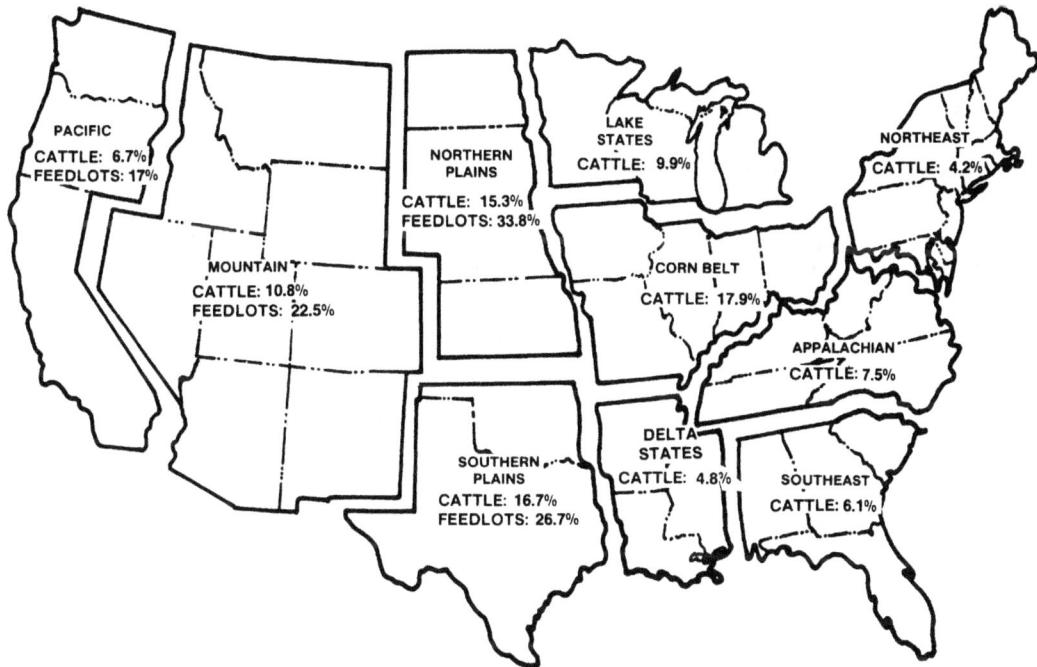

CATTLE Cattle and milk cows — Total Population About 125 millions (1978) Number is percentage of total population in the region

SOURCE U S Department of Agriculture, "Agricultural Statistics 1979", United States Government Printing Office, Washington, D C 1979

FEEDLOTS Percentage of the number of large feedlots (larger than 4,000 head) in the region
Total feedlot number 551 - Animals marketed in 1976 about 13 millions

SOURCE Schooley, F et al, "Mission Analysis for the Federal Fuels for Biomass Program", SRI International, Menlo Park, Cal , Dec 1978

Figure IV-2. Distribution of Cattle and Milk Cow Population and Large Feedlots by Regions

Exports of corn and soybeans have expanded at an average rate of about 11 and 9 percent, respectively, over the 1960 to 1977 period. Part of these exports are used for animal feed and this foreign market could absorb some coproducts such as DDG if the buyers are willing to substitute DDG for the feeds presently used. U.S. exports amounted to about 47 and 17 million tons of corn and soybeans, respectively, in 1976. The potential of the foreign markets should be investigated seriously by ethanol producers.

Geographic Distribution of the Market. As was discussed in Chapter III, the major markets for DDG and DDGS are cattle and milk cows where DDG and DDGS are a protein supplement. Other farm stock, pigs, chickens, and others, provide a significantly smaller market. Figure IV-2 shows the distribution of the cattle population and large feedlots (larger than 4,000 heads) by regions. Over 60 percent of the potential market is located in the Northern and Southern Plains, Mountain,

and Corn Belt regions. Over 80 percent of the large feedlots are located in the Mountain and Northern and Southern Plains regions. As a point of reference, a 50-million-gallon-per-year ethanol plant produces enough DDG and DDGS (see Table III-5) to feed about 120,000 steers or about thirty 4,000-head feedlots. As an example, the total cattle population of the Northern Plains region in 1978, i.e., about 18 million head, could have absorbed the DDG and DDGS production of about 120 50-million-gallon-per-year plants if DDG and DDGS were the only sources of feed. The impact of such an approach on other sources of feed would, however, be quite dramatic.

The data presented do indicate that a large potential market for DDG is available, a large fraction of the market is in the major grain-producing or neighboring regions, and a large national ethanol program may induce significant changes in the production patterns of feed-related agricultural products.

Price of Coproducts. Table III-9 shows the historical trends in the prices of corn, stillage, DDG, and soybean meal. The discussion relating to the data indicates the interdependence of the prices of these commodities as well as the range of price fluctuations to be expected.

- **Factors affecting the price of coproducts.** The price of the coproducts will be influenced by factors such as processing and transportation. Processing costs will be addressed in a later section. Transportation costs may be an important item because of the relative geographical distribution of the markets for ethanol and its coproducts and of the feedstock supply regions.

Figure IV-3 compares shipping rates for the three commodities involved in a marketing/production effort: grains, feed, and ethanol. Rates are expressed per gallon of ethanol and therefore estimate the impact of market location versus production location directly on the price of ethanol delivered. It must be stressed that the shipping rates shown in the graph are susceptible to significant variance among states. The data show that proximity of the feedstock supply is an important factor, i.e., hauling the grain rather than the products is less favorable, especially where hauling distances reach 100 miles and over. For distances less than 100 miles, it is preferable to look for proximity of the feed market rather than proximity of the ethanol market. Above that

distance, the shipping costs of both commodities are comparable.

As an example, assume that the ethanol market for a plant is the East Coast of the United States, about 500 miles away from the feedstock supply and that the DDG will be shipped to European markets. A plant located on the East Coast, at a terminal by a shipping harbor, would incur grain shipping costs of $0.232 per gallon of ethanol. A plant located in the feedstock region would incur shipping costs of only $0.135 to deliver the ethanol and DDG to their markets. Siting of the plant therefore must be carefully discussed, as it may have a major impact on overall plant economics.

- **Projected prices and uncertainties.** As was discussed in Chapter III, trends in prices may be derived from historical data and projections of crop yields and demand for crop products. These projections may be dependent on local or regional factors and a careful analysis will have to be performed once a tentative market has been identified. Fairly sophisticated computer routines have been developed by agricultural consulting firms and it is suggested that an analysis of the feedstock supply and products market picture be performed by such consulting firm as part of the feasibility study.

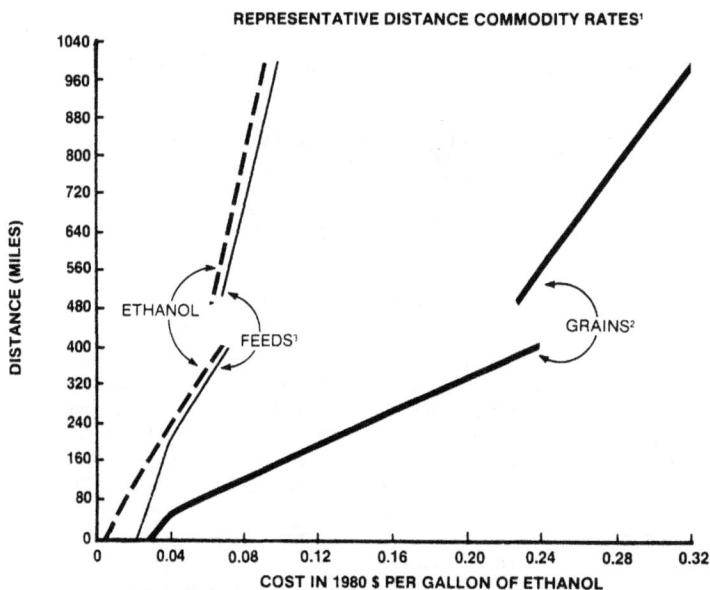

REPRESENTATIVE DISTANCE COMMODITY RATES[1]

[1]Rates for 400 miles or less are for 45.000 lb hauling trucks
Rates for more than 400 miles are for railroad carloads
[2]Assumes 56 lb per bushel. 2 5 gallons ethanol per bushel
[3] Assumes 6 6 lb DDG per gallon ethanol

Figure IV-3. Representative Distance Commodity Rates (In 1980 $ per gallon of Ethanol)

COMMERCIAL SCALE ETHANOL PRODUCTION AND FINANCING

OTHER COPRODUCTS

Carbon dioxide is a coproduct of ethanol fermentation. Recovery of carbon dioxide for soft drink production, food processing, or dry ice production is justified only if a local market for these carbon dioxide products is readily available. The value of this coproduct is very site- and demand-specific and cannot be projected on a general basis. Prices of $3 to $5/ton have been reported for uncleaned, uncompressed, raw gas. In most cases, however, no credit for this coproduct should be taken.

RELATIONSHIP BETWEEN THE ETHANOL AND COPRODUCTS MARKETS AND THE FEEDSTOCK SUPPLIES

Figure IV-4 summarizes the data of Chapters III and IV concerning the respective location of the ethanol and coproducts markets and the feedstock supplies.

The major grain feedstock regions are the Northern Plains and Corn Belt. The major market regions for DDG are the same, as well as regions to the north and west. Major markets for ethanol are the East and West Coast regions. In view of the discussion presented above on the relative transportation costs of the commodities involved, siting a plant in or close to the grain-producing regions gives direct access to the grain supply and coproduct markets and limits the marketing problem to that of distribution of ethanol in the major market areas. As was discussed, blending points for gasohol are available in all potential market areas for ethanol.

As was discussed in Chapter III, some local feedstock resources exist in other areas such as potatoes in the Mountain region. Figure IV-4 suggests that a plant located in Idaho, close to a potato or potato waste supply, would have good access to both ethanol and coproduct markets.

The conclusions derived from the data of Figure IV-4 will have to be refined to account for site- or region-specific factors such as water and auxilliary fuel resources.

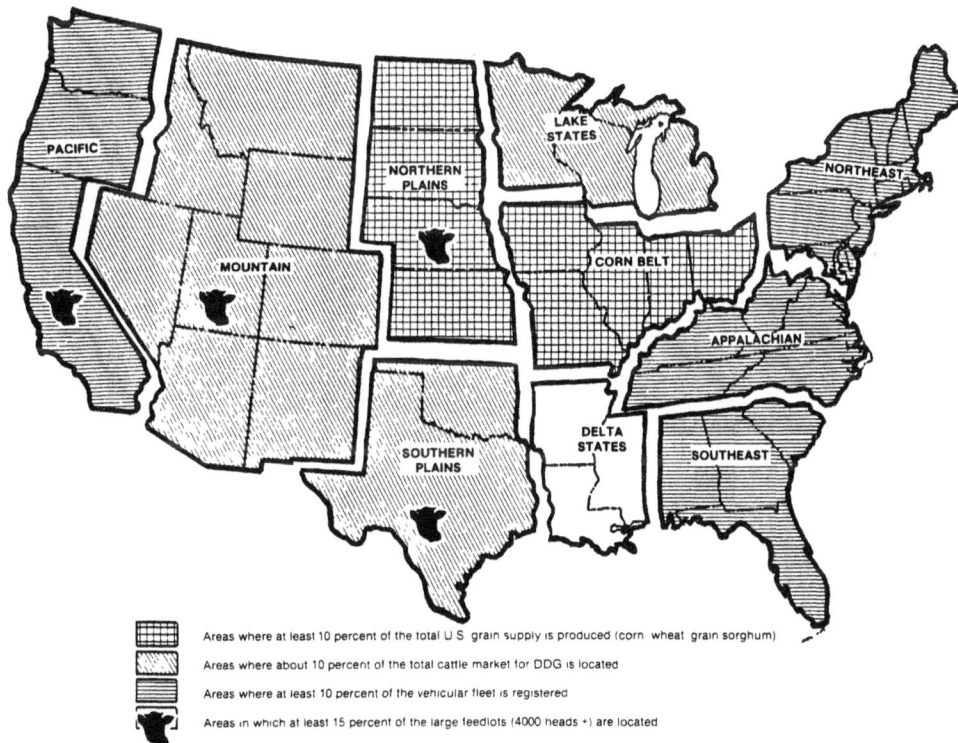

Areas where at least 10 percent of the total U S grain supply is produced (corn wheat grain sorghum)

Areas where about 10 percent of the total cattle market for DDG is located

Areas where at least 10 percent of the vehicular fleet is registered

Areas in which at least 15 percent of the large feedlots (4000 heads +) are located

SOURCE U S Department of Agriculture "Agricultural Statistics 1979, U S Government Printing Office, Washington, D C , 1979 U S Department of Commerce "Statistical Abstract of the United States 1978", 99th Annual Edition, Washington, D C , 1978

Figure IV-4. Location of Major Grain Feedstock Supplies and Potential Markets for Ethanol and Coproducts

CHAPTER V
Plant Design Concepts

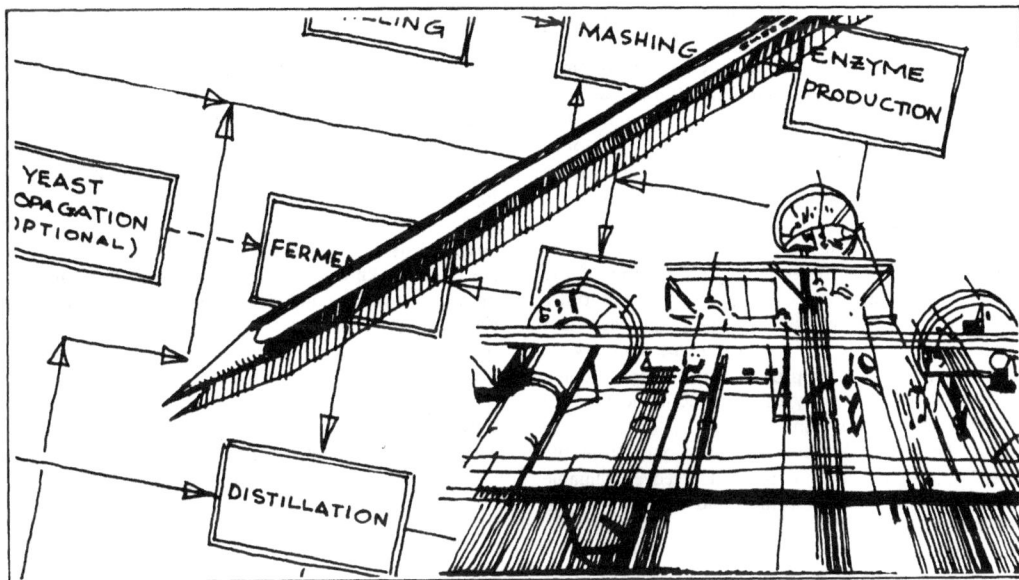

ETHANOL PRODUCTION AS A PROCESS

Ethanol production through fermentation of renewable feedstocks such as sugars, molasses, grains, or other starchy materials is well-established technology. Most of the ethanol used by the beverage industry is obtained from such feedstocks on a commercial scale. Production of ethanol for fuel uses the same technology, with modifications where necessary, to account for the end-use of the product. For example, ethanol produced for blending to gasohol must be anhydrous (water-free), while neutral spirits for beverage use may contain from 8 to 30 percent water. Such processing differences, however, are merely adaptations of existing technology, which can be incorporated in the design of reliable ethanol fuel plants.

Traditional ethanol production plants, those oriented toward beverage production, are designed to optimize the quality of the distilled spirits. As a result, energy efficiency has not been a major concern for these plants. When ethanol is produced for fuel, energy efficiency and operating economics become major concerns. As a result, plant design and operation are in a state of evolution, tending toward improved plant performance in response to the goals of the DOE and the public statutes. Enough flexibility must be included in the plant design to permit the integration of new developments as they appear, in order to maintain or improve plant performance.

A few words of caution are appropriate for the investor unfamiliar with the process of ethanol production. Although the process is well known and has been used on a commercial scale, ethanol production through fermentation is a complex process employing biochemical reactions as well as sophisticated engineering concepts. Consistent and reliable operation can be achieved, provided the prudent operator follows precautions required by biological processes and operates the facility according to established engineering practice. These include cleanliness, close control of operating conditions such as temperature, and careful storage of product. Plant operating personnel *must* have the background and experience to deal with the problems related to the common, but complex, unit processes.

Ethanol production involves hazardous materials, and plants must be designed and operated to minimize risks. The prospective investor must be cognizant of these aspects of ethanol production and assure that the engineering firm retained to perform the technical part of the feasibility study has the experience and knowledge required to advise him in the design and operation of the proposed plant.

OPTIONS FOR ETHANOL PRODUCTION

The conversion processes for producing ethanol from sugar-containing feedstocks and starchy grains are discussed in the chapter on feedstocks, Chapter III. The basic fermentation process involves the action of a living organism known as yeast, more specifically, a particular yeast such as *Saccharomyces cerevisiae,* on a fermentable sugar such as glucose. The products of fermentation are a dilute alcohol solution which contains both yeast and the unfermented portion of the feedstock, including some sugar, protein, and other carbohydrates. The mixture of unreacted feed and fermentation products is known in the industry as "beer." The fermentation process can be conducted in a single vessel and the array of other equipment which comprises the typical ethanol plant is required to prepare the feedstock, separate the alcohol from the "beer", purify the alcohol product, recover valuable coproducts, provide for storage and handling of all

feedstocks and products, and provide utility and maintenance services for the plant. The assembly of specific items of equipment into a given plant is a process which is managed by an architectural and engineering firm, in coordination with the owner and his engineering consultant. While all plants will embody the same generic types and sequence of operations, each plant is more or less unique in the adaptation to specific regional parameters and in the reflection of the "personality" of the system engineer.

OVERALL PLANT OPERATION

The plant design discussed in this Chapter is representative of all starchy grain conversions, and its major functional blocks are briefly described below (see Figure V-1).

Grain feedstock is received, unloaded, weighed, and placed in working storage. In use, it is withdrawn from on-site storage bins and cleaned by screening and/or air classification to remove tramp metal, sand, rocks, and other foreign material which could damage subsequent equipment. The grain is then milled (ground) to produce the fine meal necessary for efficient extraction of the starch which will ultimately be converted to ethanol. The meal is accumulated in "surge" storage hoppers before being weighed, mixed with liquid to form the "mash", and cooked. In the cooking process, the mash

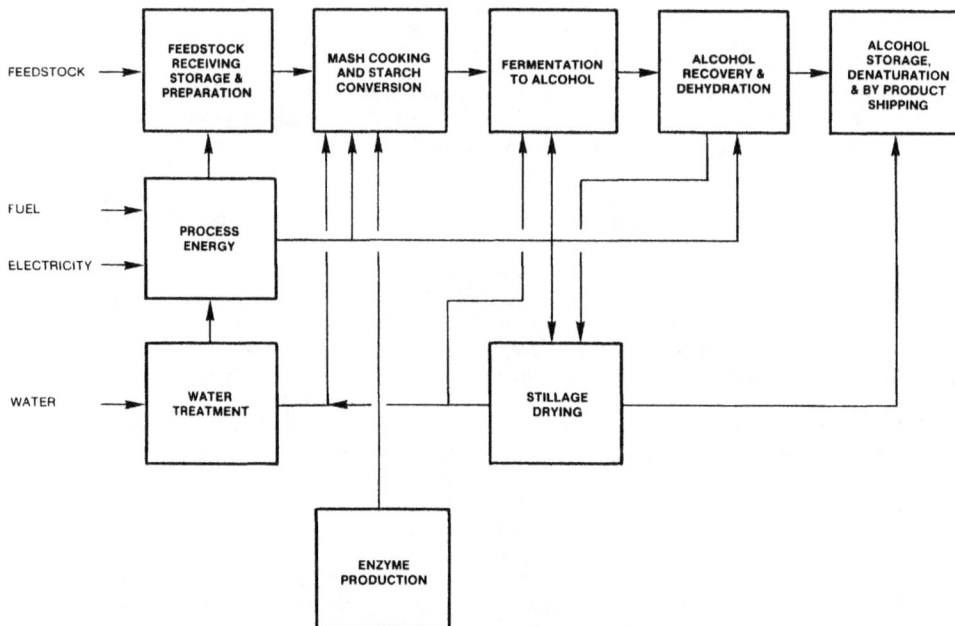

Figure V-1. Functional Diagram of Plant Design

is mixed at about 145° F and heated to a final temperature in the range of 200 to 350° F. Cooking temperature affects the required cooking time, the yield of usable starch from the grain, and the type of equipment required to conduct the cooking operation. The starch granules in the meal swell and "gelatinize" to produce a dispersion of starch in the mash. The cooking process also serves to sanitize the mash; commercial ethanol fermentation processes are not operated from a sterile mash, although the term sterilization is frequently, and erroneously, used. The cooked mash is cooled to about 150 to 170° F and mixed with an enzyme system called fungal amylase to convert the "solubilized" starch to sugars. This process is called saccharification, and the necessary enyzmes may be produced on-site or purchased commercially, depending on the economics of the specific ethanol plant.

The cooked, saccharified mash is cooled to about 80° F and pumped into the fermentation vessel. The yeast is added and, in some plant designs, a second addition of amylase enzyme completes the conversion of starch to sugar during the fermentation. The time required to complete the fermentation depends on the initial temperature and the composition of the mash. In general, batch fermentation processes follow an S-shaped curve. They start slowly as a result of the lag following yeast addition, during which time the yeast cells acclimate to the mash. The mash then passes through a period of rapid activity during which the yeast cells multiply, producing ethanol, carbon dioxide gas, and small quantities of liquid coproducts (known chemically as higher alcohols, aldehydes, etc.). As the sugar in the mash is depleted and alcohol content increases, the yeast progresses to a relatively inactive phase and, eventually, alcohol production ceases.

The fermented mash, containing perhaps 7 to 10 percent alcohol and known as dilute beer, is transferred to a holding tank, known as the beer well, from which it is pumped to the distillation towers. Distillation is a process by which liquid mixtures are separated into two or more product streams. In the case of a fermentation ethanol plant, the beer is separated into an alcohol-rich product containing up to 96 percent ethanol (and 4 percent water) and a coproduct stream known as whole stillage.

The ethanol product is further purified to anhydrous alcohol in a specialized process known as azeotropic distillation. This step is necessary to overcome a property of ethanol-water mixtures which prevents purification beyond about 96 percent ethanol with conventional distillation. The product from the azeotropic distillation section is rendered unfit for beverage use according to procedures established by the BATF. It is then stored on site, pending shipment to market.

The major coproduct of the fermentation process is the whole stillage which leaves the bottom of the beer still.

Whole stillage is separated into two fractions, processed and eventually recombined to form a dried, solid product known as DDGS. The DDGS is stored on site prior to shipment to market. The stillage processing section involves equipment for centrifuging, evaporation, and drying, all of which are common unit operations. Combined with a capital investment for equipment is an operating cost (principally for energy) associated with recovery of this valuable coproduct.

In addition to the process stages directly involved with production and purification of the ethanol product and DDGS, the fuel-ethanol plant will require ancillary operations for wastewater treatment, steam production, and interfacing with the local utility power grid. Each of these utility needs is satisfied with existing technology.

Choice of Feedstocks

Alcohol produced from starchy grain is the most common form of fermentation ethanol in the United States. The selection of a feedstock for fuel-alcohol production, however, will be guided by economic aspects associated with regional availability of specific feedstocks on a long-term basis. The choice is certainly not restricted to grains; for example, the Brazilian alcohol fuel industry is producing large quantities of fuel-grade ethanol from sugar cane. With modifications in the design of the feedstock receiving, storage, and preparation processes, fermentation ethanol may be produced from a variety of sugar- or starch-containing crops. The modifications are required to accommodate the physical properties of the feedstock, as well as the nature of the carbohydrate (i.e., starch versus sugar).

Sugar crops. In the case of sugar crops, the feedstock will be pressed and/or steamed immediately after receiving and unloading to recover the sugar juice. The juice is concentrated to avoid degradation and loss of sugar during storage. The juices are then fed directly to the fermentation section, thus eliminating the cooking and saccharification required for starchy feedstocks.

Starch crop. Minor variations may be introduced in the process flow of Figure V-1 if grains are replaced by other starch crops. For example, with potatoes as a feedstock, cleaning to eliminate dirt and slicing to expose the starchy interior may be necessary before cooking and saccharification.

When grains are the feedstock, grain preparation may involve more complex methods of products recovery. Dry or wet milling processes may be used to recover germ, oil, meal, vitamins, and other valuable components before fermentation of the residual starch fraction (processes described in Chapter III).

It must be recognized that including a dry or wet milling process prior to ethanol production amounts to two processing plants in series. Although there may be situations in which this approach is justified, the integrated system will be more complex in terms of management, labor, and marketing of the products. The capital investment will also be higher; for example, a wet milling plant having a capacity of 60,000 bushels per day—the grain capacity required by a 50-million-gallon-per-year ethanol plant—would cost about $90 million (1978 dollars), compared to about $58 million dollars for the ethanol plant alone. In addition, because of the conversion of parts of the grain to other products and the losses incurred in the extra processing, the ethanol yield per bushel of corn would be less than for the straight ethanol plant. Integration of grain milling and ethanol production will probably occur when an ethanol plant is added to an existing grain milling plant.

Multi-feedstock options. Many feedstocks, and particularly sugar feedstocks or food processing wastes, are available only on a seasonal basis. Multi-feedstock capability at the front-end of the plant is therefore desirable. However, in view of the different front-end treatment required by sugar and starch crops, multi-feedstock capability will require two separate process flows before fermentation if switching from sugar to starch feedstocks is desired.

The ability to switch between different sugar or starch crops, however, may be achieved through incorporation of flexibility in the front-end operations, which permits the adjustment of equipment operating conditions to crop-specific requirements (temperature adjustments, for instance).

Ethanol and Coproducts Options

The basic plant described in Figure V-1 assumes that anhydrous ethanol and DDG are produced. As was previously discussed, anhydrous ethanol will be easier to market in the near term than lower-proof alcohol. The gasohol market is the primary established outlet for alcohol fuel, and requires anhydrous ethanol for blending. Technically, straight, hydrated ethanol can be used in automobile engines, but its use requires engine modifications. The production of hydrated alcohol would reduce the investment and energy requirements of the plant, but would also limit the marketability of the product. This option is only justified when a captive, long-term market for the product has been secured.

Similarly, the elimination of the stillage-processing portion of the plant would offer savings in capital cost and energy-related operating cost. The option of disposing of the coproduct in the form of wet whole stillage raises enormous logistical problems of preservation, storage, and distribution for commercial-size plants. Here again,

unusual site-specific conditions will be necessary for consideration of this processing option.

Process Supplies Options

Process thermal energy, electricity, water, and chemicals are consumed during operation of the plant.

Several fuel options are available for an on-site boiler to raise process steam. Coal is the only fossil fuel acceptable for a new plant; however, the option of using renewable fuels such as wood wastes, urban or industrial solid wastes, or biomass should be considered carefully. Adoption of renewable fuels entitles the plant to tax credits which may improve the overall economics of the plant. The selection of an alternate fuel will be based on many factors: availability and price of the fuel, combustion efficiency, cost of the boiler, and environmental requirements. As an example, wood waste at $12 per green ton costs about $1.40 per million Btu, compared to $1.04 per million Btu for coal at $25 per ton (12,000 Btu/lb). The capital cost of a coal boiler is generally lower than that of a wood-fired boiler of equivalent steam capacity. However, air emission controls may be much more expensive for high- or medium-sulfur coal than for wood. Each case will have to be analyzed on an individual basis.

Electricity can be purchased or cogenerated on site. In the latter case, the possibility of selling excess electricity back to the local utility must be evaluated against the supplementary capital required by a cogeneration unit.

In the design of Figure V-1, part of the process water is recycled after treatment. A fresh water supply could be used if available and inexpensive. Environmental regulations may, however, require a high degree of treatment of the wastewater before disposal.

The design of Figure V-1 assumes that enzymes are produced on site. For small plants, it will probably be more economical to purchase the enzyme needed. In the case of sugar crop, this area of the plant would be eliminated.

Plant Size Options

Fuel-ethanol plants, like other industrial plants, benefit from economies of scale. As an example, the capital investment for a 50-million-gallon plant will be only about 2 to 3 times that of a 10-million-gallon plant. The economies resulting from plant size must, however, be balanced against the possible increased cost of feedstock mobilization and increased distribution costs for the ethanol and coproducts. Regional availability of inexpensive feedstocks such as culled potatoes or food processing wastes may justify the proportionally higher investment of a smaller plant. Generally speaking, large plants are more favorable from an economics aspect but

COMMERCIAL SCALE ETHANOL PRODUCTION AND FINANCING

specific site or regional conditions may favor small plants.

New Technologies

Several areas of the fuel-ethanol production process are under investigation to improve their efficiency and cost.

Mash cooking with electrical heating or extrusion cooking, instead of through steam injection, is claimed to increase sugar yields from starch and reduce the dilution of the mash. Batch fermentation of the mash is the present conventional method of operation in industry. Continuous fermentation, which could result in faster fermentation and reduced equipment volume, is presently under development but raises problems relating to cleaning and sterilization (sanitization). Vacuum fermentation and ethanol recovery and recycling of yeasts are also under investigation. Various methods for improving ethanol recovery, i.e., improvement or replacement of distillation, are also under investigation. Methods such as liquid extraction, selective absorption by various agents, reverse osmosis, and ultrafiltration are considered, as well as improved methods for the recovery and processing of coproducts. A long-term objective of much research is the design of an economical process by which cellulosic feedstocks could be converted to ethanol by fermentation.

The economics and technical feasibility of most of these new approaches are not fully established. It is, however, essential that the investor management team of the proposed plant keep informed of these new trends.

Plant Design Selection

Some guidelines must be kept in mind when selecting a plant design. First consideration should be given to locating near a water source such as a river. Consideration must also be given to minimizing the potential impact of unknown quantities, such as disruptions in the supply of energy feedstocks, through design for multi-feedstock or multi-fuel capability (wood and coal, for instance). Providing for the incorporation of future improvements in design should also be examined.

The analysis of some of the possible options will require additional front-end expenses that may be well justified in the long run.

TYPICAL MATERIALS AND ENERGY FLOW DIAGRAMS

The fuel ethanol process described herein is based on a publicly-available technical assessment conducted for the Department of Energy ("Grain Motor Fuel Alcohol Technical and Economic Assessment Study," Raphael Katzen Associates International, Dec., 1978). The process is the result of a detailed design using proven technologies, and represents a system which could be constructed today. No existing fuel ethanol or distilled spirits plant has been constructed to this design, however, and the choice of the Katzen design as a reference fuel alcohol plant in this document is for illustrative purposes only, and does not constitute an endorsement.

Process Description

The illustrative example used throughout this guide is a hypothetical 50-million-gallon-per-year, corn-based, coal-fired, fuel ethanol plant. The processing elements of the plant are described in this section as a means of familiarizing the reader with the elements of fuel ethanol production at a level of technical detail with which the owner/investor should be familiar in order to work closely with a consulting engineer or architectural and engineering firm. The hypothetical plant is based on a highly integrated, thermally-efficient conceptual design developed for DOE by Raphael Katzen Associates. Energy efficiency and coal-firing are very desirable aspects of this and any design.

The energy and feedstock efficiency claimed for the Katzen-engineered plant are illustrated in the following figures:

Alcohol yield = 2.57 gallons/bushel of corn

DDG yield = 18.2 pounds/bushel of corn

Coal required (Illinois #6) = 41,700 Btu/gallon of alcohol

Electrical energy required = 1.32 kWh/gallon of alcohol

The physical plant layout is illustrated in Figure V-2.

Receiving, Storage, and Milling

Shelled corn is delivered to the plant by railroad hopper cars or grain trucks. A single railroad unloading station and two truck unloading stations have been provided. The unloading arrangement has been planned so that a railroad hopper car and a truck can be unloaded simultaneously. (See Figure V-3.) Railroad hopper cars hold about 52.5 tons of grain (1,875 bu of corn). The fully loaded hopper car weighs about 67.4 tons. Grain trucks come in various dimensions, but a typical truck will hold about 800 cu ft or 643 bu of corn. Unloading conveyors have been specified to accommodate a total unloading rate of 7,500 bu/hr, enough grain in a single 8-hour shift to operate the plant for a full day.

Trucks delivering grain to the plant are lifted by means of truck dump hoists and are weighed on one of two

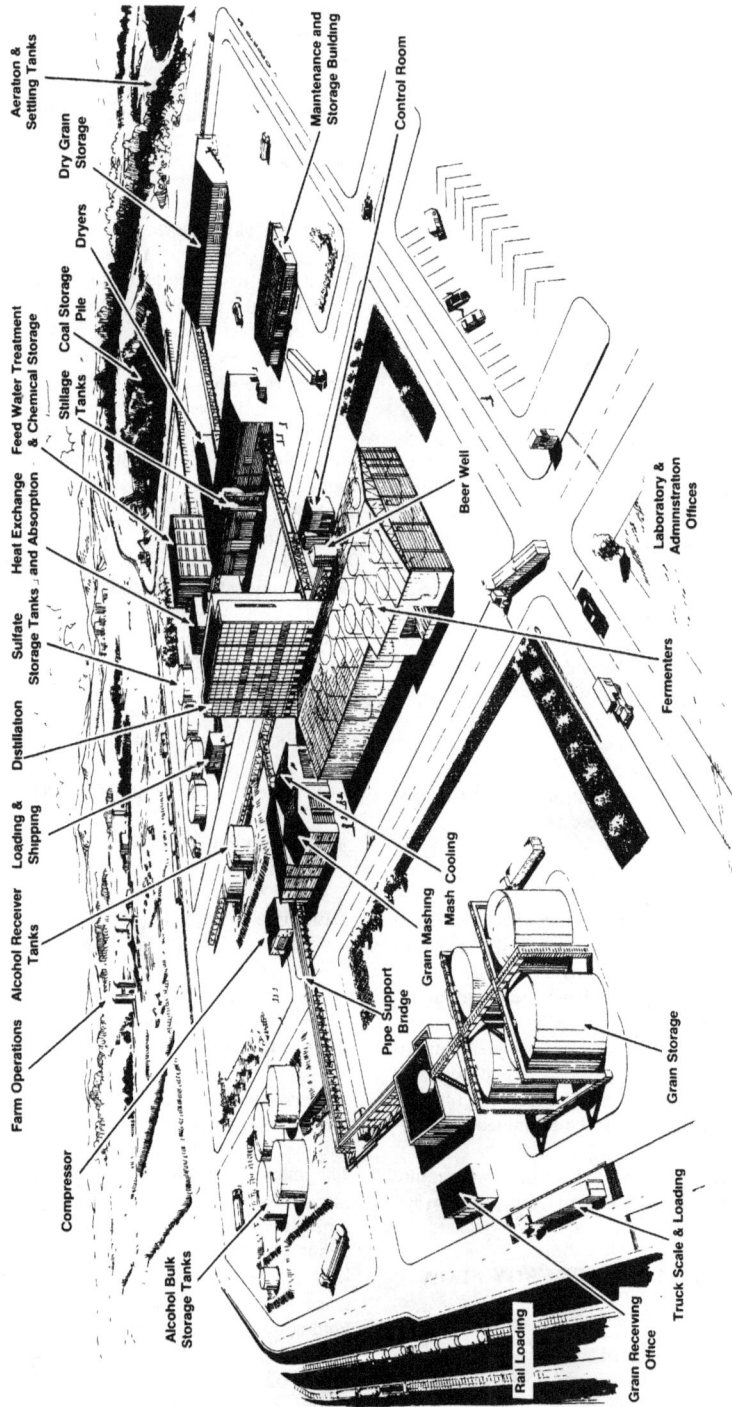

Aeration &
Settling Tanks

Maintenance and
Storage Building

Control Room

Dry Grain
Storage

Dryers

Coal Storage
Pile

Stillage
Tanks

Feed Water Treatment
& Chemical Storage

Heat Exchange
and Absorption

Beer Well

Sulfate
Storage Tanks

Distillation

Loading &
Shipping

Alcohol Receiver
Tanks

Farm Operations

Fermenters

Mash Cooling

Grain Mashing

Laboratory &
Administration
Offices

Compressor

Pipe Support
Bridge

Grain Storage

Alcohol Bulk
Storage Tanks

Rail Loading

Grain Receiving
Office

Truck Scale & Loading

Figure V-2. Katzen Base Case Plant

COMMERCIAL SCALE ETHANOL PRODUCTION AND FINANCING

truck dump scales. The grain passes from the truck into a bin housed in a pit. Grain is discharged from these bins through star valves into either of two truck unloading conveyors.

In the case of grain delivered to the plant by railroad hopper cars, the car is weighed on a rail car scale and the grain is then dumped from the car through a bin and star valve to rail car unloading conveyor. The grain passes into either of two rail car unloading cross conveyors. A truck unloading conveyor, coupled with a rail car unloading cross conveyor, then can be delivered into bucket elevator. The grain is lifted to a position above the grain storage bins, and passes from the bucket elevator into one of two distributing conveyors. These conveyors are arranged to deliver to the storage bins or may convey their grain directly into storage by-pass conveyors which deliver directly into a surge hopper.

The surge hopper has been sized to hold 7,500 bu of grain. This provides a nominal holdup time of three hours. When the surge hopper is full, grain can be diverted into storage in any of the grain storage bins. The total grain storage capacity is equivalent to grain usage for one week. When grain is being received, operation could have grain passing directly to the surge hopper. When it is filled, grain then would be diverted to the storage bins. When grain is not being received, it would pass from the storage bins to the surge hopper through the individual storage bin bottom conveyors, through a collecting conveyor, and into a bucket elevator. The grain is thus lifted and discharged into the surge hopper.

Grain discharges from the surge hopper at the rate of 2,453 bu/hr into the grain cleaner, which separates materials in the grain which are foreign to the process, including sand, tramp metal, etc. Light materials in the grain are picked up from the screens and air transported through a blower, to the bag house in the coproduct recovery section, where they become part of DDG. Tramp metal and other oversize materials are rejected from the grain cleaner and periodically removed from a collecting bin.

Grain, suitable for processing, passes into the hammer mills which deliver into a surge bin. The ground grain then passes through a star valve at the base of the surge bin and is pneumatically conveyed to the process section for mash cooking and saccharification.

The grain receiving, storage, and milling area has been separated from other plant processing areas because of the dust problem associated with these front-end operations.

Mash Cooking and Saccharification

Corn meal is received from the milling area in the surge tank. (See Figure V-4.) This tank is sized to allow con-tinuous meal input while the output to the batch weigh tank is shut off when the batch tank is being emptied into the continuous weigh tank. The batch weigh tank provides an accurate record of the total grain used, and the continuous weigh tank provides a reading of how much grain is used within any given period.

The continuous weigh tank feeds the mash mixing tank where the other mashing ingredients are added. This tank is sized for a 2.5-minute residence time and is fitted with an agitator to promote thorough mixing. The other main ingredients are recycled thin stillage (backset) and water. The water comes from recycled condensates and from makeup fresh water. The condensates are hot, and their use is regulated to maintain a tank temperature of 145° F. The total water input to this tank is controlled to produce about 22 gallons of mash per bushel of grain input (56 lb/bu basis). The thin stillage is added in an amount about 10 percent of the final mash volume going to the fermenters.

The mash is transferred from the mixing tank to the mash pre-cooker. This tank has provision for adding live steam in case insufficient condensate is available to attain the 145° F pre-cooking temperature. This tank is sized for a residence time of about 7 minutes.

The mash is further heated in the mash heater located downstream of the pre-cooker. This heater uses 15 psig steam from the pressure flash to heat the mash to 229° F. Final cooking of the mash takes place in the mash cookers by injection of live steam to attain a temperature of 350° F. The cookers consist of several 20-foot lengths of 10-inch diameter pipe connected with 180 return bends, and they are sized to provide a cooking time of 1.5 minutes.

The cooked mash is flashed to 15 psig in the pressure flash tanks. Some of the steam from this flash is used to preheat the mash as discussed above, and the remaining flash steam is used for beer heating in the distillation section. Additional water is added at this point in an amount to provide a final mash volume of 30 gallons per bushel. The mash is then further flash-cooled in the vacuum flash tanks. The temperature when leaving these tanks is 145° F, which corresponds to a pressure of about 3.3 psia. This vacuum is maintained by the flash vapor condenser, its associated steam ejector, and the ejector condenser. The condensate from all of these heaters and condensers goes to a hot well, from where it is pumped back to the mash mixing tank as discussed above.

After the vacuum flash, the mash discharges directly into the fungal amylase mixers, where the fungal amylase is added. From the mixer the mash is pumped to the pipeline saccharifier, where the starch is converted to fermentable sugars. Part of this mash (approximately 15 percent) is delivered to the fungal amylase section.

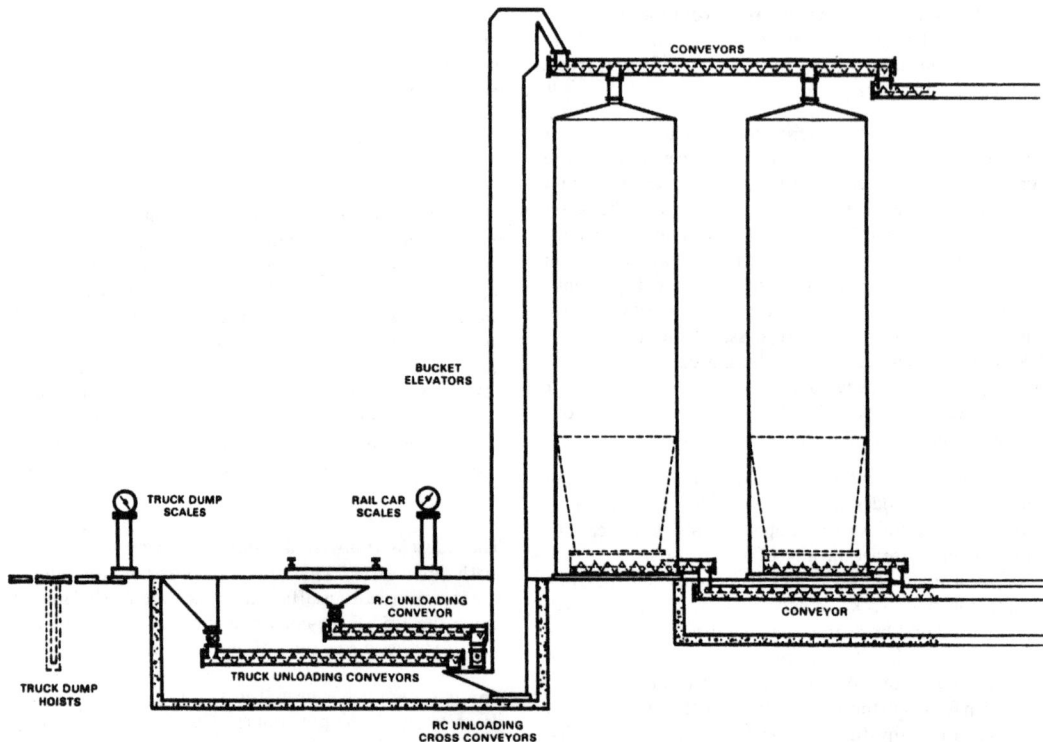

The pipeline saccharifier is sized for a 2-minute residence time. From there, the converted mash is fed thrrough the mash coolers to the fermenters. The first six mash coolers use cooling tower water at 85° F to reduce the mash temperature to 100° F. Then, in the remaining four coolers, well water at 60° F is used to complete the cooldown to 80° F before the mash enters the fermenters.

Fungal Amylase Production

The system consists of seven seed tanks, seven batch fermenters, a system for delivering sterile compressed air, and a pump for delivering the product to the cooking and saccharification section. (See Figure V-5.)

Fungal amylase is prepared batch-wise using a seed tank to grow the inoculum which is initially started in the laboratory. One seed tank is used as a starter for a fermenter tank which is sized at 33,750 gallons to provide fungal amylase for one day of operation (27,000 gallons plus 25 percent freeboard). The total batch cycle is one week, consisting of one day for tank cleaning, charging, and sterilizing; five days fermentation; and one day for usage. Twelve thousand five hundred scfm of compressed air is supplied at 25 psig, which corresponds to 0.5 scfm per cu ft of fermenter volume for five fermenters at a time.

Prior to filling the fermenters with mash, they are thoroughly cleaned, and after filling, the mash is heated to 250° F using steam in the tank jacket in order to sterilize the tank and its contents. After sterilization, the mash is cooled to 90° F using well water in the tank jacket. A small amount of additional cooling will be required throughout the fermentation period in order to remove the heat added by the agitators.

The contents of one seed tank are added to the fermenter after it has been cooled down. A period of about five days is then required for completion of the batch. During this time the tank is agitated and aerated with compressed air.

Moyno pumps are used to transfer the finished fungal amylase to the saccharification section.

Fermentation (Batch)

The fermenters receive mash continuously from the mash converter located in the mashing and saccharification section. The fermenters are batch-operated and consist of sixteen 250,000-gallon vessels which are arranged in sets of four with one heat exchanger and circulation pump for each fermenter set. (See Figure V-6). The fermenters are designed for a liquid loading of 80 percent of maximum capacity. Since cooling is needed

COMMERCIAL SCALE ETHANOL PRODUCTION AND FINANCING

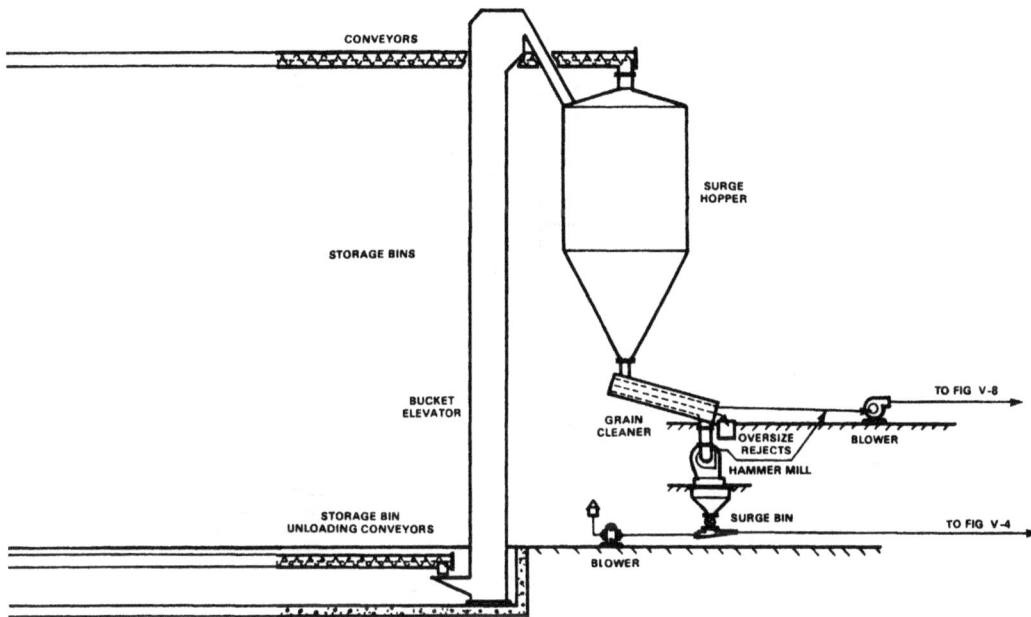

Figure V-3. Grain Receiving, Storage, and Milling

for only about 8 or 10 hours out of the 48-hour fermentation cycle, one exchanger will service the needs of four fermenters. The fermenters are filled on a three-hour cycle.

The yeast, *Saccharomyces cerevisiae,* is manually added to a yeast mix-tank and then transferred to the fermenter as it is being filled (about 300 lb of yeast per batch). The yeast is purchased rather than manufactured on location. The inlet mash temperature is about 80° F, and the temperature gradually rises to a maximum of about 95° F during the fermentation period. Cooling is provided during peak period by recirculation of the mash through the fermenter cooler; well water at 60° F is the cooling medium. Each fermenter requires a flow of approximately 1,200 gal/min during the peak period. The recirculation for cooling also serves to agitate the tank. At the end of the fermentation period, the fermenter contents are transferred to the beer well, from which they are pumped continuously to the distillation section.

The fermenters are cleaned and sterilized by means of automatic spraying machines installed in each fermenter tank. Each tank has two such spraying machines. After each fermentation cycle, the tank is washed with a cleaning solution, sterilized with an iodine solution, and rinsed with clean sterile water in preparation for the next cycle.

Distillation

Dilute beer feed from the fermentation section of the plant is collected in the beer well, from which it passes continuously to a heat exchanger in the distillation section of the plant. (See Figure V-7).

The dilute beer feed, amounting to approximately 1,150 gallons per minute, contains 7.1 weight percent alcohol and 6.92 percent solids. Solids consist of both dissolved solids and suspended solids, in approximately equal amounts. The beer leaves the beer well at a temperature of 90° F and undergoes a series of preheating steps before it enters the first stage of distillation. The dilute beer first passes into the tube side of a condenser-preheater. In this unit, approximately 23 percent of the total preheating is accomplished. This first preheating step utilizes a portion of the vapors condensed from the dehydration tower. These vapors are condensed to supply this first-stage preheating. The warmed dilute beer feed next passes to condenser-preheater, where additional preheat amounting to about 8.5 percent of the total is added. In this condenser-preheater, a portion of the overhead vapors from the pressure stripper-rectifier (PSR) is condensed to supply second-stage preheat. The dilute beer feed next passes through two stages of feed preheating, wherein a portion of the heat in the bottoms stream from the PSR tower is utilized in a two-stage flash operation. These stages add about 21.5 percent of

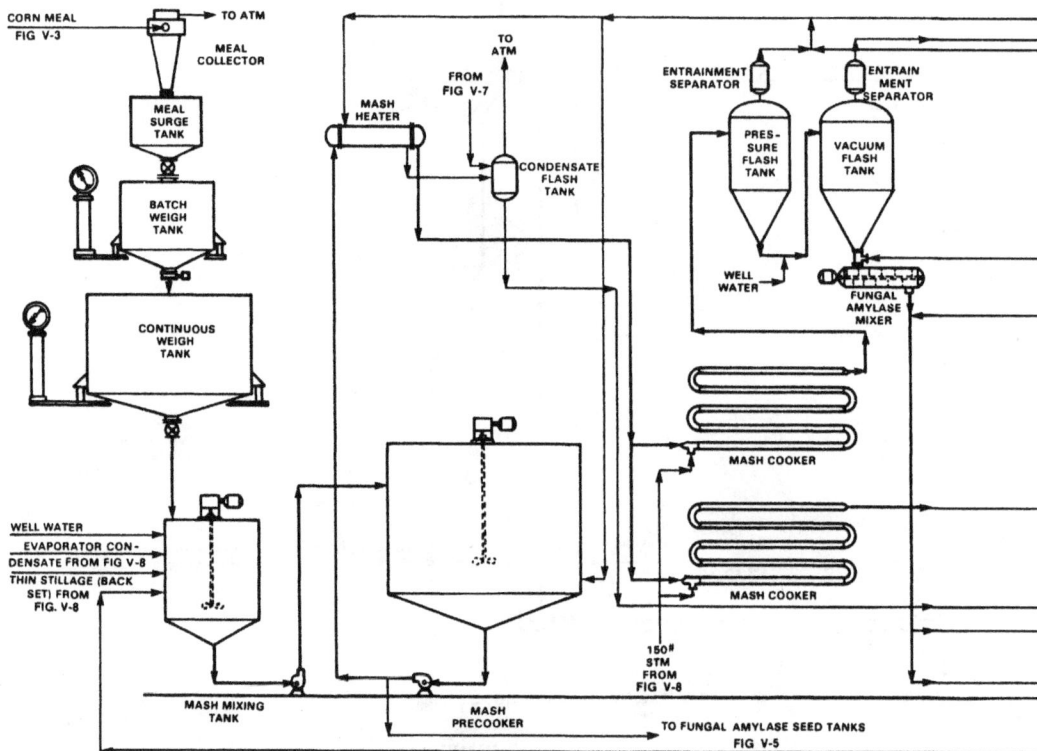

the total feed preheating. The warm dilute beer feed next passes into a steam condenser where low-pressure steam is used to accomplish additional preheating. Approximately 23 percent of the total feed preheat is added in the steam condenser. The heating medium, in this case, consists of low-pressure steam taken from other parts of the plant. The feed is finally preheated, approximately to saturation temperature, in an additional two-stage heating step using flash heat taken from the bottoms stream out of the PSR. Approximately 24 percent of the total feed preheat is accomplished there.

The hot, saturated, dilute beer feed next passes into the degassing drum, where dissolved carbon dioxide is flashed off. This represents one of the products of the fermentation reaction. It is not recovered. Any alcohol or water vapor, accompanying the vented carbon dioxide, is condensed in a vent condenser from which it drains back to the flash drum.

The saturated dilute beer feed enters the midsection of the PSR tower. Because of the high suspended solids content of the beer feed, the lower section of the PSR has been designed as a disc-and-donut type tower. This represents an effective contacting device which tends

to be self-purging and does not allow the buildup of solids which would block ordinary distillation trays. The PSR operates with a head pressure of 50 psig. The non-volatile dissolved solids and suspended solids in the dilute beer feed wash down through the stripping section of the PSR and a very dilute alcohol steam, containing less than 0.02 weight percent alcohol, is removed from the bottom of the tower. The dilute stillage containing the dissolved and suspended solids leaves the base of the PSR tower at about 304° F. In the bottom section of the PSR tower, alcohol is effectively stripped from the dilute beer. The aqueous bottoms stream then passes through a series of flash stages. These stillage bottoms are subjected to progressive reductions in pressure through four flash stages. The flash vapor that develops in these stages is utilized to accomplish a portion of the feed preheating as described previously. In these four flash stages, the temperature of the hot stillage is reduced from 304° F to approximately 212° F. The whole stillage, containing about 7.5 percent total solids, is next pumped to the coproduct recovery section of the plant where the solids in the stillage are recovered as an animal feed coproduct.

Heat is supplied to the base of the PSR by means of condensing 150 psig steam on the shell side of parallel

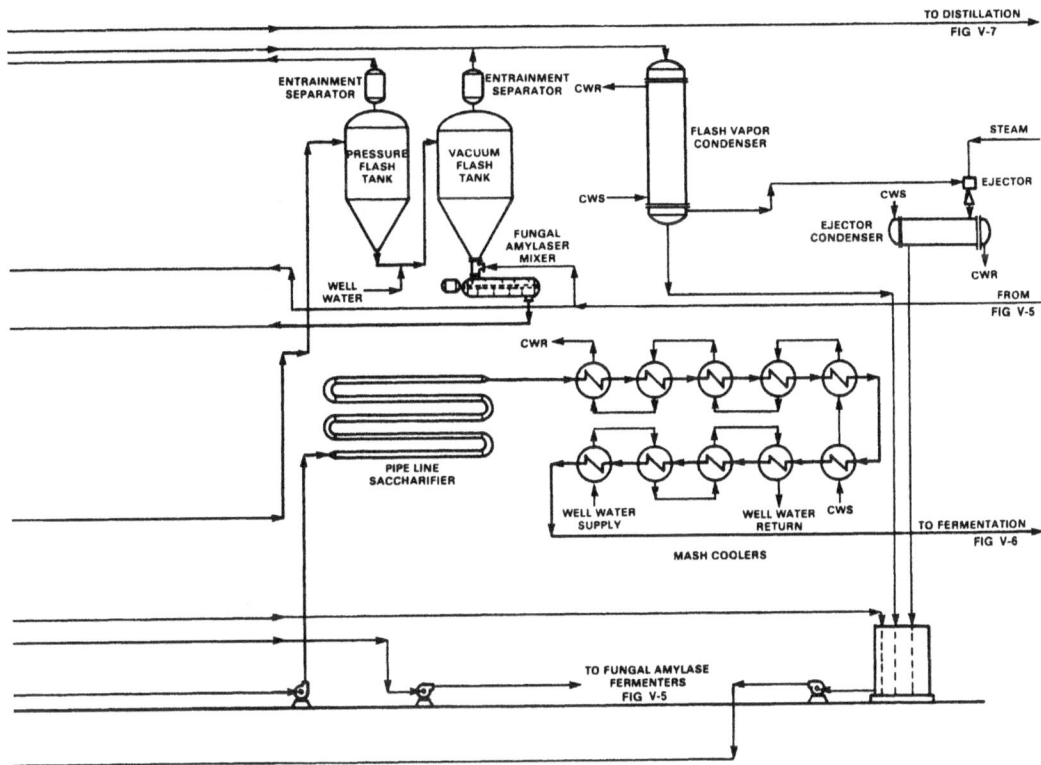

Figure V-4. Cooking and Saccharification

forced-circulation reboilers. Total steam supplied to the base of the PSR tower through the shell sides of the reboilers is 110,000 pounds per hour.

The upper portion of the PSR contains perforated trays and has a reduced diameter compared to the stripping section. The lower section of the PSR tower is 138 inches in diameter while the top section of the tower, containing 28 perforated trays, has a diameter of 102 inches.

Alcohol-rich vapors generated in the PSR pass overhead from the tower at a temperature of about 250° F and a pressure of 50 psig. These vapors may be utilized as a source of heat by condensing in the reboilers which are attached to the base of the dehydration tower and the hydrocarbon stripper. Sufficient vapor is generated in the PSR to allow a portion of the total overhead vapor to be utilized in a condenser-preheater to do some of the feed preheating which has been described previously. Of the total overhead vapor generated in the PSR, 10.9 percent is utilized for feed preheating, 81.6 percent is used to supply heat to the dehydration tower, and 7.5 percent is employed to supply heat to the hydrocarbon stripper. The upper five trays of the PSR operate in a total reflux condition. That is, all of the overhead vapor

is condensed and returned to the tap tray of the PSR. The liquid product from the PSR is removed as a liquid side draw stream about five trays from the top of the tower. From there, it passes to the midsection of the dehydration tower.

The dehydration tower is 138 inches in diameter and contains 50 perforated trays. The tower operates at essentially atmospheric pressure. The bottoms stream represents the anhydrous motor fuel-grade alcohol and has a concentration of 99.5 volume percent ethanol (199° proof); the balance is water. The bottoms stream from the dehydration tower is pumped through a product cooler which utilizes cooling water to reduce the temperature of the product alcohol to about 100° F. The cooled product next passes to product storage.

Heat is supplied to the base of the dehydration tower through parallel force circulation reboilers. The overhead product from the dehydration tower is a ternary, minimum-boiling azeotrope consisting of hydrocarbon, alcohol, and water. A portion of these overhead vapors is utilized for feed preheating and the balance condensed in the primary condenser, which utilizes cooling water to remove the heat of condensation in

FROM FIG V-4
STEAM

FROM FIG V-4

AIR
FILTER

PRE-FILTERS

AIR
COOLER
CWS

CWR

AIR

AIR COMPRESSOR

FUNGAL
AMYLASE
PUMP

TO FUNGAL AMYLASE MIXERS #220A&B
WELL WATER SUPPLY
WELL WATER RETURN

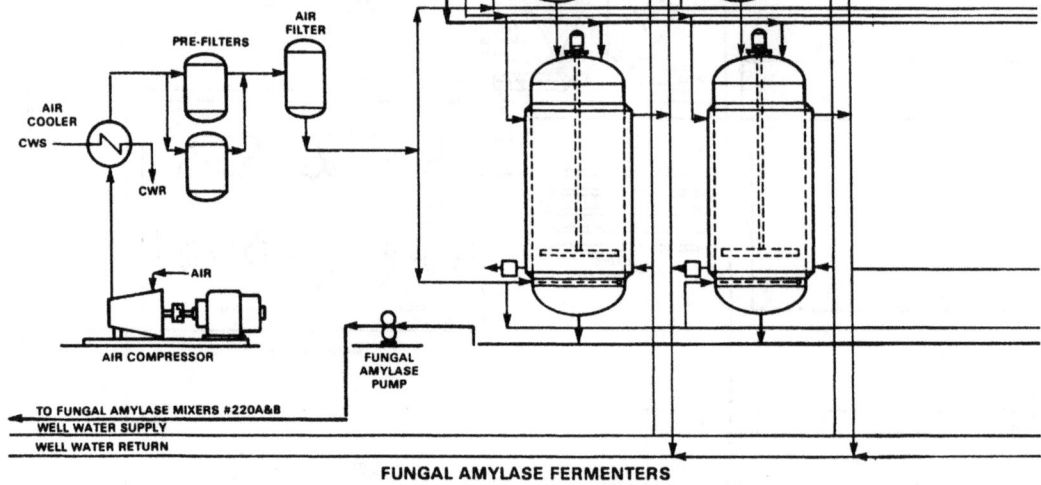

FUNGAL AMYLASE FERMENTERS

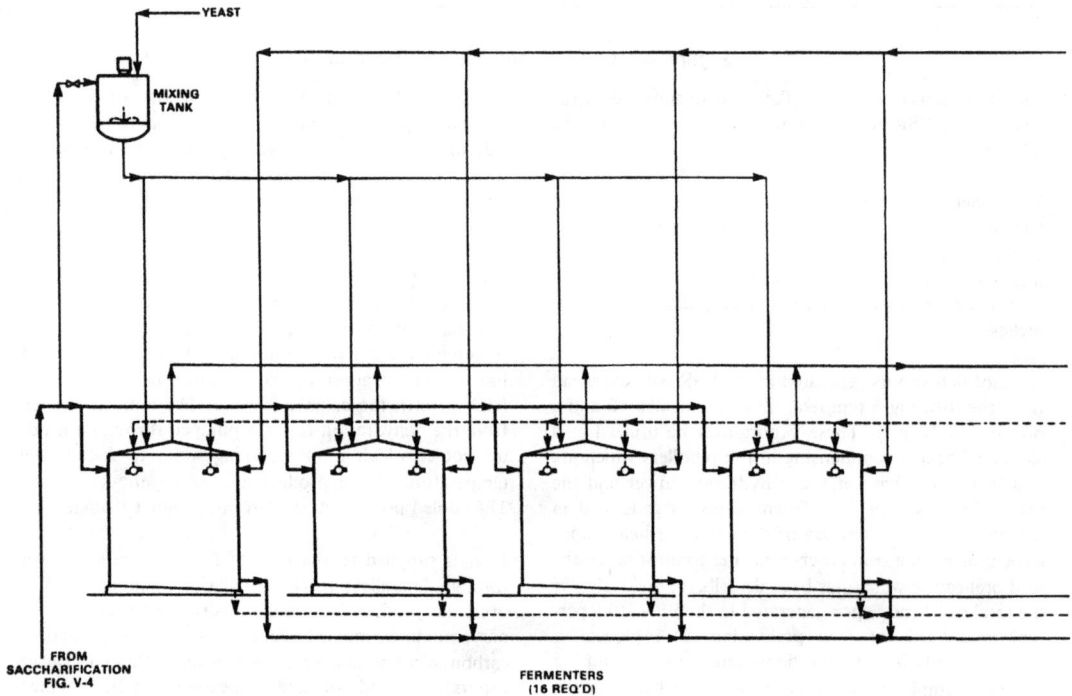

YEAST

MIXING
TANK

FROM
SACCHARIFICATION
FIG. V-4

FERMENTERS
(16 REQ'D)

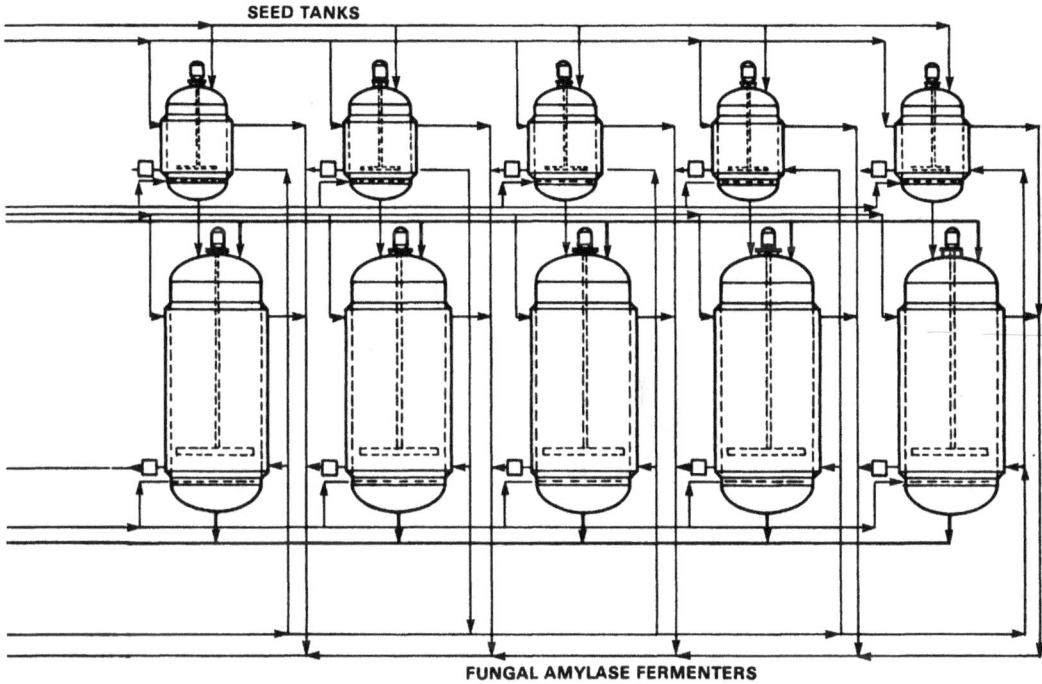

Figure V-5. Fungal Amylase Production

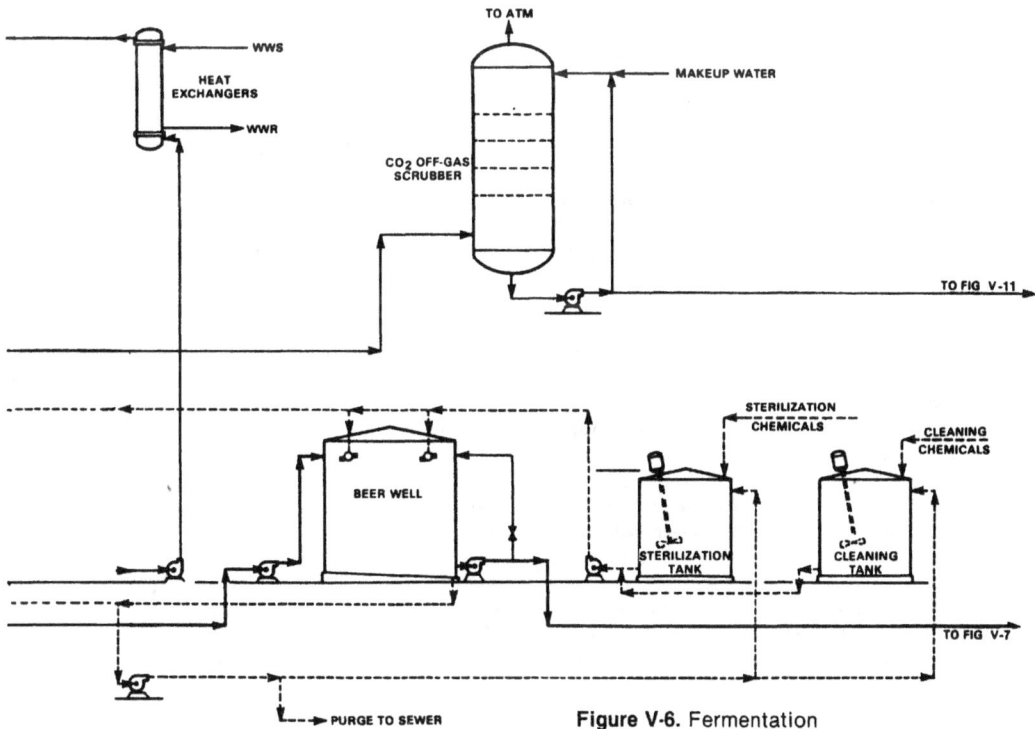

Figure V-6. Fermentation

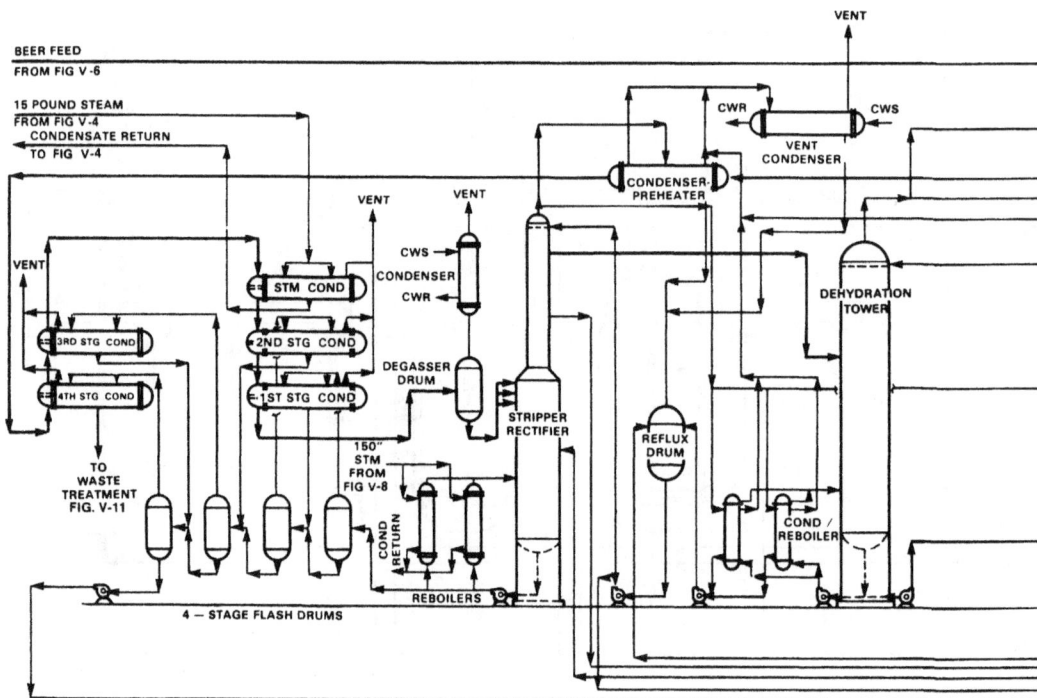

BEER FEED
FROM FIG V-6

15 POUND STEAM
FROM FIG V-4
CONDENSATE RETURN
TO FIG V-4

these vapors. The condensed vapors pass to a reflux cooler where they are further subcooled by cooling water prior to being fed to the decanter. The subcooled liquid entering the decanter separates into two layers. The upper layer is the larger in volume, and represents the hydrocarbon-rich layer. The lower, which separates, is a water layer containing some alcohol and hydrocarbon. The upper layer from the decanter is pumped back to the top tray of the dehydration tower. The lower layer from the decanter is pumped to the top tray of the hydrocarbon stripper which removes the remnants of hydrocarbon and alcohol contained in the feed to the top tray. The bottoms stream of the hydrocarbon stripper is essentially aqueous, and is removed and sent to the waste treatment plant. Thermal energy is supplied to the base of hydrocarbon stripper via alcohol-rich vapor condensing on the shell-side of a reboiler. The condensed alcohol-rich vapor is passed by to a reflux drum where it joins other vapor condensate before being returned to the top tray of the PSR. Overhead vapors containing alcohol, hydrocarbon, and water from the atmospheric pressure hydrocarbon stripper pass to a condenser-preheater, where they are condensed. The condensate is returned through the reflux cooler to the decanter. The aqueous stream passing from the bottom of the hydrocarbon stripper contains less than 0.02 weight percent alcohol.

This distillation system is covered by a patent allowed, but not yet issued, to Raphael Katzen Associates, International, Inc.

Fusel Oil and Head Removal

In the yeast fermentation process, certain extraneous products, in addition to ethyl alcohol, are formed. These are generally higher alcohols, i.e., higher molecular weight alcohols known as fusel oils, and light ends which include such materials as aldehydes, etc.

The distillation system provides for the removal of these extraneous components in the following manner. (See Figure V-7). The fusel oils have the property of being more volatile than alcohol in dilute aqueous solution, but are less so than alcohol in concentrated alcohol solution. For this reason, they tend to concentrate on some tray in the rectifying section of the PSR. These fusel oils, thus having concentrated, can be removed as a liquid side drawstream from the PSR. They are removed and passed through a fusel oil cooler and to a fusel oil washer, a water washing extraction column in which the alcohol content of the fusel oils is washed from them, under reduced temperature, by counter-currently contacting the cooled fusel oil side stream with a stream of cold water. The heavy aqueous stream, containing the

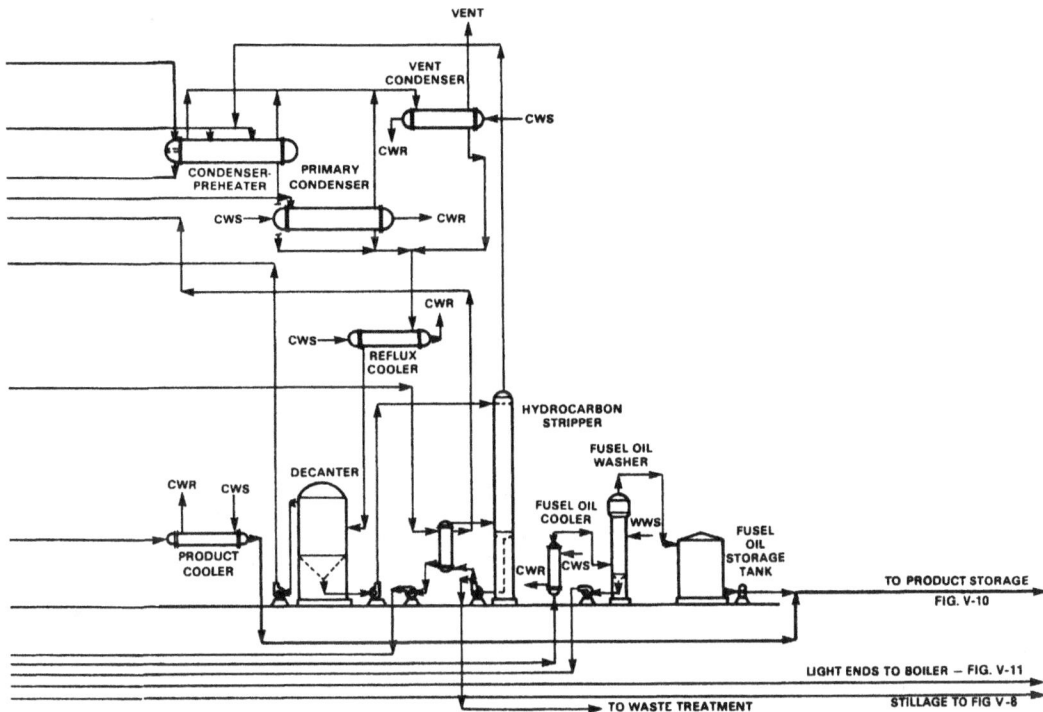

Figure V-7. Distillation

extracted ethyl alcohol, is removed from the base of the fuel oil washer. This stream is returned to the lower section of the PSR for alcohol recovery. The light fuel oil stream is decanted from the top of the fuel oil washer and passes into the fuel oil storage tank.

In general, the fermentation process, when utilizing corn, will produce about 4 to 5 gallons of fuel oil for every 1,000 gallons of anhydrous alcohol product. These fuel oils do have a heating value and can be reblended into the product. If this reblending operation is not desired, then the fuel oils may be passed to the plant boiler where they are used as fuel. Fuel oils should have no harmful effect upon motor fuel-grade alcohol.

Light extraneous fermentation products such as aldehydes are effectively removed in this distillation system by withdrawing a very small purge from the total reflux stream passing back to the top tray of the PSR. This light component purge, in general, cannot be reblended into the alcohol to be used for motor fuel blending, because these light products would tend to cause vapor lock when the ethanol is blended with gasoline to produce gasohol. Therefore, these materials are removed and sent to the plant boiler where their fuel value is recovered.

The fuel oil and light ends must be removed because their presence would upset the equilibrium associated with the dehydration step, and could cause problems in the decantation step.

The distillation scheme* for producing motor fuel-grade alcohol, as described here, utilizes only 17.5 pounds of process steam per gallon of anhydrous motor fuel-grade alcohol product. This great reduction in energy use is accomplished by optimizing the feed preheating scheme and by utilizing the heat content of high-pressure vapors produced in the PSR to supply the reboil heat for both the dehydration step and the hydrocarbon-alcohol stripping.

Evaporation and Drying of Stillage Residue

Stillage from the distillation area is delivered to the whole stillage tank, where it is pumped to the solid bowl centrifugals that operate on a continuous basis. (See Figure V-8). These centrifugals separate the whole stillage into two fractions: thin stillage containing 6.5 to 10 percent total solids and thick stillage containing about 35 percent total solids. Part of the thin stillage, referred to as backset, (corresponding to 10 percent of

---

*Patents allowed but not yet issued.

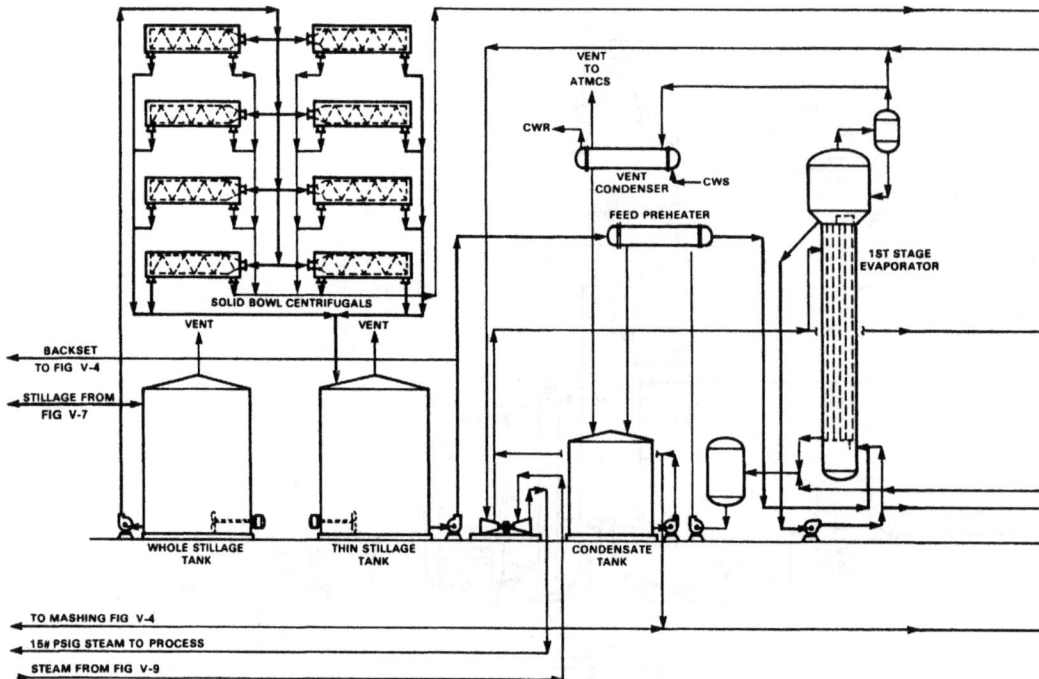

the total mash) is recycled to the mash mixing tank in the cooking and saccharification section.

The remaining thin stillage is evaporated in a vapor recompression evaporator to about 55 percent solids. Because of a cooling effect in the centrifugal separators (caused by evaporation of the stillage in contact with air), the thin stillage must be reheated from about 165 to 208° F before it enters the evaporator. Heating is accomplished by using evaporator condensate cooled from 230 to 185° F. Power for driving the evaporator's vapor compressor (approximately 6,200 hp) is provided by a steam turbine which uses 580 psig steam and exhausts at about 160 psig. This exhaust steam is used for distillation, mash cooking, and heating supplemental air for spent grains drying. The vapor compressor operates at approximately atmospheric pressure at the inlet and compresses the vapor to about 21 psia. The compressor outlet steam is superheated, but before it enters the evaporator bodies it is desuperheated by injection of condensate. This is done to get the best heat transfer possible and to prevent "baking" of solids on the evaporator surfaces. The water vapor from the evaporator bodies passes through entrainment separators and then to the vapor compressor. However, before entering the compressor, it must be superheated by mixing with a recycle flow of superheated vapor from the compressor outlet, to maintain dry, non-corrosive conditions in the vapor compressor.

The thick stillage is mixed with the concentrated thin stillage and recycled dry grains in the wet grains minglers. (See Figure V-9). The amount of dry grains recycle is regulated to maintain a wet grains moisture content of 30 percent to minimize stickiness.

The wet grains are then fed to rotary dryers where they are tumbled in contact with hot flue gas from the power boiler. Supplemental hot air for drying is provided by an air heater using 150 psig steam for heating. The hot gas enters the dryers at about 600° F with a wet bulb temperature of about 145° F. The gas and dry grains leave the dryers at about 190° F. The flue gas goes through cyclone collectors to remove most entrained solids before it is delivered to the flue gas scrubber.

The dry grains, at about 10 percent moisture, are transferred to the dry grains hopper. About 75 percent of the dry grains is recycled in order to regulate the moisture content of the wet grains. The remaining net make of dry grains is ground in a hammer mill and then cooled in the product cooler, an auger-type heat exchanger that uses well water to cool the grains to about 100°F. The cooled coproduct is then transferred pneumatically to storage and shipping.

COMMERCIAL SCALE ETHANOL PRODUCTION AND FINANCING

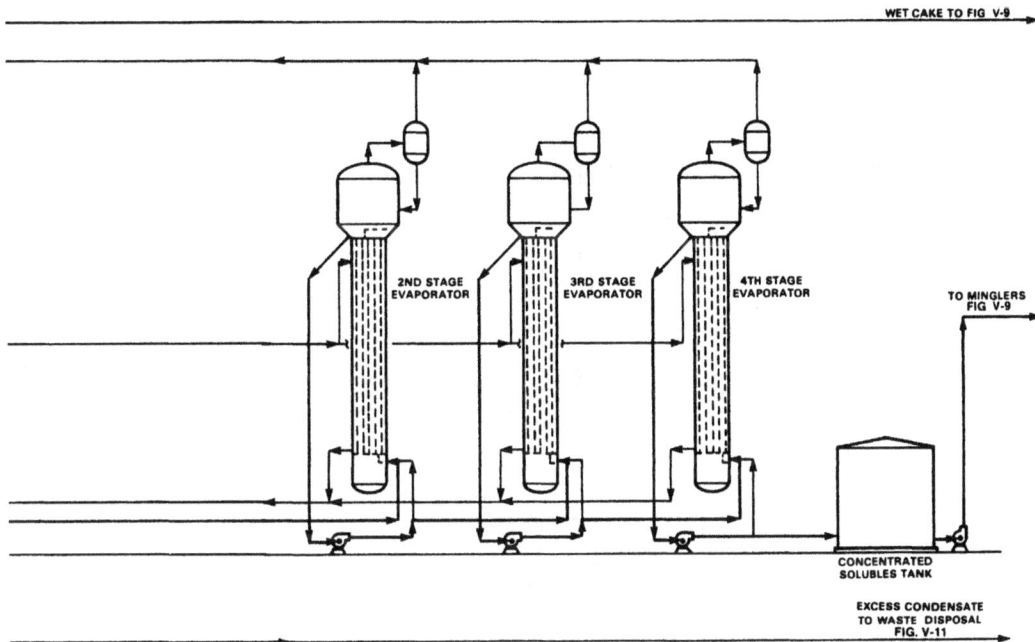

Figure V-8. Residue Feed Processing

Alcohol, Ammonium Sulfate, and Dry Grains Storage and Shipping

The alcohol is received from the distillation section and stored in receiver tanks, each of which is sized for one day of production. (See Figure V-10). While one tank is being filled, the other is checked for quality and quantity by the government inspector. After inspection, its contents are sent to long-term storage. The four storage tanks hold a total of about 4.2 million gallons—about 28 days of production. Upon transferring the product from storage to a tank car or truck, denaturant (gasoline) is added at a rate of 1 gallon per 100 gallons of alcohol. The denaturant tank holds about 50,000 gallons. Accurate metering is provided by positive displacement meters.

A water solution, containing 40 percent ammonium sulfate, is produced by the flue gas scrubbing system. This solution is to be sold as field fertilizer and is stored in four tanks having a capacity of 1 million gallons each. This corresponds to storage of the total ammonium sulfate production for about 9 months, which allows for storage during the fertilizer off-season.

Dry grains are stored in an A-frame-type building with a storage capacity of about 295,000 cu ft, equivalent to

about one week's production. (See Figure V-10). Shipping of the dry grains from storage is done on a first in/first out basis and utilizes a front-end loader to load the pneumatic conveyor system which transfers the grains to the live bottom surge bin at a rate of 88,000 lbs per hour. This rate is based on shipping out the dry grains for an average of 12 hours each day. Shipment may be made by either truck or rail, and shipping scales are provided for weighing the shipments.

Coal-Fired Boiler

Steam for the plant is provided by a coal-fired boiler rated at 250,000 lb/hr of steam at 600 psig and 600° F. (See Figure V-11). Calculated plant usage is about 200,000 lb/hr of steam. The firing rate is 12.6 T/hr of coal, having a gross heating value of 10,630 Btu/lb (Illinois No. 6 coal). The coal contains approximately 3.8 percent sulfur (moisture-free basis). Small fuel inputs are also provided by light ends from distillation of the alcohol and by de-watered sludge from waste treatment.

The coal unloading facility provides for direct transfer to the coal bunker or to a storage pile. A front-end loader is used to transfer coal from the pile back to the unloading area where it can be transferred to the coal bunker.

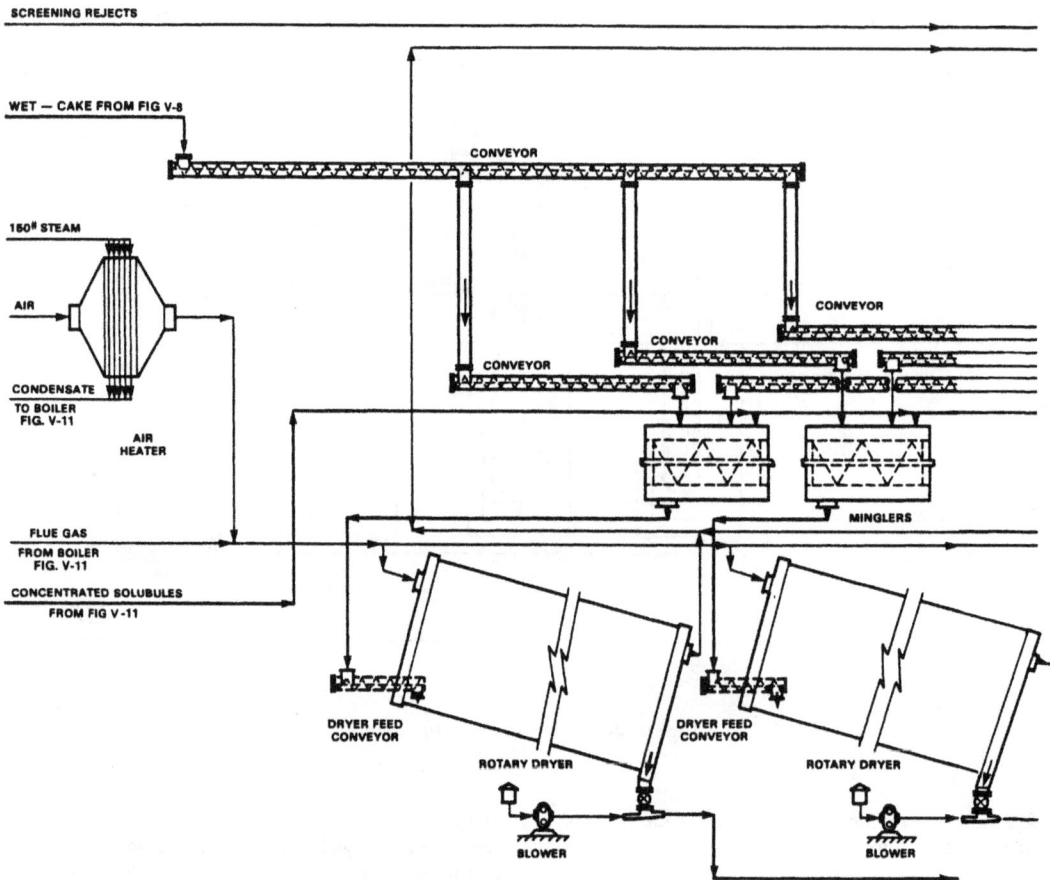

WET — CAKE FROM FIG V-8

CONVEYOR

150# STEAM

AIR

CONVEYOR

CONDENSATE
TO BOILER
FIG. V-11

AIR
HEATER

CONVEYOR

CONVEYOR

MINGLERS

FLUE GAS
FROM BOILER
FIG. V-11

CONCENTRATED SOLUBULES
FROM FIG V-11

DRYER FEED
CONVEYOR

DRYER FEED
CONVEYOR

ROTARY DRYER

ROTARY DRYER

BLOWER

BLOWER

Boiler feedwater is provided by condensate return from the process, where possible, and by makeup water that has been filtered and conditioned in a conventional boiler feedwater treatment system. About two-thirds of the boiler feedwater is condensate return.

The flue gas passes through the cyclone collector for particulate removal. The collector consists of numerous small cyclones (multi-clones) housed in a single chamber. The recovered particulate goes to the boiler at a temperature of about 725° F. This is a high flue gas temperature by normal standards, but in this case, does not adversely affect the overall plant thermal efficiency since the hot flue gas, tempered with air to 600° F, goes to the stillage drying section where its heat is used to dry the distiller's grains. For this reason, no flue gas heat economizer or tack is required with the boiler.

The coal is fed to the boiler by four stoker-spreader units. The spreader feeding system was selected (rather than pulverized blown coal) because of the small boiler size and because the boiler inefficiency, due to excess air, does not affect the process thermal efficiency.

Water Supply

The water supply system provides well water for meeting process makeup requirements, process cooling where cooling tower water is not cool enough, and maintaining a supply of fire protection water. (See Figure V-12). Well water is also used to provide makeup to the cooling tower. Three wells are provided with a capacity of 1,800 gpm each. The well water storage tank has a capacity of 100,000 gallons and provides for surges in demand and allows short-term shutdowns of the wells, as required for maintenance or repairs, without interrupting the supply of water to the plant.

The fire protection system consists of the fire protection tank which has a capacity of 300,000 gallons, four fire water pumps (2 electric and 2 diesel), and an underground fire water distribution system, along with the appropriate fire hydrants and spray headers. Each pump has a capacity of 2,000 gpm. The diesel-powered pumps are used only in the event of an electric power failure.

COMMERCIAL SCALE ETHANOL PRODUCTION AND FINANCING

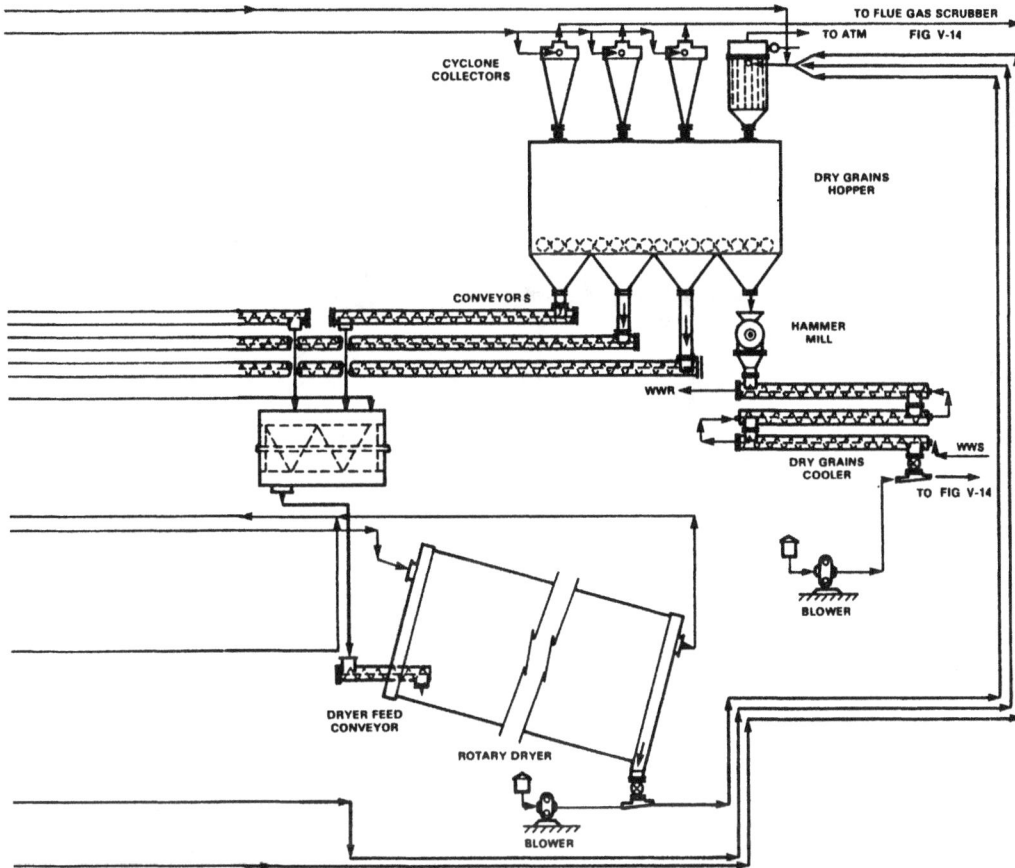

Figure V-9. Stillage Mixing

The cooling tower is designed to provide 16,000 gpm of water at 85° F from a warm water return temperature of 115° F, and an ambient design wet-bulb temperature of 75° F. The tower consists of two cells with a two-speed fan for each cell. The three cooling tower pumps are rated at 7,500 gpm each.

Wastewater Treatment

The wastewater flows include process wastewater and sanitary wastewater. (See Figure V-13). The wastewater is collected at a lift station and pumped to the treatment plant.

The wastewater treatment plant is designed for secondary treatment with two extended aeration tanks and two settling tanks. The aeration tanks are 95 ft in diameter. The sludge thickening tank is 20 ft in diameter.

The influent stream passes through a bar screen and grinder prior to entering the first aeration tank. Water

from the first stage is split; the major portion is recycled to the first aeration stage, while the remainder is sent to the thickening tank. Clarified water from the first settling stage overflows to the second aeration tank, whereby additional biochemical oxygen demand (BOD) is removed. Nutrients may be added to either aeration tank, but due to the nature of the wastewater flows, should not be needed. The water from the second aeration tank overflows to the second-stage settling tank.

The sludge from the second-stage settling tank is recycled to either the first- or second-stage aeration tank. A stream is sent to the thickener. The two-stage aeration system, coupled with the flexibility to recycle sludge from either stage, allows good control of effluent BOD.

The clarified water from the second-stage settling tank flows by gravity to the chlorine contact tank. Chlorine is added to the contact tank for destruction of final traces of impurities. The effluent from the chlorine tank flows to a nearby river.

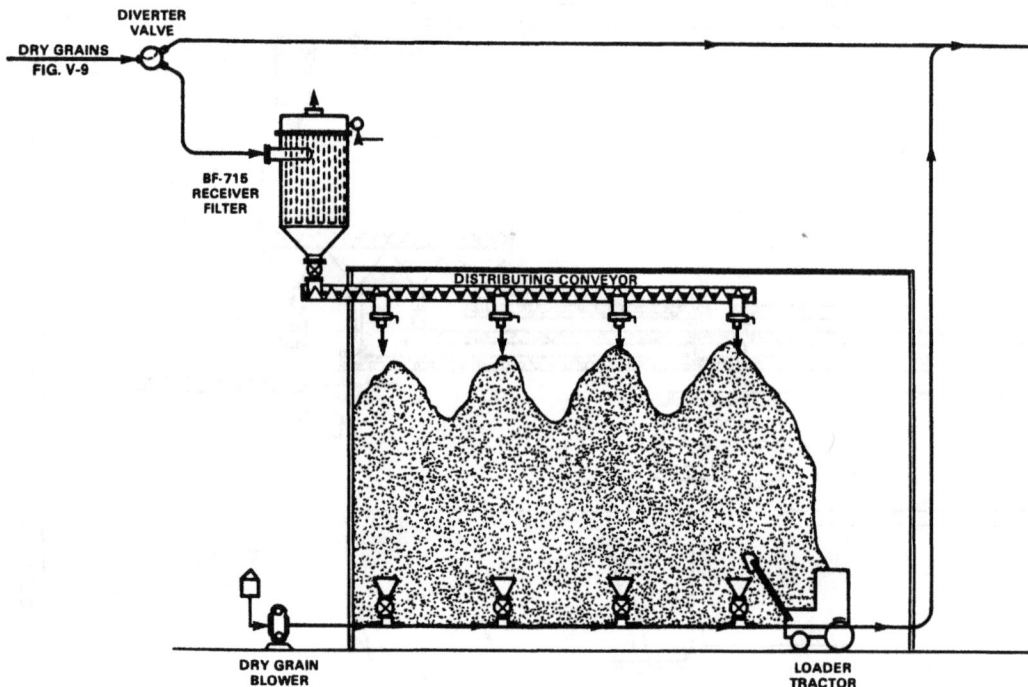

DIVERTER VALVE

DRY GRAINS
FIG. V-9

BF-715
RECEIVER
FILTER

DISTRIBUTING CONVEYOR

DRY GRAIN
BLOWER

LOADER
TRACTOR

Sludge from both the first- and second-stage aeration tanks is collected in the sludge thickener and pumped to the dewatering press. The sludge is mixed with primary sludge collected from the flue gas desulfurization system and dewatered to about 25 percent solids. The solids from the dewatering press are chopped up in a flaker and conveyed to the boiler for burning.

The treatment plant is designed to remove 95 percent of the effluent BOD.

Flue Gas Scrubber

The hot flue gas from the coal-fired boiler is sent to dry grains recovery where it mixes with dilution air for grains drying prior to being sent on to the flue gas scrubbing system. (See Figure V-14). Because of the high sulfur content of Illinois No. 6 coal (and also the high purchase price of low- versus high-sulfur coal) a flue gas desulfurization system is required. The desulfurization system recovers the SO_2 as ammonium sulfate and differs from the more conventional limestone scrubbing systems in that no calcareous sludge is produced in the system. The only coproduct is the ammonium sulfate.

In the flue gas scrubbing system,* water sprays cool the gas and remove particulate in three stages. The first stage removes particulate only. The next two stages cool the gas and remove additional particulate. Ammonia is

used in a two-stage absorption section for removal of sulfur dioxide from the flue gas. The ammonium sulfite/bisulfite solution from the scrubbers is neutralized to ammonium sulfite and oxidized to ammonium sulfate. The ammonium sulfate is suitable for sale as agricultural fertilizer. A more detailed description of the system is given below.

The hot gas enters the spray quench section where it is quenched to within 5° of its wet bulb temperature, 143° F.

The saturated gas enters the first section of the flue gas scrubber where it is washed with a very high rate of water flow from spray nozzles. The gas is then cooled in two successive spray cooling sections to 110° F. Part of the heat removed in the first cooling section is used to reheat the exit flue gas in the heat exchanger, to improve the flue gas buoyancy. The remaining heat from the cooling sections is removed by cooling water using plate-and-frame-type heat exchangers. Particulate matter, removed in the quench and cooling sections, is removed by the cyclone cleaner. This material is concentrated further in a clarifier and then combined with sludge from the waste treatment system. The mixture is

---

*U S Patent 3,957,951 and patents pending, licensed by Raphael Katzen Associates International

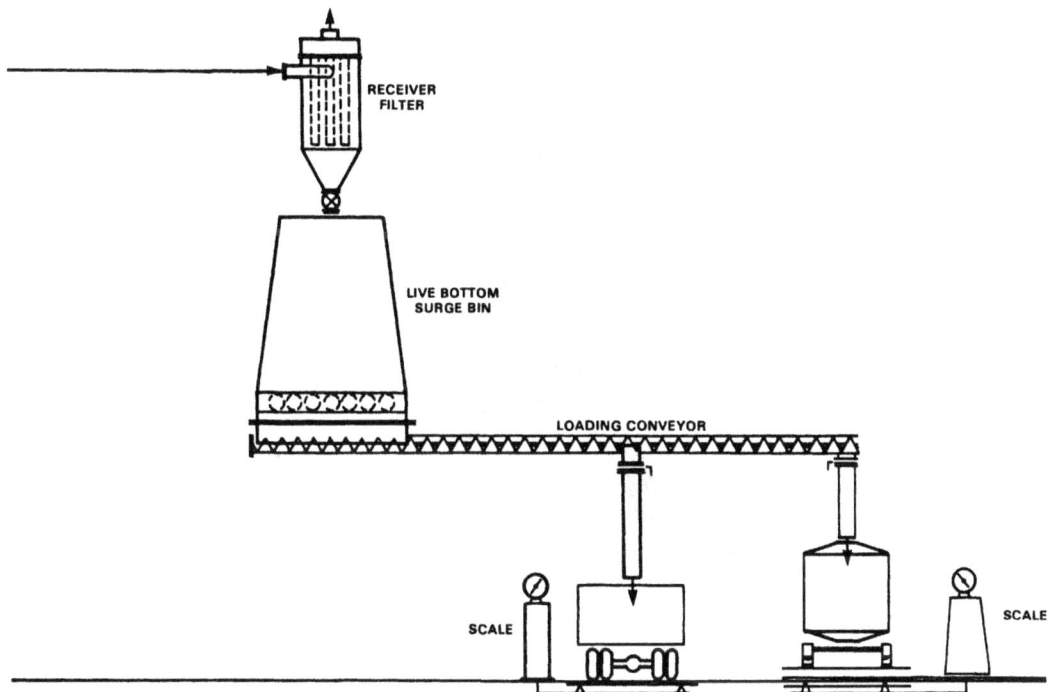

Figure V-10. Storage and Shipping of Dry Grains

then dewatered in the dewatering press and sent to the coal-fired boiler. The boiler is equipped with special feeder/spreaders to handle the sludge.

In the absorption stages, the pH of the liquid is carefully controlled in order to minimize the gas phase relation between ammonia vapor and sulfur dioxide gas. The pH is controlled by adjusting the absorption solution circulation rate and the rate of ammonia addition to each stage. The heat of reaction from SO_2 absorption is removed by another plate-and-frame exchanger.

The flue gas draft for the scrubbing system, as well as for the spent grains driers and interconnecting ducting, is supplied by the flue gas fan located at the scrubber outlet. The fan is driven with a 1,000-hp motor and will handle 150,000 acfm with a 30'' water guage pressure increase to the flue gas.

The product ammonium sulfite/bisulfite solution from the scrubber is taken from the lower absorption loop downstream of the first-stage absorption loop pump. This solution is neutralized with aqueous ammonia to ammonium sulfite and then oxidized to ammonium sulfate in the oxidizer reactor. This solution is then cooled to 100° F in the oxidizer exchanger. The water makeup to the scrubber will be controlled to produce a final oxi-

dized solution strength of 40 percent ammonium sulfate, which can be used for direct application to fields for fertilizer or for a blending material by fertilizer manufacturers.

Materials Flow

Table V-1 summarizes the daily materials flow through the system. Most noticeable is the large amount of water required by the process, i.e., about 7.25 gallons of water per gallon of ethanol produced.

Energy Flows

Table V-2 summarizes the energy flows through the system by major operation in the process. Distillation requires over half the steam input to the process, followed by cooking and saccharification, which require about 30 percent of the total. Enzyme production, DDG recovery, and utilities are the major components of electricity demand.

The efficiency of the system (ratio of energy out to energy in) may be expressed in several ways:

- If only the fossil fuel inputs are considered (electricity being based on a 10,000 Btu/kwh equiv-

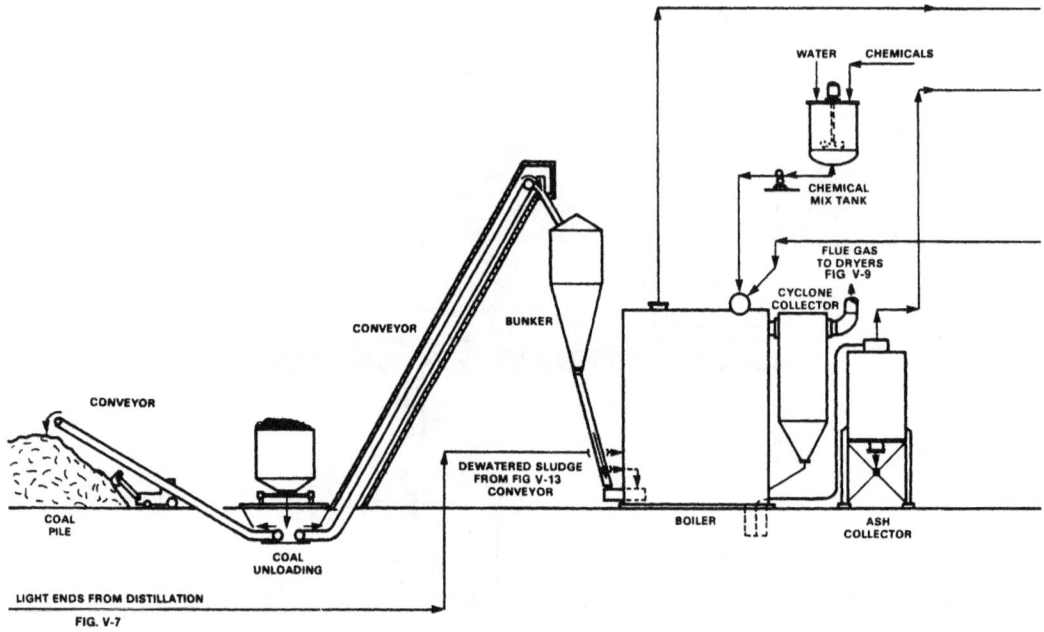

CONVEYOR

CONVEYOR

BUNKER

WATER   CHEMICALS

CHEMICAL
MIX TANK

FLUE GAS
TO DRYERS
FIG V-9

CYCLONE
COLLECTOR

CONVEYOR

COAL
PILE

COAL
UNLOADING

DEWATERED SLUDGE
FROM FIG V-13
CONVEYOR

BOILER

ASH
COLLECTOR

LIGHT ENDS FROM DISTILLATION

FIG. V-7

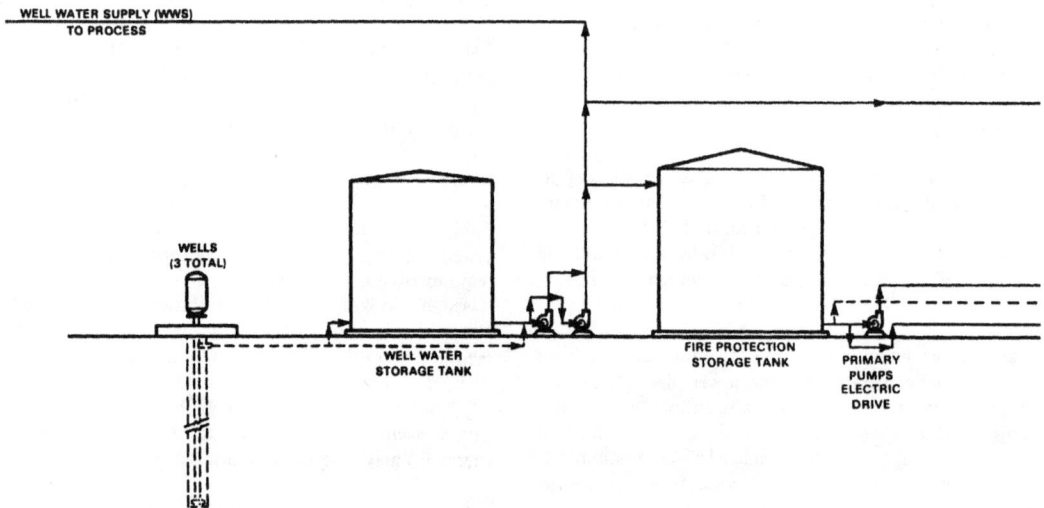

WELL WATER SUPPLY (WWS)
TO PROCESS

WELLS
(3 TOTAL)

WELL WATER
STORAGE TANK

FIRE PROTECTION
STORAGE TANK

PRIMARY
PUMPS
ELECTRIC
DRIVE

COMMERCIAL SCALE ETHANOL PRODUCTION AND FINANCING

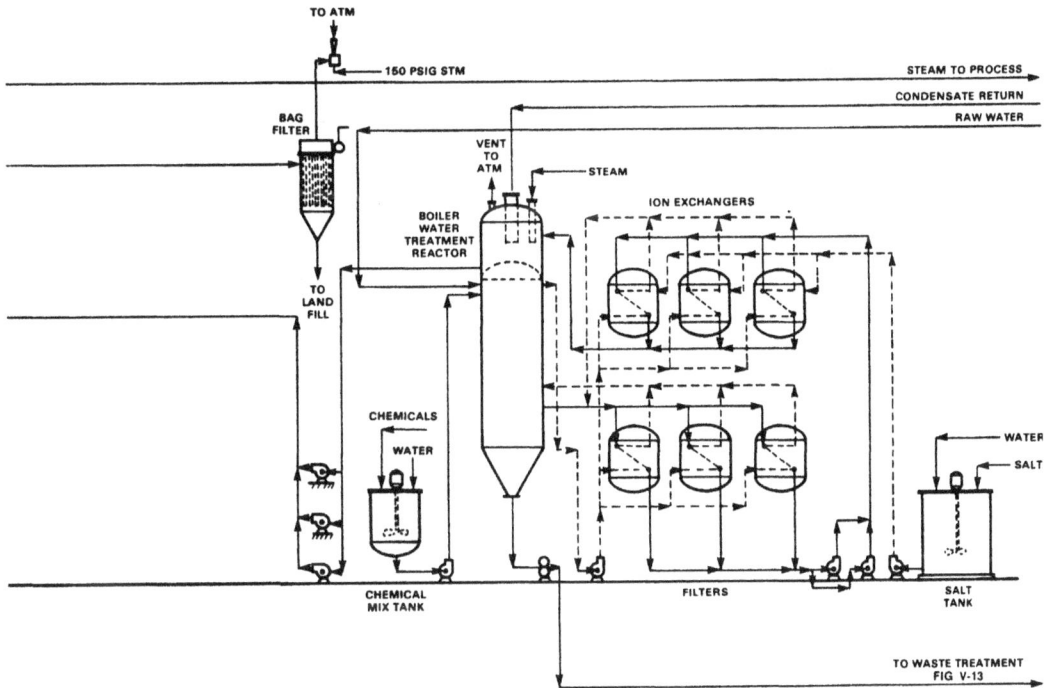

Figure V-11. Utilities, Boiler

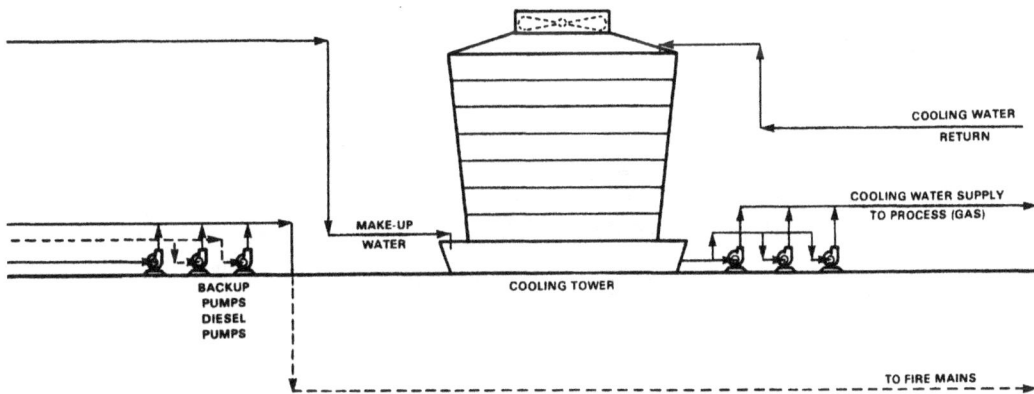

Figure V-12. Utilities, Water Supply

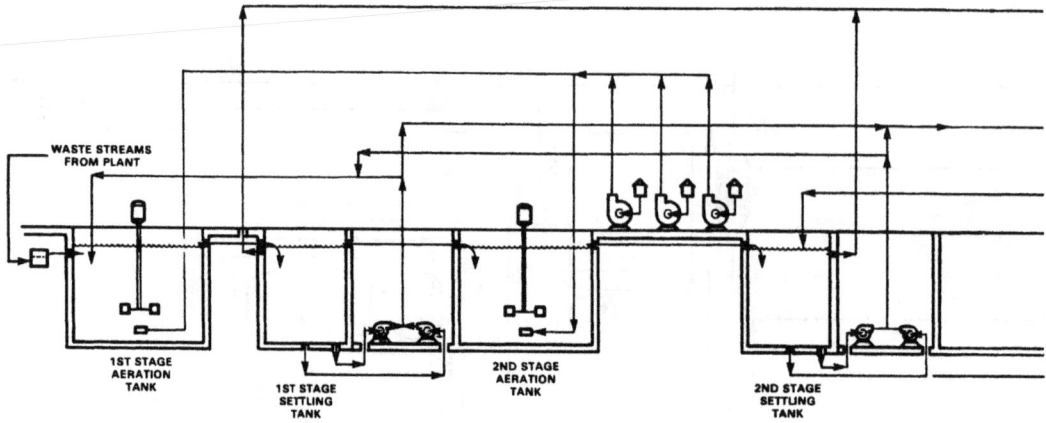

WASTE STREAMS FROM PLANT

1ST STAGE AERATION TANK

1ST STAGE SETTLING TANK

2ND STAGE AERATION TANK

2ND STAGE SETTLING TANK

SULFATE

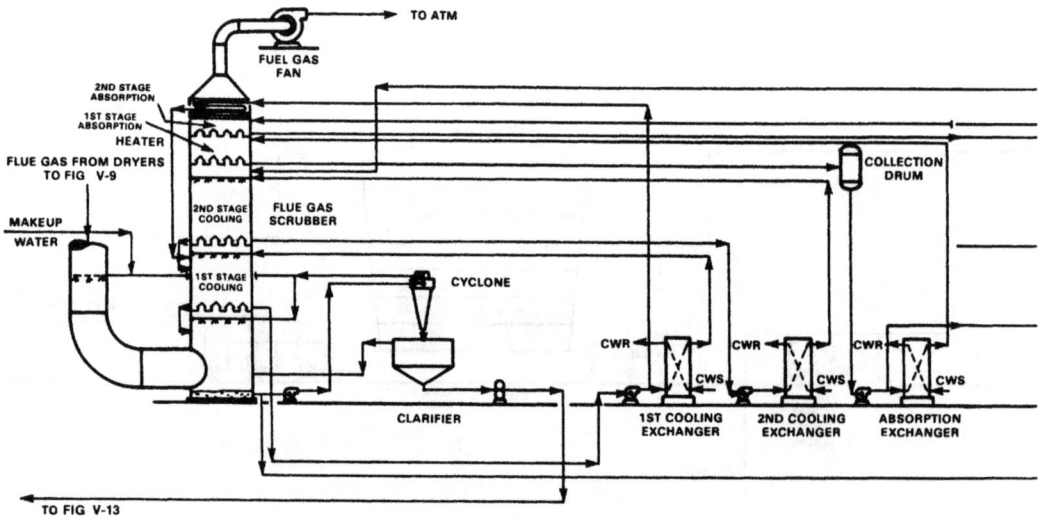

TO ATM

FUEL GAS FAN

2ND STAGE ABSORPTION

1ST STAGE ABSORPTION

HEATER

FLUE GAS FROM DRYERS TO FIG V-9

MAKEUP WATER

2ND STAGE COOLING

FLUE GAS SCRUBBER

1ST STAGE COOLING

CYCLONE

COLLECTION DRUM

CWR   CWR   CWR

CWS   CWS   CWS

CLARIFIER

1ST COOLING EXCHANGER

2ND COOLING EXCHANGER

ABSORPTION EXCHANGER

TO FIG V-13

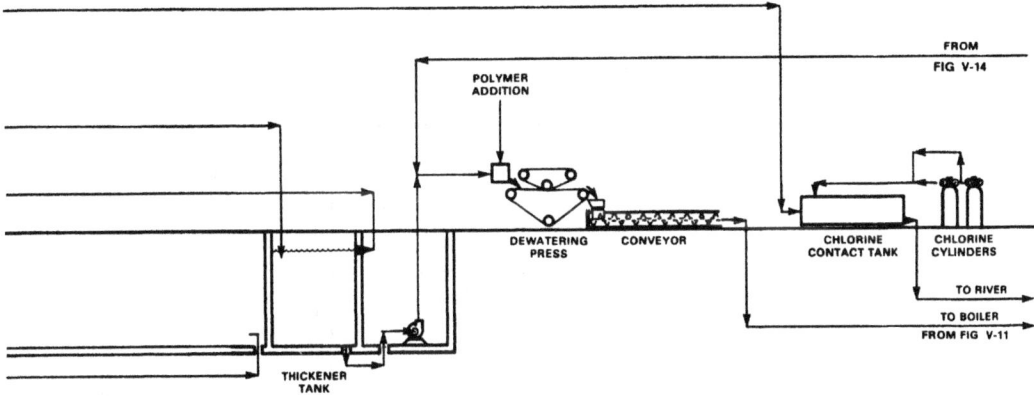

Figure V-13. Utilities, Wastewater Treatment

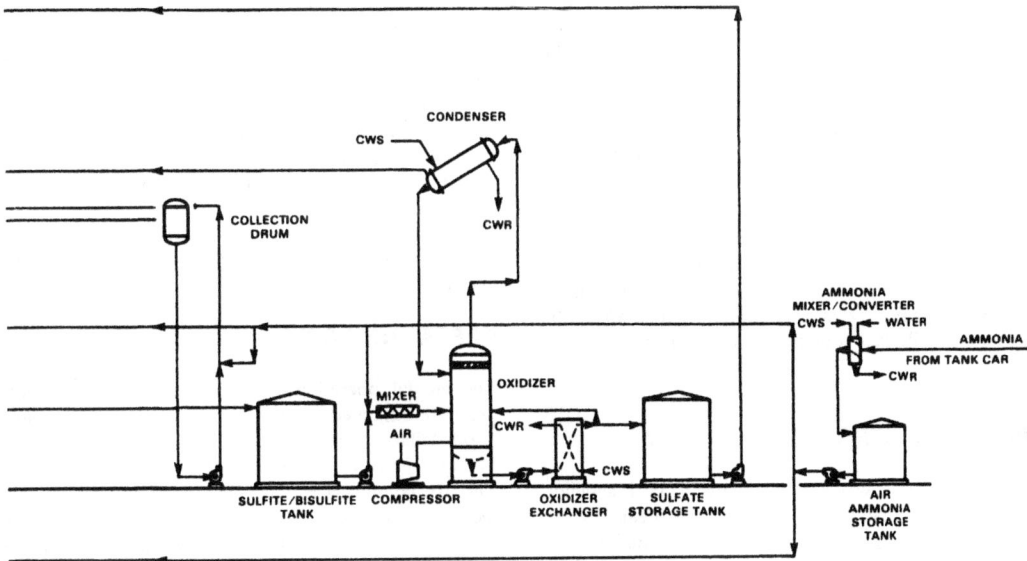

Figure V-14. Utilities, Flue Gas Scrubber

Table V-1. Materials Flow for a 50-Million-Gallon-Per-Year
Ethanol Plant

ITEM	UNITS	DAILY QUANTITY
INPUTS		
Corn	Bushel	58,900
Coal	Tons	296.7
Yeast	Tons	1.2
Denaturant	Gallons	1,500
Ammonia	Tons	9.2
Water	Gallons	1,250,000
OUTPUTS		
Ethanol	Gallons	151,515
Distillers Dried Grains and Solubles	Tons	536.7
Ammonium Sulfate	Tons	31.6
Wastewater	Gallons	1,100,000

SOURCE Katzen, R , "Grain Motor Fuel Alcohol Technical and Economic Assessment Study" Raphael Katzen
Associates for the U S DOE, Dec , 1978

alent): alcohol/(coal + electricity) = 154.4% or about 55,000 Btu per gallon of ethanol.

- If the energy used to produce the corn and a credit for DDG production are included: (alcohol + DDG)/(coal + corn + electricity) = 105.1% (corn production requires about 95,000 Btu/bushel)

- If the total thermal energy in the corn and DDG are considered: (alcohol + DDG)/(coal + corn + electricity) = 68.9%.

Efficiency is not really a concern to the investor except as it affects economics. It is more a matter of national policy to determine what cost in fossil fuel (coal) can be tolerated to displace a barrel of imported oil.

PLANT INVESTMENT

The total plant investment is estimated to be $58 million (end of 1978 dollars). Table V-3 shows a breakdown of the investment by plant section. The two major items are the utilities and the DDG recovery sections, which together account for over half the total investment. For a plant in the range of 50 million gallons per year, the investment is on the order of $1.16 per gallon capacity (1978 dollars).

For plants of 10 million gallons capacity, the investment was estimated at about $25.2 million (1978 dollars) or about $2.50 per gallon capacity. The investment for a 100-million-gallon plant was estimated at $100 million or about $1.00 per gallon capacity. Figure V-15 summarizes the plant investment data generated by Katzen and others. It is apparent that large investment economies are achieved when the capacity increases from 10 to about 50 million gallons per year. Above this range, the economies of scale are much less significant, since many pieces of equipment in the larger plants are multiples of those used at the 50-million-gallon capacity and some of the large pieces of equipment such as the distillation towers must be field- rather than factory-assembled.

OPERATING COSTS

The annual operating costs for the 50-million-gallon plant are estimated at $44.51 million or $0.89 per gallon (1978 dollars). The costs include straight line depreciation over 20 years. Table V-4 summarizes the cost items. The first column gives the 1978 equivalent cost and the second, the operating cost for the first year. The major cost item is corn and the second is the credit for the DDG coproduct. The net cost of corn (corn cost minus credit) amounts to about 58 percent of the annual operating cost. Therefore, ethanol production will be very sensitive to fluctuations in feedstock and coproducts market prices.

COMMERCIAL SCALE ETHANOL PRODUCTION AND FINANCING

Table V-2. Energy Flow for a 50-Million-Gallon-Per-Year
Ethanol Plant

ITEMS	PROCESS STEAM		ELECTRICITY	
	LB/HR	%	kW	%
Receiving, Storage, Milling			507	6.1
Mash Cooking & Saccharification	61,000	30.5	216	2.6
Enzyme Production	1,400	0.7	1,696	20.4
Fermentation	400	0.2	333	4.0
Distillation	117,000	58.5	133	1.6
DDG Recovery	12,800	6.4	2,253	27.1
Storage & Denaturing			58	0.7
Utilities	5,400	2.7	3,076	37.0
Buildings	2,000	1.0	42	0.5
TOTAL	200,000	100.0	8,314	100.0

SOURCE Katzen, R , "Grain Motor Fuel Alcohol Technical and Economic Assessment Study" Raphael Katzen
Associates for the U S DOE, Dec , 1978

Table V-3. 50 Million Gallon Per Year Plant Investment
Base Case

IDENTIFICATION	DECEMBER 1978 COST (DOLLARS)	PERCENT OF INVESTMENT
Receiving, Storage & Milling	$ 2,086,800	3.96
Cooking and Saccharification	2,824,300	5.36
Fungal Amylase Production	3,485,900	6.61
Fermentation	4,195,600	7.96
Distillation	5,123,800	9.72
Dried Grain Recovery	13,018,400	24.69
Alcohol Storage, Denaturing & Coproduct Storage	4,399,900	8.34
Utilities	15,090,000	28.62
Building, General Services, and Land	2,494,000	4.73
Subtotal	52,719,000	100.00
+ 10% Contingency	5,272,000	
TOTAL PLANT COST	$57,991,000	

SOURCE Katzen, R , "Grain Motor Fuel Alcohol Technical and Economic Assessment Study" Raphael Katzen
Associates for the U S DOE, Dec , 1978

Under the same assumptions as those used in Table V-4 (20 years straight depreciation), the operating costs for 10- and 100-million-gallon plants are $1.22 and $0.84 per gallon, respectively.

ECONOMIC ANALYSES

Economic analyses were performed for the 50-million-gallon plant described above to determine the price of ethanol which will cover the production cost and pro-vide a suitable return on equity. The return on equity is based on a Discounted Cash Flow-Interest Rate of Return (DCF-IROR) analysis.

The schedule of plant life and financial assumptions used in the analysis are summarized in Table V-5.

Figure V-16 summarizes the results of the analysis for three plant sizes. Ethanol price is expected to range from about $1.40 to about $2.20 per gallon in 1983.

PLANT DESIGN CONCEPTS

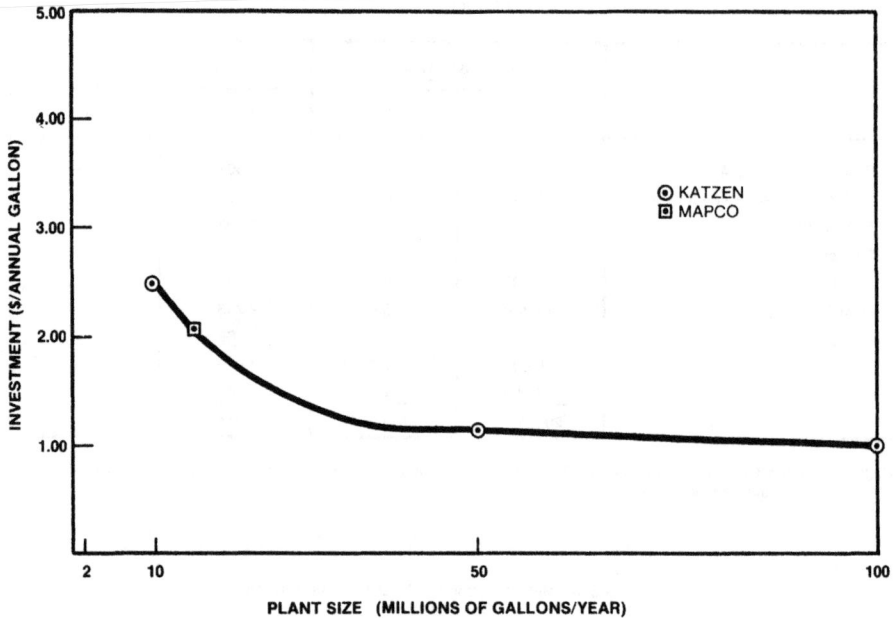

SOURCE Katzen, R , "Grain Motor Fuel Alcohol Technical and Economic Assessment Study" Raphael Katzen Associates for the U S DOE, Dec , 1978

Figure V-15. Ethyl Alcohol Plant Costs in $ Per Annual Gallon

SENSITIVITY OF ETHANOL PRICE

The sensitivity of ethanol price to various parameters is discussed below.

Sensitivity to Feedstock Price

The base case discussed above assumed corn price at $2.30/bushel. Figure V-17 shows the dependence of ethanol price on that of corn for two levels of DCF-IROR. As is expected from the previous analysis, ethanol price is very sensitive to corn prices: an increase of 10 cents per bushel in corn price results in an increase of about 5 cents per gallon in ethanol price.

Sensitivity to Coproduct (DDG) Price

Figure V-18 shows the impact of variations in DDG prices on the price of ethanol. An increase of about 10 percent in DDG price results in a decrease of about 3 percent in ethanol price.

Sensitivity to Changes in Feedstock and Fuel

An analysis similar to that performed for corn was performed for wheat, milo, and sweet sorghum feedstocks. In the latter case, there is some uncertainty in the results because a complete plant design was not available. Table V-6 summarizes the results of the analysis for the base case of a 50-million-gallon plant.

The similarity in design of the corn, wheat, and milo plants is reflected by equivalent investments for the plants. Sweet sorghum is a sugar crop which requires a front-end processing unit different from corn (juice extraction and clarification). Sweet sorghum is only available for about 6 months of the year. To avoid extensive storage facilities, it was assumed that the plant operates half of the year on sweet sorghum and half of the year on corn. As was pointed out in an earlier section, the double feedstock capability results in higher investment costs because of the need for two separate front-end feedstock processing trains.

COMMERCIAL SCALE ETHANOL PRODUCTION AND FINANCING

Table V-4. Plant Operating Costs for 50 Million Gal/Yr
- Base Case

		EQUIVALENT 1978 COST		1st YEAR OPERATION* 1983 COST	
		ANNUAL, $ MILLION	$/GAL	ANNUAL, $ MILLION	$/GAL
FIXED CHARGES					
Depreciation	20 Years	2 900	0 058	3 200	0 064
	10 Years	5 8	0 116	6 4	0 128
Licenses Fees		0 029	0 001	0 040	0 001
Maintenance		1 829	0 036	2 560	0 051
Tax & Insurance		0 914	0 019	1 280	0 026
SUBTOTAL	20 Years	5 672	0 114	7 080	0 142
	10 Years	8 6	0 172	10 3	0 206
RAW MATERIALS					
Yeast		0 320	0 006	0 449	0 009
NH_3		0 373	0 007	0 522	0 010
Corn		44 770	0 896	62 679	1 254
Coal		2 410	0 048	2 273	0 067
Miscellaneous Chemicals		0 180	0 004	0 252	0 005
SUBTOTAL		48 053	0 961	67 276	1 346
UTILITIES					
Electric Power		1 646	0 033	2 305	0 046
Diesel Fuel		0 012	0 000	0 017	0 000
Steam (from plant)		0 000	0 000	0 000	0 000
C W (from plant)		0 000	0 000	0 000	0 000
SUBTOTAL		1 658	0 033	2 322	0 046
LABOR					
Management		0 240	0 005	0 337	0 007
Supervisors/Operators		2 194	0 044	3 072	0 061
Office & Laborers		1 202	0 024	1 683	0 034
SUBTOTAL		3 636	0 073	5 091	0 548
TOTAL PRODUCTION					
COST, TPC	20 Years	59.019	1.181	81.769	1.640
	10 Years	61.9	1.24	85.0	1.70
COPRODUCTS					
Dried Grains		19 175	0 384	26 845	0 537
Ammonium Sulfate		0 413	0 009	0 578	0 012
SUBTOTAL		19 588	0 393	27 423	0 548
MISCELLANEOUS EXPENSES					
Freight		2 504	0 050	3 506	0 070
Sales		1 930	0 039	2 705	0 054
G&AO		0 644	0 013	0 901	0 018
SUBTOTAL		5 078	0 102	7 102	0 142
TOTAL OPERATING	20 Years	44.509	0.890	61.456	1.230
COST	10 Years	47.4	0.948	64.7	1.29

*Assumed 7-percent inflation rate

SOURCE Katzen, R , "Grain Motor Fuel Alcohol Technical and Economic Assessment Study" Raphael Katzen Associates for the U S DOE, Dec , 1978

Table V-5. Financial Assumptions

Engineering Period	1979-80
Plant Construction	1980-83
Start-up	1982-83
Operationg Period	10 or 20 Years
DCF-IROR	15%
Working Capital	10% of operating costs
Inflation	7%
Depreciation	10 or 20 Years
Taxes (federal/local)	50% of profits after expenses
Investment Tax Credit	10%
Equity	100%

SOURCE Katzen. R , "Grain Motor Fuel Alcohol Technical and Economic Assessment Study" Raphael Katzen Associates for the U S DOE, Dec , 1978

Table V-6. Ethanol Prices for Various Feedstocks and Fuels
(50-Million-Gallon Plant · 1978 dollars)

FEEDSTOCK	FEEDSTOCK PRICE ($/bu)	INVESTMENT ($ millions)	OPERATING COSTS ($ millions/yr)	ETHANOL PRICE ($/gallon)
Corn	2 30	58 0	44 5	1 05
Wheat	3 15	58 0	57 1	1 31
Milo (grain sorghum)	2 20	58 0	42 6	1 02
Sweet				
Sorghum*	14 42/ton	91 6	58 8	1 40
Fuel				
Corn Stover	25 0/ton**	57 0	46 3	1 09

*Half corn, half sweet sorghum feedstock

**Dry basis, 8000 Btu/lb

SOURCE Katzen, R , "Grain Motor Fuel Alcohol Technical and Economic Assessment Study" Raphael Katzen Associates for the U S DOE, Dec , 1978

COMMERCIAL SCALE ETHANOL PRODUCTION AND FINANCING

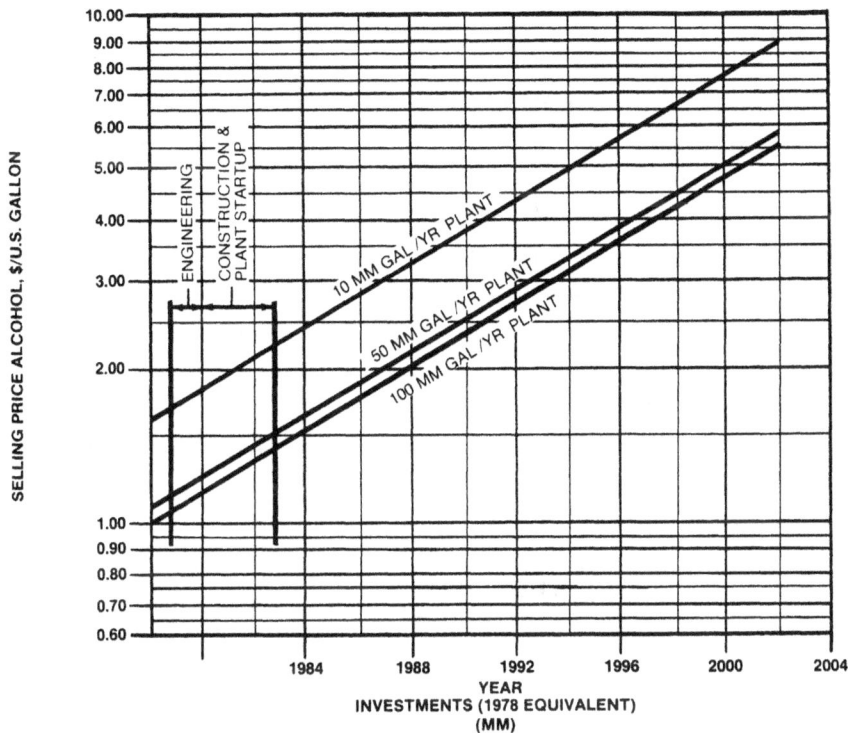

SELLING PRICE ALCOHOL, $/U.S. GALLON

ENGINEERING

CONSTRUCTION & PLANT STARTUP

10 MM GAL/YR PLANT

50 MM GAL/YR PLANT

100 MM GAL/YR PLANT

YEAR

INVESTMENTS (1978 EQUIVALENT)
(MM)

CAPACITY	FIXED	W.C.	TOTAL
10 MM	25 2	1 8	27 0
50 MM	58 0	5 6	63 6
100 MM	100 0	10 2	110 2

SOURCE Katzen, R , "Grain Motor Fuel Alcohol Technical and Economic Assessment Study" Raphael Katzen
Associates for the U S DOE, Dec , 1978

Figure V-16. Plant Capacities Other Than Base Case - Selling Price vs. Production Year

Of the feedstocks analyzed, grain sorghum shows a slight advantage over corn under the conditions of the analysis.

Switching from coal to corn stover for fuel results in a slight increase in ethanol price. This option could result in environmental problems as the removal of stover from the fields could increase the risk of erosion.

Sensitivity to Financial Parameters

The sensitivity to parameters such as debt-to-equity ratio, DCF-IROR, and Investment Tax Credit is shown in Table V-7. The data were derived from the Katzen report, to which the reader is referred for details. In most cases, the impact of variations in financial parameters is smaller than that resulting from variations in feedstock or DDG prices.

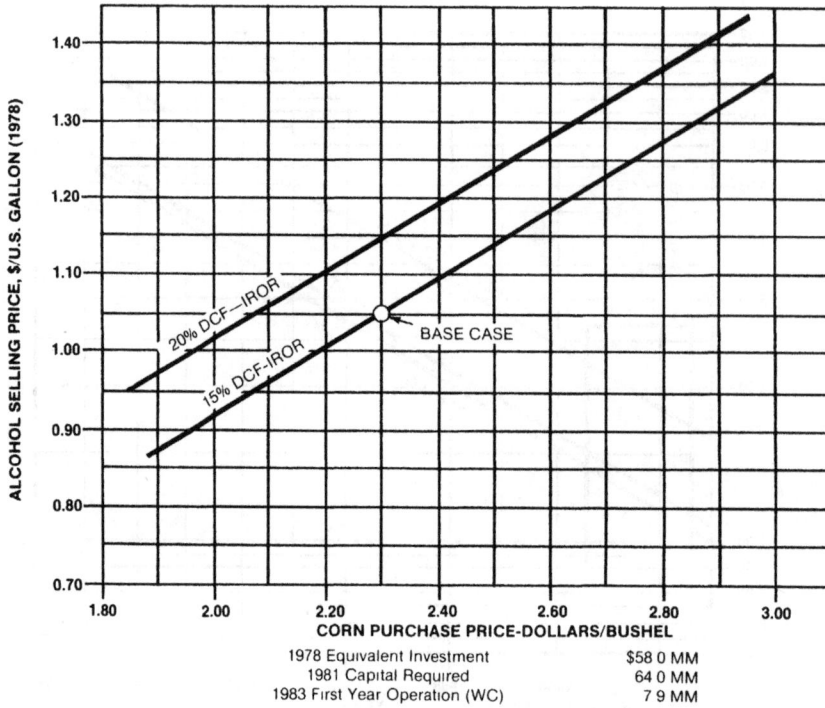

1978 Equivalent Investment	$58 0 MM
1981 Capital Required	64 0 MM
1983 First Year Operation (WC)	7 9 MM

SOURCE Katzen, R , "Grain Motor Fuel Alcohol Technical and Economic Assessment Study " Raphael Katzen Associates for the U S DOE, Dec , 1978

Figure V-17. Corn Purchase Price Sensitivity Analysis

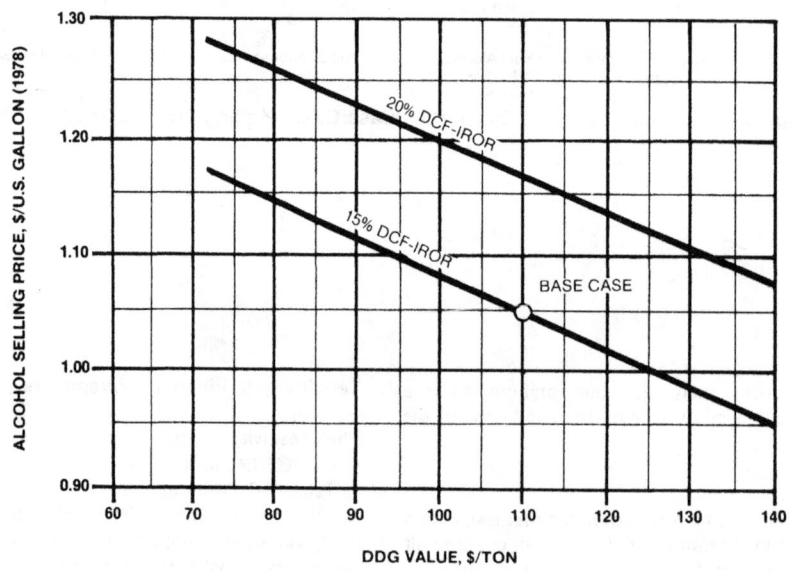

SOURCE Katzen, R , "Grain Motor Fuel Alcohol Technical and Economic Assessment Study" Raphael Katzen Associates for the U S DOE, Dec , 1978

Figure V-18. Alcohol Selling Price (1978) vs. DCF-IROR

COMMERCIAL SCALE ETHANOL PRODUCTION AND FINANCING

Table V-7. Sensitivity of Ethanol Price to Financial Parameters

PARAMETERS	VARIATION OF PARAMETER (%)	IMPACT ON ETHANOL PRICE (%)
DCF-IROR	+33[1]	+10
Investment Tax Credit	+300[2]	-2
Plant Depreciation Period	+100[3]	+2
Debt/Equity Ratio	+80[4]	-10

[1]DCF-IROR increased from 15 to 20%
[2]Investment Tax Credit increased from 10 to 30%
[3]Depreciation Period increased from 10 to 20 years
[4]Debt increased from 0 to 80 percent of capital

SOURCE Katzen, R , "Grain Motor Fuel Alcohol Technical and Economic Assessment Study" Raphael Katzen
Associates for the U S DOE, Dec , 1978

CHAPTER VI
Safety Aspects

Safety is of primary importance in a system producing a flammable product. In addition to the main product (ethanol), other hazardous compounds are consumed and generated in the process system. The chemicals considered hazardous are: sulfuric acid (H_2SO_4); calcium oxide, (lime, CaO); hexane ($n\text{-}C_6H_{14}$); calcium hydroxide (hydrated lime, $Ca(OH)_2$); carbon dioxide (CO_2); and ethanol (grain alcohol, C_2H_5OH or EtOH). The hazards of these chemicals are briefly described in Table VI-1.

The codes and standards which are applicable to the production of fuel-grade ethanol are listed in Table VI-2.

SAFETY CONSIDERATIONS

Ethanol, 99.9 percent pure, is stored at ambient temperature and slight positive pressure. The flash point is 55° F and flammability limits are 3.3 to 19.0 vol%. The saturated vapor/air mixture above the liquid ethanol is flammable between 50 and 110° F. Hence, a CO_2 blanket should be used in the ethanol storage tanks.

N-hexane exists as a superheated vapor at 450° F and 10 psig. The flash point is 0° F. The flammability limits are 1.2 to 7.5 vol%. The saturated vapor/air mixture above the liquid hexane is flammable between − 20 and 40° F.

Benzene has flammability limits of 1.35 to 6.75 vol% in air, and its flash point is 12° F. The saturated vapor/air mixture above the liquid is flammable between − 10 and 64° F.

HAZARDS

Fermentation ethanol plants are often located in rural areas, to be close to their source of feedstock. Remoteness from city water supplies and fire departments places responsibility for fire protection almost entirely on the plant itself. Safety also depends on good construction and proper arrangement and safeguards for processes.

Because of the fire and explosion hazards inherent in handling large quantities of flammable liquids and also the potential for grain dust explosions in the grain

Table VI-1. Hazardous Liquids and Gases

HAZARDOUS MATERIAL	DESCRIPTION	FIRE AND EXPLOSION HAZARD	LIFE HAZARD	PERSONAL PROTECTION	EMERGENCY PROCEDURES	HANDLING TECHNIQUES
Sulfuric Acid (H_2SO_4)	Colorless (pure) to dark brown oily dense liquid	Not flammable but highly reactive Reacts violently with water and organic materials with evolution of heat Attacks many metals releasing hydrogen	Causes severe deep burns to tissue when contacting liquid Vapors are extremely irritating to eyes and mucous membranes (nose and throat) AVOID CONTACT!	Wear rubber gloves rubber apron and indirectly ventilated liquid-tight chemical goggles when transferring acid to storage tank or manually adding acid to any vessels	In case of contact immediately flush skin or eyes (affected area) with large quantities of water for at least 15 minutes FOR EYES GET MEDICAL ATTENTION!!	Transfer acid by pump or gravity flow Never use a compressed gas or air to pressurize an acid container Always add acid to water never water to acid a violent reaction will occur causing acid to be ejected from container Make addition slowly to minimize heating Use proper carboy truck and tilter if acid is received in carboys Small glass containers of acid shall be handled in impact-resistant chemical carriers
Calcium Oxide (CaO)	Colorless crystal, also known as unslaked, quick or burnt lime	Noncombustible but reactive When wetted it swells, gets hot and becomes calcium hydroxide $Ca(OH)_2$ (slaked lime caustic lime)	Causes skin burns Less corrosive than caustic soda (NaOH or sodium hydroxide) or caustic potash (KOH or potassium hydroxide) Dust is highly irritative to the eyes and mucous membranes and prolonged contact with skin can cause dermatitis Avoid contact	Wear rubber gloves, chemical goggles and long-sleeved shirt or jacket	In case of contact, immediately flush skin or eyes (affected area) with large quantities of water for at least 15 minutes For eyes get medical attention	Transfer crystals from container to rubber pail Pour required quantity into empty addition vessel Close vessel cover and add required quantity of water to dissolve crystals
Ethyl Alcohol (EtOH)	Clear colorless, fragrant liquid, burning taste Also, known as ethanol or grain alcohol	Flammable liquid Vapors form flammable mixtures with air BURNS WITH INVISIBLE FLAME *	Exposure to concentrations above 1000 ppm may cause headache and irritation of the eyes nose and throat If continued for prolonged time will cause drowsiness and stupor Contact with liquid can cause defatting of skin No known cumulative effect as is common with methyl alcohol	Wear standard safety glasses When breaking lines for maintenance wear neoprene gloves and liquid-tight chemical goggles	In case of body splash flush with large quantities of water Dilute liquid spills with large volumes of water Attack small spill fires with ABC dry chemical extinguishers	Product will normally be handled in a closed system When transferring to transport vehicle proper bonding and grounding procedures will be used
Carbon Dioxide (CO_2)	A colorless odorless, tasteless gas	None Is a fire extinguishing agent	Is a simple asphyxiant Symptoms include dizziness headache, shortness of breath muscular weakness drowsiness, and ringing in ears OSHA standards require oxygen concentration of 19 5% or greater before entry is made into tanks or vessels All vessel entry is to be made under the Safe Work Permit system Skin contact with CO_2 snow will cause frost burns		Remove victim from oxygen-deficient atmosphere Rescuers must wear self-contained breathing apparatus if vessel entry is necessary If breathing has ceased start mouth-to-mouth resuscitation Call for medical assistance	

*NOTE Flammability Characteristics of Ethanol — Flash point, 55°F Flammable limits 3 3-19% Ignition temperature 685°F Vapor density (air = 1) 0 8 (water soluble) Specific gravity (water = 1) 0 8 (water soluble) Boiling point 173°F Odor detectable at 5-10 ppm Flash point of 5% alcohol/water solution 144°F Flash point of 10% alcohol/water solution, 120°F Flash point of 15% alcohol/water solution 110°F Flash point of 20% alcohol/water solution 97°F

COMMERCIAL SCALE ETHANOL PRODUCTION AND FINANCING

Table VI-2. Codes and Standards for the Production of Fuel-Grade Ethanol

Title	Code*
Prevention of Dust Explosion in Industrial Plants	NFPA63
Basic Classification of Flammable and Combustible	NFPA321
Liquids Static Electricity	NFPA77
Flammable and Combustible Liquids Code	NFPA30
Occupational Noise Exposure	OSHA191094
Machinery and Machine Guarding	OSHA Subpart 0
Power Piping	ANSI B31,1
Standard for Steel Aboveground Tanks for Flammable and Combustible Liquids	UL142
Boiler and Pressure Vessel Code (B & PV)	ASME Code Section IV & VII Division I
All electrical instrumentation	NFPA70-1978
National Electric Code	Class II Division I°

*Abbreviations: NFPA —National Fire Protection Association

    OSHA -Occupational Safety and Health Administration

    UL   —Underwriters Laboratory

    ASME —American Society of Mechanical Engineers

    ANSI —American National Standards Institute

storage areas, safety depends on supervision by well-trained operators, good maintenance, and process equipment safeguards.

Grain handling, milling, and feed preparation at distilleries present dust explosion hazards. Although grains and feeds are slow burning, fires in these materials may be deep-seated and difficult to extinguish. Wet grains will heat and sour if not dried promptly.

Process fire and explosion hazards are present during distilling, but are considered negligible during mashing and fermenting. Strict government regulations that require seals on every pipe joint, valve, and spigot reduce the probability of flammable liquid or vapor being released during distilling operations.

Flammable liquid hazards are also present in varying degrees in the various distilled-alcohol handling areas.

Because of alcohol's lower heat of combustion, radiant heat energy, and complete miscibility with water, lower sprinkler system demands are required than with other flammable liquids of equivalent flashpoint.

The quantity of water needed to extinguish fires in alcohol/water mixtures depends upon the temperature of the liquid above its fire point and the effectiveness of mixing. The amount of water can be estimated from the following formula, assuming perfect mixing:

$$\text{Volumes of water needed per volume of burning liquids} = \frac{\%\ \text{alcohol in solution before fire}}{\%\ \text{alcohol at point of fire extinguishment}} - 1$$

Assume that a solution will be extinguished when the alcohol concentration is reduced to 20 percent. Applying the formula, a mixture containing 95 percent alcohol would require 3.75 volumes of water to extinguish each volume of burning liquid. A mixture containing 50 percent alcohol would require 1.5 volumes.

INSURANCE LOSS EXPERIENCE

A survey of industry losses for the years 1933 to 1972 indicated that approximately 75 percent of all property damage resulted from fires in 12 alcohol warehouses without sprinkler systems. However, several serious fire and explosion losses occurred in still-buildings. The most serious losses in still-buildings involved explosions with ensuing fires where sprinkler systems were damaged by the explosion. Several fires also occurred in driers which were processing dried grains from spent stillage or slops.

PLANT SAFETY CONSIDERATIONS

General

Grain handling, milling, and feed preparation facilities should be designed, arranged, and safeguarded in accordance with safety standards for grain storage and milling.

Construction and Location

Mashing and Fermenting. Mashing and fermenting areas should preferably be of fire-resistive or noncombustible construction.

Distilling.

1. Distilling operations should be separated from other buildings by at least 100 ft (30 m). Existing still-buildings that adjoin other buildings should be completely cut off by blank fire walls, parapeted above adjoining buildings. Avoid basements, pipe trenches, and other spaces beneath still-buildings.

2. Preferably locate distilling equipment with a minimum of enclosing structure. Structures should be of damage-limiting construction. Load-bearing steel members and exposed steel equipment supports should be fire-proofed with material having a minimum two-hour fire-resistance rating. For existing buildings of substantial construction, provide explosion venting capacity through venting windows and roof panels in as high a ratio as practical.

3. Floor cutoffs are advisable at operating levels in high, enclosed buildings. If complete floor cutoffs are not practical, provide solid noncombustible mezzanines with curbs at levels supporting receivers or other equipment containing appreciable quantities of flammable liquids.

4. Unless the maximum possible spill can be extinguished by dilution while confined, provide emergency drainage facilities for the distilling area of buildings to prevent escaping liquids from exposing other areas or buildings.

Distilled Alcohol Handling.

1. Alcohol handling areas should preferably be of fire-resistive or noncombustible construction.

2. Distilled-alcohol handling areas should be cut off from surrounding occupancies. Vertical cutoffs should be provided in multi-story buildings. Cutoffs should have at least a one-hour fire-resistance rating.

3. Provide curbs, ramps, or trapped floor drains at doorways and other openings to prevent the spread of flammable liquids to other departments. Floor drains in each distilled-alcohol handling area should be designed to handle expected sprinkler discharge unless the maximum possible spill can be extinguished by dilution while confined.

Occupancy

Mashing and Fermenting. Grain meal should be discharged to precookers only through tight connections to prevent liberation of dust.

Distilling.

1. Pressure vessels should be designed and constructed in accordance with applicable codes, standards, State and local laws, and regulations.

2. Stills should be equipped with vacuum and pressure relief devices piped to outdoors. Any condenser vents also should be piped to outdoors. Vents should be sized to discharge the maximum vapor generation possible at zero feed and maximum heating within the pressure limitations of the protected equipment. Vents should terminate at least 20 ft (6.1 m) above the ground and preferably at least 6 ft (1.8 m) above roof level and should be located so that vapor will not re-enter the building. Vent terminals should be equipped with flame arresters.

3. Equipment should be designed and maintained to eliminate or at least minimize any liquid and vapor leaks.

4. Where gauges are needed, use Factory Mutual-approved gauging devices. If ordinary gauge glasses are used, both connections normally should be kept closed and provided with weight-operated, quick-closing valves. Protect the glass from mechanical injury. Where possible, tail boxes should be replaced with armored rotameters and specific gravity indicators, or with other instrumentation not subject to accidental breakage or leakage.

5. The steam supply for distillation should be thermostatically controlled and interlocked to shut down and sound an audible alarm on cooling-water failure. Alternately, powered standby pumps or gravity supplies of cooling water should be provided.

6. Stills and other large equipment containing flammable liquids should be purged with steam or an inert gas (steam will be most generally available) before they are open for inspection or repair. Equipment should be washed with water following steaming.

7. Ventilation, designed and installed to ensure air movement throughout the entire structure, should be provided to prevent accumulation of explosive vapor-air concentrations within the

building. The stack effect (i.e., natural ventilation) may suffice if: the building is high; permanent openings are provided at grade and roof elevations; the equipment can be drained and cleared of vapors during shutdowns; and heat losses from the equipment maintain a temperature above that of the outdoors during all operating periods. If these operating conditions cannot be satisfied, or if blank walls or solid floors interfere with natural ventilation, mechanical exhaust ventilation should be designed to provide 1 cfm/ft^2 (0.3 m^3/min/m^2) of floor area. Locate suction intakes near floor level to ensure a sweep of air across the area.

8. Electrical equipment, including wiring and lights, should be suitable for Class 1, Group D locations. Still-buildings should be considered Division 2 locations.

Distilled Alcohol Handling.

1. Noncombustible, vapor-tight construction should be used for all tanks containing flammable concentrations of alcohol. Tanks should be kept tightly closed except when taking samples.

2. Tanks should be equipped with vents of adequate size terminating outdoors. Vents should be equipped with Factory Mutual-approved flame arresters if the flashpoint of the contents is less than 100° F (38° C).

3. Factory Mutual-approved liquid-level gauges should be installed on all tanks. If ordinary gauge glasses must be used, weight-operated, normally closed valves should be installed at both tank connections and the glass should be protected against physical damage. Wherever possible, top tank connections should be provided and liquids transferred by pumping through the top rather than by gravity flow. If draw-off stations are located in the same area as the supply tank, automatically operated emergency shutoff valves should be provided in gravity-feed lines. Flexible metallic hose should be used on all connections to scale tanks where fire exposure would release the tank contents or expose its vapor space.

4. Mechanical exhaust ventilation should be provided as needed, and arranged with suction near floor level to ensure air movement throughout the building. At dump troughs and similar installations, localized intakes are desirable.

Careful attention should be given to below-grade installations, windowless buildings, sumps, pipe trenches, and similar installations. Usually, 0.25 cfm of air/ft^2 (0.075 m^3/min/m^2) of floor area will be adequate. The use of factory approved, portable flammable vapor indicators is recommended.

5. Electrical equipment, including wiring and lights, should be suitable for Class 1, Group D locations. Tank storage areas should be treated as Division 2 locations.

Fire Protection.

1. Provide automatic sprinkler protection for distilleries, preferably of a type designed to flood the area.

2. Sprinkler control valves, dry-pipe valves, and riser drains should be readily accessible at all times to plant personnel. This is particularly important for areas under direct government supervision that may be locked during non-operating periods.

3. Small hoses with combination shutoff nozzles should be provided throughout the distillery. Hose stream demand is a minimum of 500 gpm (190 dm^3/min) for at least 60 minutes.

4. Suitable portable fire extinguishers should be provided throughout the distillery.

The implementation of adequate safety procedures impacts the number and kinds of personnel needed to run an ethanol plant. It is estimated that a typical 50-million-gallon-per-year ethanol plant would require the services of a medical doctor half-time, assisted by four full-time nurses. Such a plant would also need at least one safety engineer to oversee safety procedures. Thus, ensuring adequate plant safety adds to the operating costs through the addition of 5.5 personnel for a 50-million-gallon-per-year plant.

Training/Operating Procedures

1. Prior to plant acceptance, detailed operating procedures (including safety procedures) should be documented in a technical operation manual. Detailed process diagrams, with color-coded piping illustrations, must be provided.

2. Operators must be trained to wear hardhats and be aware of plant safety features and location of fire equipment.

CHAPTER VII
Environmental Concerns

The discussion of environmental considerations will focus on the conversion of feedstock to ethanol, but the major issues associated with feedstock production and end-use of the ethanol in blends with gasoline will be summarized briefly. The role of environmental assessment in business decisions also will be addressed.

The major environmental concerns associated with ethanol production from grain are: emissions from the process heat source; and two multi-source environmental problems, distillery wastewaters and occupational exposure to process and byproduct chemicals.

Distillery wastewaters are acidic and high in suspended solids, BOD, and chemical oxygen demand and may present a potential health hazard if not properly treated or disposed. Stillage is particularly high in BOD and mineral salts (especially alkali salts), and both surface and groundwaters would be affected by improper or inadequate treatment or disposal practices. Standard wastewater treatment systems are commercially available.

Toxic and corrosive process chemicals employed in the conversion of starch feedstocks to ethanol or arising as coproducts in the conversion series include acids, dehydrating agents, denaturants, fuel oil and aldehyde byproducts, and chemicals used for equipment maintenance. Environmental concerns attendant with the use of such substances include spills and other accidental emissions and potential occupational health hazards resulting from long-term exposure. The practice of good industrial hygiene should minimize danger to workers.

Impacts due to air emissions and solid wastes generated during combustion of fossil fuel, especially coal, to meet process heat requirements need to be considered, with special attention to commercial-scale alcohol conversion facilities with on-site, industrial-size boilers. Both the environmental problems and the control measures are generic to coal-fired boilers, and thus represent no new problems peculiar to the fuel alcohol industry.

Environmental issues arising in the production of feedstock grains are erosion and use of scarce resources such as water and land. Erosion serves not only to reduce soil fertility and productivity, but also as a transport mechanism whereby entry of pollutants into waterways and the food chain as a whole is facilitated.

The utilization of ethanol/gasoline blends as a motor fuel could have environmental consequences for both users of the fuel and the general population. Health and safety risks occur for individual users, and combustion of the fuel entails potential air quality degradation and ecosystem effects.

BIOCONVERSION OF GRAIN FEEDSTOCKS TO ETHANOL

Overview of Emission and Effluent Sources

For the discussion of environmental impacts during fermentation of biomass to ethanol and recovery/purification of the product, the reference plant will be a 50-million-gallon-per-year facility, employing the energy-efficient Katzen design. Dry-milled corn is the biomass feedstock and coal is the fuel for generation of process steam requirements. For such a reference plant, Table VII-1 indicates the major sources of emissions and effluents during the conversion process.

In the feedstock storage and preliminary processing sections of the operation, the major atmospheric emission is particulates from the physical preparation of the feedstock. Liquid effluents include various wash waters and flash cooling condensate. The only solid of concern is grain dust, which may be contaminated by pesticides or fungi. Aflatoxin contamination of midwestern corn is rare.

During hydrolysis and fermentation, the major air emissions arise from fermenter vents; CO_2 and accompanying volatile organic fermentation products and byproducts escape to the atmosphere. Wash waters are

Table VIII-1. Bioconversion of Grain to Ethanol - Emission and Effluent Sources.

OPERATION	EMISSIONS TO ATMOSPHERE	LIQUID EFFLUENTS	SOLID WASTES
Feedstock Storage	Transfer operations (fine dust)		Grain, dirt
Milling and Cooking	Mechanical collectors for milling operations (particulates)	Wash water (dissolved and suspended solids, organics, pesticides, alkali) Flash cooling condensate (dissolved and suspended solids, organics)	Grain dust (from mechanical collectors)
Hydrolysis and Fermentation	Fermentation vents (CO_2, hydrocarbons)	Wash water (dissolved and suspended solids, organics, alkali)	
Distillation and Dehydration	Condenser vents on columns (volatile organics)	Rectifier bottoms (organics), dehydration bottoms (organics)	
Storage	Storage tanks (hydrocarbons)		
Coproduct Recovery	Dryer flue gases (NO_x, SO_x, CO, particulates, hydrocarbons). Evaporator condenser vents (hydrocarbons)	Evaporator condensate (dissolved and suspended solids, organics)	Grain dust (from direct-contact dryers)
Steam Production (coal-fired)	Flue gases (NO_x, SO_x, CO, particulates)	Boiler blowdown (inorganics), cooling water blowdown (dissolved and suspended solids, organics)	Coal dust, ash (bottom ash and fly ash)
Environmental Control Systems	Evaporation from biological treatment ponds (organics)	Scrubber blowdown (dissolved and suspended solids, organics)	Sludge (from flue gas treatment), biological sludge (from wastewater treatment)

Source R. M. Scarberry and M. P. Papai, "Source Test and Evaluation Report· Alcohol Synthesis Facility for Gasohol Production," Radian Corporation, January, 1980.

COMMERCIAL SCALE ETHANOL PRODUCTION AND FINANCING

high in BOD, dissolved solids, and suspended solids; in addition, they may contain traces of chemicals used as nutrients or used to control the growth of undesired organisms during the fermentation. For starch feedstocks, the associated pesticide residues are thought to be destroyed during the cooking process.

In the alcohol purification (distillation and dehydration) and denaturation steps, the major sources of atmospheric emissions are condenser vents on columns. Evaporator condensates may contain materials that require treatment before disposal. No solid wastes are obtained.

Evolution of criteria pollutants occurs in byproduct processing through the exhaust of flue gas used in drying. In addition, evaporator condensate contains dissolved and suspended solids and various organic materials. The direct-contact dryers generate grain dust.

Steam production using coal-fired boilers causes emissions of criteria pollutants. Aqueous streams of environmental concern include boiler blow-down and cooling water blowdown. Solid wastes include coal dust, fly ash, and bottom ash.

Environmental control systems may produce secondary wastes. For example, biological treatment ponds yield evaporative emissions. Scrubber blowdown contains dissolved and suspended solids, as well as various organics. Biological sludge from wastewater treatment may contain pesticides, benzene, ammonia, and various metals.

Impacts Associated with Specific Process Steps

Specific impacts in the three major categories of air quality, water quality, and solid wastes are discussed in more detail for each stage of the conversion process. A separate discussion of multi-source environmental problems, specifically wastewater treatment and occupational safety and health, is also presented.

- **Feedstock Storage**

  Feedstock storage facilities may represent a potential health hazard by serving as breeding grounds for rodents and other pests. Spoilage of grain due to natural fermentation and other decay processes gives rise to a variety of substances which, depending on the degree of containment of the storage area, could be discharged to waterways or the atmosphere. The use of fungicides to retard spoilage could affect worker health, as well as groundwater and surface water quality. Finally, the generation of large amounts of fine dust in the transfer of grain into and out of the storage area results in poten-

tial respiratory health hazards to workers, as well as an explosion hazard.

Experience in controlling dust in the grain industry should be transferable to similar operations associated with fuel alcohol plants.

- **Milling and Cooking**

  If corn is processed by hammer mills, particulates in the form of grain dust can be a problem, along with high levels of noise (85 to 88 dBa) generated by the processing equipment. Controls for dust from grain are similar to those used for fly ash particulates. Grain dust, chaff, and impurities collected in cyclones are not considered solid wastes, as they are cycled to the dryer for inclusion in the DDG byproduct. During the cooking process, a beneficial side effect is the apparent destruction of pesticides, many of which are vulnerable to heat and decompose to simpler compounds during the heat treatment. In a recent sampling program conducted at the Midwest Solvents plant in Atchison, KS, pesticides were detected in the feedstock grains, but no traces were found in the solid wastes or wastewater effluents downstream of the gelatinization step.

  Flash cooling of the mash in multi-effect evaporators yields a condensate (from a surface condenser employed at the last stage) which is high in dissolved solids and suspended solids, and volatile organic materials.

- **Hydrolysis and Fermentation**

  Nearly equal weights of ethanol and CO_2 are produced in the fermentation reaction. Other byproducts include a myriad of oxygenated organic compounds, primarily acids, alcohols, and aldehydes. The CO_2 can be trapped or merely vented to the atmosphere. If the CO_2 is recovered, the gaseous stream is passed through a water scrubber to condense water and volatile organic materials (including ethanol) before the cryogenic recovery of CO_2. Alternatively, if CO_2 is vented, the water scrubber may still be employed. The gaseous stream is accompanied by volatile organic byproducts, as well as some of the product ethanol. Temperature is carefully controlled during the fermentation reaction to minimize evaporative losses of ethanol, which increase by a factor of 1.5 for every 5° C increase in temperature. Closed fermenters with off-gas condensation and scrubbing may be used to minimize ethanol losses to the atmosphere.

  A variety of chemicals added to the fermenting mash include acids for pH control, yeast nutrients (ammonium salts, urea, phosphates), and chemicals or antibiotics to control the growth of undesired organisms without affecting the yeast. Because of the variety of byproducts and nutrients/chemicals used in the fermentation process, it is not surprising that the fermentation wash water contains high levels of dissolved solids, suspended solids, organics, and alkali.

Care must be taken, when personnel enter the fermenters during cleaning, to avoid the possibility of asphyxiation in an atmosphere with a high concentration of CO_2.

- **Distillation and Dehydration**

During alcohol concentration and purification, the major atmospheric emissions are from the vent condensers employed at distillation column openings to retard (but not entirely eliminate) the escape of organic materials to the environment. As the vent streams contain volatile organic materials too dilute for economic recovery, flares are often used to control emissions. Alternatively, the vent streams could be routed to the process burners as air feed. The evaporator condensates, rich in dissolved solids, suspended solids, and organic materials, are sent to wastewater treatment. Distillation and dehydration bottoms contain benzene (if used as the dehydrating agent) and other organics. Because benzene is a known leukemogen, strict precautions must be taken in work areas during its use. The Federal standard (29 CFR 1910.1000) for exposure to benzene is 10 ppm (8-hour, time-weighted average), with a ceiling concentration of 25 ppm and a peak of 50 ppm allowable for 10 minutes. A recent sampling program conducted at the Midwest Solvents plant by the Radian Corporation revealed benzene concentrations in the low range of 2.7 to 59.4 ppm for the liquid streams. As these values are near the detection limit for benzene, they are assumed to have limited accuracy; nevertheless, it is known that 27 gallons of benzene (approximately 1 gallon for every 5,500 gallons of product ethanol) would be lost daily somewhere in the process for a 50-million-gallon-per-year plant.

- **Hydrocarbon Storage**

Evaporative emissions during storage can be controlled through the use of tanks with floating roofs or internal floating covers to reduce the air space above the stored liquids. Another option is the inclusion of a vapor recovery system.

- **Byproduct Recovery**

If stillage is not dried or otherwise treated before disposal, water quality could be degraded by this waste stream. Stillage is relatively high in protein content, as well as unconverted starches and sugars, various fermentation products, and yeast. Stillage drying operations may affect air quality. Particulates, primarily in the form of grain dust generated during the drying process, are emitted, and dryer flue gases, which use boiler flue gases from the plant's heat source, contain criteria air pollutants, including SO_2, CO, and NO_x. Grain dust which is collected may be added to the DDG byproduct. Aqueous wastes, such as evaporator condensate from coproduct recovery operations, are sent to the wastewater treatment system.

Aflatoxin contamination of midwestern corn, the primary source of corn for fuel ethanol production, is rare. Although more than 50 percent of southeastern corn is contaminated, it represents only 8 percent of the national crop, and most of it is confined to intrastate use. In the rural Southeast, contaminated corn is directly ingested, and epidemiological information indicates no higher human liver cancer incidence than in other areas of the United States. As aflatoxins are not distillable, any which survive the fermentation process will be concentrated in the DDG byproduct. It has, however, been demonstrated that mature animals rapidly metabolize and dispose of aflatoxins. The only potential sensitivity could arise from feeding contaminated DDG to dairy cattle, as one type of aflatoxin remains in the milk.

- **Steam Production**

Combustion of coal in industrial-size boilers results in atmospheric emissions, aqueous effluents, and solid wastes. To fulfill process steam requirements for the reference-size ethanol plant, 296.7 tons/day of coal is consumed, and 27.4 tons/day of ash is generated. Ash from a typical plant would be hauled to a nearby landfill for disposal. Coal dust collected during storage and transfer of the fuel can be routed to the boiler.

The atmospheric emissions from a coal-fired boiler at a fuel alcohol plant that are most likely to require controls are sulfur dioxide and particulates. Control of nitrogen oxides is required in some states, but emission levels are expected to be within legislated limits in most.

Several standard options are available for control of particulates, including fly ash and coal dust entrained in the flue gas. For systems of the size required for a commercial-scale fuel alcohol plant, fabric filters are usually the best alternative for particulate control.

For sulfur dioxide emissions, the simplest control in many cases is the use of low-sulfur coal, which could obviate the need for sophisticated control systems downstream. Several flue gas desulfurization systems are commercially available, but are expensive in terms of both capital costs and operating expenses. To remove both sulfur dioxide and particulates, a flue gas scrubbing system may be employed (at an initial cost of $4.2 million in 1978). Ammonia is used to remove sulfur dioxide, followed by neutralization of the ammonium sulfite/bisulfite solution and its oxidation to ammonium sulfate, which can be sold as a fertilizer. This type of chemical recovery desulfurization system has the advantage of producing no calcareous sludge, the secondary waste generated in conventional limestone scrubbing systems.

Water quality could be affected by leachate from coal or ash storage or ash disposal piles (although coal ash is generally disposed of in off-site landfills), as well as by cooling tower blowdown. Depending on the soil and groundwater characteristics in the area, leachate collection and treatment may become necessary. Boiler and cooling water blowdown are routed to the wastewater treatment section of the plant.

Because a fossil fuel-fired steam generator represents one of the major emission sources at an ethanol production facility, an obvious all-around environmental control measure is the reduction of the amount of process steam required. Any improvement in plant efficiency will result in a corresponding decrease in fossil fuel combustion and its associated atmospheric emissions, and aqueous and solid waste disposal problems.

Besides improvements in engineering design, substantial gains in this area can be obtained by using waste heat from other industries or by siting new plants imaginatively.

- **Environmental Control Systems**

  Environmental control systems may generate secondary pollutants, such as wastewater from various scrubbers. As mentioned above, certain types of flue-gas treating systems give rise to sludge, as does the wastewater treatment system. The only secondary atmospheric pollutant is the evaporative loss from biological treatment ponds; this emission consists of an assortment of volatile organic compounds.

Multi-Source Environmental Concerns

Wastewater streams arise from numerous separate sources in a fuel ethanol plant and potential health and safety hazards pervade the occupational environment. These two multi-source problems are discussed briefly.

- **Wastewater**

  Wastewaters from distilleries (approximately 1.1 million gallons per day for a 50-million-gallon-per-year plant, if extensive efforts are made to recycle water) are acidic and contain high levels of total solids (25,000 pounds per day for the reference-size plant); suspended solids (3,000 pounds per day); and BOD (7,300 pounds per day). Experience in the beverage alcohol industry indicates that levels of dissolved organics are below concentrations that require treatment before discharge to surface waters. Moreover, no significant problems are expected with regard to acidity levels in wastewaters, as the pH can be adjusted before discharge. High BOD loadings, however, will necessitate treatment. Suspended solids can be removed by preliminary screening and sedimentation in a holding tank before the biological oxidation step. Several biological oxidation systems are commercially available; spare aeration or equalization basins are recommended for added safety. Treatment can occur at the alcohol plant site or at publicly owned treatment works (POTW), if the latter have sufficient spare capacity to treat the large volumes of wastewaters associated with ethanol production. At many potential Farm Belt locations for ethanol plants, the local POTW may have insufficient capacity to receive distillery effluents, and treatment systems will thus become necessary parts of fuel alcohol facilities.

- **Occupational Safety and Health**

  In a fuel alcohol plant, the major potential threats to worker health and safety are explosion, fire hazards, and various modes of exposure to toxic and corrosive chemicals. Adequate controls or mitigating measures are currently available to cope with the problems.

  Chemical Exposures Ethanol itself can cause mild irritation of the eye and nose and can defat the skin, causing dermatitis. Prolonged inhalation produces irritation of the eyes and upper respiratory tract, headache, drowsiness, tremors, and fatigue. It may increase the toxicity of other inhaled, absorbed, or ingested chemical agents. Moreover, since ethanol and some common prescription drugs interact unfavorably when ingested, the possibility of synergism between these drugs and inhaled or absorbed ethanol should be considered. The Federal standard for workplace exposure to ethanol is 1000 ppm (1,900 mg/m^3). Work areas should be well ventilated, and normal safety precautions should be taken in handling the liquid.

The same general considerations apply to exposures to other organic compounds employed in or generated during the ethanol production process. The primary entry routes to the body are inhalation and dermal absorption. Workers should be educated regarding the proper handling of chemicals; protective gloves and aprons should be worn when appropriate; and emergency spill containment procedures should be well established in the occupational environment.

Fire/Explosion/Burn Hazards Whenever any volatile organic compounds are in use, standard precautions must be taken to prevent ignition of leaks or fumes. Explosion-proof motors, for example, along with other specially protected electrical equipment, should be routinely used. Equipment should be available and emergency procedures in place to deal with chemical fires.

In any industry that employs large quantities of process heat, particularly when it is in the form of steam, precautions need to be taken against burns. Certain routine preparations and minor equipment modifications, such as using baffles to direct steam gasket leaks away from the work area, can be made. Prevention of contact burns from steam lines can be accomplished by making the lines conspicuous or by insulation.

Environmental Control Costs

As indicated throughout the preceding discussion, hardware for adequate control of various pollutants generated in the production of fuel alcohol from biomass is commercially available and is similar to equipment used in other industries (e.g., food processing, drug manufacture, chemical manufacture, and beverage alcohol production). The major capital items include a flue gas scrubbing system, $4.2 million; wastewater treatment system, $2.0 million for the reference 50-million-gallon-per-year fuel ethanol plant; fire protection system, $0.60 million; ash collection package, $0.17 million; vapor-controlled storage tanks, $0.16 million (incremental cost over cone roof tanks); and assorted vent condensers, $0.13 million. The cost of

environmental controls, $7.5 million, accounts for approximately 13 percent of the total fixed investment for a fuel ethanol plant. This estimate is for direct controls only (i.e., physical hardware) and does not include installation, maintenance and operating costs, or the cost of implementing procedures (such as conducting seminars for employees on appropriate actions during spills and other potential workplace emergencies). It should be mentioned that an alternative to the capital-intensive, on-site wastewater treatment system is the discharge of distillery wastewater to POTW, provided that the latter has the excess capacity to treat the industrial wastes. In a comparison of self-treatment costs to costs of treating at a POTW (including Industrial Cost Recovery charges), it has been shown that for a POTW with a 15-million-gallon-per-day or greater capacity, it would generally be less expensive for distilleries of the reference size to discharge to the POTW. Actual charges for treating distillery effluent at a specific POTW, however, vary widely, because they depend not only on financial parameters such as interest rate, but also on the assimilative capacity of the stream into which the treated water is discharged.

Regulatory Constraints on Development of the Fuel Ethanol Industry

There are no current major Federal environmental regulatory obstacles associated with biomass conversion to ethanol; moreover, no roadblocks are anticipated. The major effects of environmental considerations on commercial-scale fuel alcohol development are in the interrelated areas of cost (discussed earlier), siting limitations, and uncertainties for investors.

The major legislation influencing fuel alcohol development is: the Clean Air Act and amendments of 1977, the Federal Water Pollution Control Act of 1972 (Clean Water Act), and the Resource Conservation and Recovery Act (RCRA) of 1976. The provisions of these acts which pertain to the alcohol industry are summarized briefly below. It should be borne in mind that more stringent state environmental regulations could be superimposed on any of the existing Federal regulations. The investment climate thus could be adversely affected by the uncertainties in environmental legislation at the State and Federal levels and the attendant uncertainties in investment costs.

- **Clean Air Act**

    All of the criteria pollutants, except lead, subject to the National Ambient Air Quality Standards (NAAQS) are emitted during the preparation and conversion of feedstock to ethanol. Under the Clean Air Act, major existing and new sources of air pollutants within an area presently attaining the primary and secondary NAAQS may be

Table VII-2. Nonattainment Areas of 48 Contiguous States*

STATE	NO. OF NONATTAINMENT COUNTIES (PARISHES)	TOTAL NO. OF COUNTIES (PARISHES)	% OF THE STATE'S COUNTIES IN NONATTAINMENT STATUS
Alabama	9	67	13
Arizona	13	14	93
Arkansas	1	75	1
California	43	58	74
Colorado	11	63	17
Connecticut	8	8	100
Delaware	1	3	33
Florida	7	67	10
Georgia	15	159	9
Idaho	5	44	11
Illinois**	34	102	33
Indiana**	14	92	15
Iowa**	13	99	13
Kansas	5	105	5
Kentucky	19	120	16
Louisiana	19	64	30
Maine	15	16	94
Maryland	10 + Baltimore	23	43
Massachusetts	14	14	100
Michigan	40	83	48
Minnesota	18	87	21
Mississippi	1	82	1
Missouri**	15 + St Louis	114	13
Montana	8	57	14
Nebraska	4	93	4
Nevada	14 + Carson City	16	88
New Hampshire	10	10	100
New Jersey	21	21	100
New Mexico	7	32	22
New York	62	62	100
North Carolina	4	100	4
North Dakota	0	53	0
Ohio*	75	88	85
Oklahoma	4	77	5
Oregon	6	36	17
Pennsylvania	67	67	100
Rhode Island	5	5	100
South Carolina	6	46	13
South Dakota	1	67	1
Tennessee	17	95	18
Texas	18	254	7
Utah	6	29	21
Vermont	14	14	100
Virginia	12	96	13
Washington	9	39	23
West Virginia	9	55	16
Wisconsin	17	72	24
Wyoming	1	23	4

*EPA Data As of December, 1979
**Included in the U S Department of Agriculture's 'Corn Belt' Growing Region

subject to emission limitations and permitting requirements to ensure the Prevention of Significant Deterioration (PSD) of air quality in the region. Sources thus controlled include coal-fired steam boilers of more than 250 million Btu per hour heat input. (For comparison, the Katzen-designed 50-million-gallon-per-year plant requires heat input of 263.3 million Btu per hour.) PSD limitations regulate criteria pollutant emissions so that the secondary NAAQS are not violated within the Air Quality Control Regions (AQCR) concerned. Although all of the criteria pollutants, as well as certain other pollutants not associated with alcohol production, are subject to PSD limitations, to date the permissible emission increments have been announced only for particulates and sulfur dioxide. It is, therefore, possible that standards yet to be proposed by the EPA could have an impact on the siting of energy facilities such as ethanol-from-biomass plants.

Certain new, modified, and reconstructed stationary sources, including large fossil fuel-fired steam generators such as those used in distilleries, are subject to EPA's New Source Performance Standards (NSPS). Limitations are placed on particulates, opacity, nitrogen oxides, and sulfur dioxide. Smaller industrial boilers are expected to be covered by revised NSPS standards in 1980 or 1981.

Siting of new fuel alcohol plants could be barred in so-called nonattainment areas. In order to begin construction or major modification of a major source in a nonattainment area, a construction permit must be obtained. Before the permit is granted, the source must obtain an emissions offset from existing sources in the region. In addition, the source must limit its air pollutant emissions to the lowest achievable emission rate or the lowest obtainable in practice, regardless of energy and economic considerations. The extent to which the fuel alcohol industry could be affected is demonstrated by noting the number of counties in each of the 48 contiguous states which have been designated as nonattainment areas, as shown in Table VII-2. Although the number of counties affected may not be proportional to their land area, the figures convey at least a general indication of the extent of the problem in a given state. In 31 states, 30 percent or less of the state's counties had been classified as nonattainment areas for one or more pollutants as of December, 1979, and in 20 of these states, 15 percent or less of the counties were affected. At the other extreme, in 13 states, 71 percent or more of the counties had been designated as nonattainment areas (see Figure VII-1), and all counties in Connecticut, Massachusetts, New Hampshire, New Jersey, New York, Pennsylvania, Rhode Island, and Vermont were in nonattainment status.

Because of proximity to feedstock supplies, the states included in the USDA's "Corn Belt" growing region (Iowa, Missouri, Illinois, Indiana, and Ohio) are likely targets for development of the fuel ethanol industry. Of the Corn Belt states, only in Ohio should siting limitations for ethanol plants be severe.

- **Federal Water Pollution Control Act**

    It is the intent of the Clean Water Act to attain zero discharge of regulated pollutants such as BOD, solids, oil, and grease by 1985. Technology-based effluent limitations on certain existing sources are predicated on the effectiveness of the best available control technology economically achievable by 1983.

    Aqueous discharges from distilleries are relatively easy to monitor and regulate, as they generally involve point sources. There are no current technology-based effluent guidelines for fuel ethanol facilities or for beverage distilleries. Fuel ethanol plants are under consideration by EPA, and a decision on the need for wastewater emission regulations for these facilities will be made in one to two years. The National Pollutant Discharge Elimination System (NPDES) distinguishes between major and minor dischargers, with a major source consisting of one discharging 100,000 gallons or more per day of wastewater. (Thus, commercial-scale ethanol plants would be classified as major sources.) Wastewater pretreatment regulations may have an impact on fuel ethanol production. Final pretreatment regulations have been set governing water quality standards for nondomestic waste that is introduced to POTW with design flows of 5 million gallons per day or more. Between now and July 1, 1983, the provision for a pretreatment program will become a condition of all new or reissued NPDES permits and industrial dischargers must be in compliance by that date.

- **Resource Conservation and Recovery Act**

    The primary areas of potential solid waste regulatory impacts applicable to ethanol-from-biomass activities are coal ash from boiler operation, grain elevator dust, particulates from the drying of distillers grains (trapped during emission control efforts and disposed of as solid waste, unless recycled), and stillage, if it is applied to soil as fertilizer. If recycling of

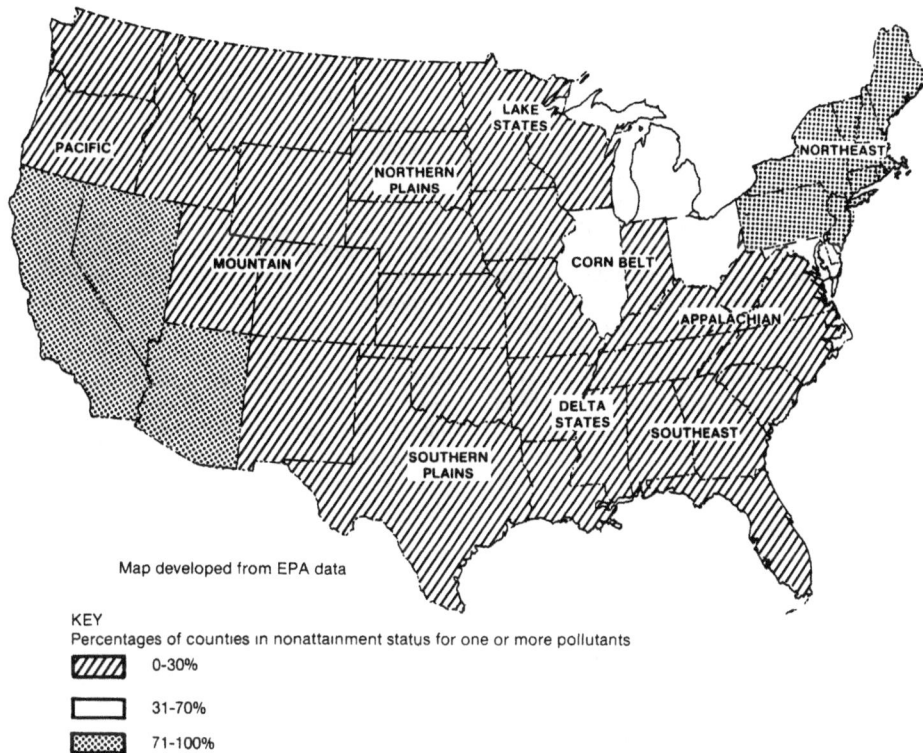

KEY

Percentages of counties in nonattainment status for one or more pollutants

/////	0-30%
☐	31-70%
▓▓▓	71-100%

Figure VII-1. Extent of Nonattainment of National Ambient Air Quality Standards in 48 Contiguous States

trapped grain and recovery of DDG from stillage are assumed, then the major Federal legislation bearing on solid wastes from the biomass preparation and conversion steps is the RCRA. The extent of coverage and methods of implementation of this Act are still being interpreted and developed by EPA and the courts. EPA has considered designating some energy-related hazardous wastes as "special waste" under RCRA. Interim standards would be established for special waste. Coal ash may be assigned to the special waste category until extensive analysis is completed.

Environmental Assessment

Environmental assessments or impact statements may be required prior to construction of fuel ethanol plants under certain conditions. The need for such assessment is triggered by Federal participation in funding, whether by direct cost-sharing or by loan guarantee. For a large project, which may produce significant environmental effects, formal assessments will be required. The Office of Environmental Compliance and Overview within DOE rules on the need for such assessments.

Even in the absence of a Federal requirement for an environmental assessment, it may be prudent for an investor to gather the pertinent environmental information, as such foresight may expedite the permitting process and facilitate public acceptance of a project. Although a Federal agency must evaluate the environmental evidence, the fact-finding activity may be delegated to the applicant or its independent contractor.

FEEDSTOCK PRODUCTION

In the production of grains as feedstocks for ethanol production, the single most critical environmental issue is erosion and its effects on water quality. Erosion serves as a transport mechanism for carrying nutrients, sediments, and pesticides to water resources, and it directly affects crop productivity and soil fertility.

Land and water use are also important issues. The need for irrigation in the western states puts increasing pressure on existing surface and groundwater resources, causing water tables to drop in some areas and the salinity content to increase in return flows to surface waters. The land use issue has two major components: 1) the loss of farmland to urban sprawl, and 2) the increasing pressure to put marginal farmland into use and convert fallow or natural areas to intensively cultivated farmlands.

Erosion

The rate of erosion is dependent on several factors, including type of soil, topography, amount and intensity of rainfall, ground cover, and control methods. It is, therefore, highly variable from site to site and over a period of time. The effects on soil are the direct loss of fertility and enhanced potential for water quality degradation. The latter arises as the biological content of the soil is decreased, in turn reducing the rate at which toxic substances applied to the soil are broken down. Hazardous substances thus are more likely to reach waterways.

The loss of sediment to waterways results in silting. Runoff from agricultural lands can cause additional problems, since fertilizer and pesticide residues are carried to the waterways. The major soil nutrients, nitrogen and phosphates, promote algal growth and thereby accelerate eutrophication of surface waters. Corn requires relatively high fertilizer application rates; indeed, although corn comprises only 20 percent of the United States' cropland, it requires approximately 40 percent of the fertilizer used nationwide. Herbicide applications for corn are also higher than the national average.

Water Usage

The quantity of water used for irrigation can be substantial, and represents only one of several competing uses for a limited resource. Groundwater currently supplies approximately 40 percent of total United States irrigation needs, primarily in the western States. The importance of this issue for corn production is not as great as for other feedstocks, as minimal amounts of land are under irrigation in the major corn-producing States. Five of the six largest corn-producing States, for example, have less than one percent of total cropland under irrigation.

In addition to the use of water for agriculture, the quality of the returned water is of importance. Irrigation water picks up mineral salts, and the resulting increase in salinity of the water can have adverse effects on downstream users.

Land Use

Increased demand for grain crops as ethanol feedstocks would result in the need to increase the amount of land in agricultural production. If marginal land is brought into production, more fertilizer and pesticides would be necessary to attain the average level of productivity for a given crop. These increases in chemical applications would intensify the water quality impacts already associated with agricultural runoff; i.e., erosion from such marginal land would create more environmental impacts per acre, or per unit of output, than the average farmland in use today.

Another potential source of farmland is the conversion of natural areas such as forests to intensively managed crop production. Wildlife habitat would thus be destroyed, and erosion rates would be substantially increased.

The land use issue is further complicated by the loss of farmland to urban development. Annually, 3 million acres of farmland are lost to urban sprawl, and one-third of this total is considered prime farmland. As more of the existing farmland is lost to urban development, a greater proportion of marginal lands may be required for crop production.

A recent USDA cropland availability study reports that 78 million acres with high potential as cropland remain in this country, with much of it in the Corn Belt and Northern Plains regions. Of this high-potential land, 15 million acres have essentially no limitations on development. The latter acreage, if planted entirely in corn and used exclusively as ethanol feedstock, corresponds to approximately 4 billion gallons of fuel ethanol per year, or 80 of the reference-size, commercial-scale plants.

USE OF ETHANOL/GASOLINE BLENDS IN HIGHWAY VEHICLES

Environmental impacts which could be associated with the use of ethanol/gasoline blends as fuel for highway vehicles are discussed briefly below. Not only users of the fuel but also the general population could be affected. Although potential impacts on the latter group are largely undefined, no barriers to the use of gasohol are anticipated.

Impacts on Users

- **Health**

  Ethanol is not highly toxic. Its major routes of entry into the human body are ingestion, inhalation, and absorption through the skin. Effects of exposure to the substance are primarily ones of discomfort, as described in an earlier section. Ingestion of ethanol/gasoline blends is unlikely, but the ethanol could be separated from the denaturing agents in some formulations. (For example, ethanol can be separated from gasoline denaturant.) A variety of physical symptoms can also arise from exposure to denaturants used with fuel ethanol.

- **Safety**

  Ethanol is a flammable liquid, with flammability limits of 3.3 to 19 percent by volume in air. Its flash point is above that of gasoline. Typical ethanol/gasoline blends pose explosion or fire hazards comparable to those of gasoline. In storage, however, ethanol presents one additional safety hazard—neat alcohols burn with invisible flames, thus making it possible for ethanol fires to remain undetected and/or to be approached too closely by workers.

Impacts on General Population

- **Air Quality**

  Both evaporative and exhaust emissions from ethanol/gasoline blends will differ slightly from those associated with unblended gasoline. Carbon monoxide and hydrocarbon emissions are not changed significantly. The exhaust is sensitive to the original air/fuel setting of the carburetor, whether or not the carburetor has been adjusted for the new fuel, the type of emission control system, and the engine configuration. A reduction of nitrogen oxides emissions is obtained with ethanol/gasoline blends, but aldehyde (currently unregulated) and evaporative emissions increase for the blend relative to gasoline. As ethanol contains no sulfur, sulfur dioxide is absent from the combustion products. Moreover, ethanol yields no particulates on combustion.

- **Ecosystems**

  Little information exists in the realm of potential effects of utilizing ethanol/gasoline blends in highway vehicles except for the above information on combustion. The major area of potential concern appears to be the possibility of introduction of ethanol to waterways through spills or other accidents. Ethanol dissolves readily in water and could, therefore, be transported easily in an aqueous medium.

SUMMARY OF ENVIRONMENTAL ISSUES

Three major environmental problem areas associated with commercial-scale fuel ethanol production are: 1) safe disposal and treatment of distillery wastewaters; 2) exposure of workers to toxic and/or corrosive chemicals; and 3) atmospheric emissions and solid wastes resulting from combustion of coal to satisfy process steam requirements. Controls, in the form of hardware and procedures, are available commercially. The cost of direct controls amounts to approximately 13 percent of the capital investment for a 50-million-gallon-per-year fuel ethanol plant.

The environmental regulatory climate can influence investment decisions. Current regulations may place limitations on siting of ethanol plants, particularly if they include coal-fired boilers. Besides the siting restrictions arising from air quality considerations, the assimilative capacity of streams will play a role in siting decisions. Beyond the current generation of Federal environmental regulations, investors need to consider the possibility of more stringent ones, as well as existing State and local regulations which may be more restrictive than their Federal counterparts. Anticipated regulations could influence both the economics and the siting of future plants.

Availability of feedstocks for fuel ethanol production is essentially unfettered by Federal environmental laws. Environmental impacts associated with agricultural production tend to be non-point-source problems and are, therefore, difficult to monitor and regulate. The major concerns of increased agricultural production to meet fuel ethanol feedstock requirements are erosion, which impacts water quality, and use of limited land and water resources.

When ethanol is blended with gasoline for use as a highway vehicle fuel, there are potential environmental implications involving both the user group and the general population. Ethanol is not highly toxic, but it does produce definite physical symptoms when inhaled or ingested. The same precautions need to be taken in handling ethanol as with any other explosive and flammable substance. The potential air quality and ecosystem impacts of ethanol combustion in highway vehicles are not fully defined, but no barriers to deployment of ethanol for fuel use are currently foreseen.

CHAPTER VIII
Business Plan

Entry into the alcohol industry requires equal planning from both business and technological perspectives. Many technologically sound projects have never gotten off the ground for lack of business acumen, adequate financing, and a cohesive team capable of transforming ideas and strategies into operational results. These concerns are of particular significance in the alternative energy field where there is insufficient financial and management history to ensure an adequate return on investment. The purpose of a business plan is to blend innovative ideas into a detailed plan of action. The principal objectives of a business plan include:

- The establishment of a project's goals and objectives, such as a monetary return, investment criteria, protection of supply, image, diversification, public or national spirit, etc. As a general rule, the more owners or investors in a project, the more diverse a project's objectives become.

- The development of a decision matrix for "Go Ahead", "No Go", and modification of plan decisions.

- The assembly of a task force team including both internal and external members capable of accomplishing the project's objectives, and likewise having the ability to advise the investors when to abort an undesirable project. This team should include:

  - Investors, whose function is to establish the objectives of the project.

  - Internal management and staff personnel, whose function is to carry out the investors' objectives.

  - External consultants and vendors, who will usually include:
    Attorneys
    Certified Public Accountants
    Engineering Consultants
    Investment Bankers
    Commercial Bankers
    Contractors
    Insurance Agents and Underwriters

Table VIII-1. Project Costs and Possible Funding for an Ethanol Plant

COST ITEM	FUNDING SOURCE		
	SEED MONEY OR EQUITY	MAY BE DEBT FINANCED	POSSIBLE GRANT
Prefeasibility Cost	X		
Underwriting the Cost of Raising Equity	X		
Feasibility Study	X		X
Exploring Sources of Financing	X		
Preliminary Engineering	X		X
Final Engineering		X	
Site Options	X		
Acquiring a Site		X	
Plant and Equipment Cost (Construction Period)		X	X
Plant and Equipment Cost (Permanent)		X	X
Working Capital Requirement for the Plant	X	X	

Once the project team is assembled, the goals have been defined, and a decision matrix established, a detailed schedule or work program should be developed for project control. This detailed schedule or work program should include, as a minimum, the following:

- Designation of a project manager or director, and description of all project members' functions.

- Specific assignment of tasks to project members with definite dates for completion.

- Definition of the project's objectives and goals.

- A method for interim reporting on the status of each assignment.

- A calendar of expected completion dates for key areas in the project which will require decisions.

The following discussion addresses the key requirements of studying, organizing, financing, constructing and operating an ethanol plant.

SEED MONEY STAGE

"Seed money" is usually the most difficult to raise for any investment. By definition, seed money pays for expenses incurred during the planning stage of a project, before decisions are made on how best to proceed; consequently, seed money is considered high risk by investors. Because of their greater risk, investors who enter the venture at this time will receive an ownership position in the project greater than their proportional dollar share of investment. Seed money is particularly

important to alcohol fuel projects to finance the prefeasibility study. The financing and economic feasibility study requires the accumulation of marketing, legal, economical, technical, and financial information by engineers, attorneys, and C.P.A.'s, to evaluate the viability of the project.

In planning an ethanol plant, one must obtain funds to cover the various project costs. These costs and possible funding sources are listed in Table VIII-1.

Once a project team is assembled, seed money acquired, and a financing plan developed, the organizational form of the entity must be considered.

ORGANIZING THE VENTURE

Alcohol plants may be established in various organizational structures depending on the objectives of investors and applicable regulations. The different entity forms which may be considered include:

- Proprietorship
- General partnership
- Limited partnership
- Joint venture
- New subsidiary of an existing corporation
- New division of an existing corporation
- New corporation
- Acquisition or merger of an existing corporation

Because of the complicated legal, accounting, financial, and tax considerations involved in choosing an organization structure, a decision should not be made until detailed discussions are held with the project's attorney and C.P.A. to determine which entity format is best suited to a given situation.

The key factors and pros and cons of the alternative entity forms are shown in Table VIII-2.

OTHER CONSIDERATIONS

Establishing a project office within an existing corporation has the following advantages:

- Ability to select from a large, tried and proven management staff
- Lower administrative burden
- Established sources of internal funds
- Existing credit rating
- Existing credit sources

Reliance on internal development has been a rule for many firms, yet the costs and risks involved in such action should be weighed against the opportunities created by acquisition which can often provide more rapid and predictable results. The addition of new corporate capabilities by this method should be investigated as early as possible for any conceivable antitrust implications. The public interest, as reflected in the antitrust laws, prohibits any acquisition or merger that substantially lessens competition or creates a monopoly. In favor of this alternative is the immediate acquisition of trained management, a new source of working capital, and an expanded credit base.

A new corporation is an obvious organizational alternative. A charter defines its purpose and scope, and formation of a new corporation has significant debt and tax advantages. Financing may be obtained by issuing new shares and company assets may be pledged as security for loans and bonds.

A joint venture is a special form of a partnership or corporation created by two or more existing legal entities to accomplish a specific, limited objective. The necessary seed money, management, labor, and board of directors are provided by the sponsoring organizations but their operational expenses thereafter are borne by the new group.

Once the legal and management structures have been determined, the organization should set regular business sessions and stockholders' meetings. It is not important

Table VIII-2. Alternative Entity Options

Factor of Comparison	Individual/Partnership or Consolidated Corporate Subsidiary	New Non-Consolidated Corporate Ownership
Start up cost, depreciation or losses may be deducted when incurred by investor	Usually passed out to the investor for a current deduction	Carried forward to offset against future income
Tax credit such as investment tax credit and energy tax credit	Usually passed out to the investor for a current deduction	Carried forward to offset against future income
Rates of taxation	The personal tax rate of the individual or partners; the corporate rate of corporate parent	From 17% of the first $25,000 to 46% on income in excess of $100,000.
Liability of investor	Unlimited to proprietor or general partner, others usually limited	Usually limited
Taxation	To the individual or corporate owner when incurred; no taxation of dividends to individuals and an 85% exemption for corporate parents	Taxed at corporate level when earned; dividends also taxed to the individual when distributed

at this point to fine-tune the organization; the most pressing detail is building the project team which is capable of structuring the financial arrangements for the project.

CRITERIA IN SELECTION OF THE PROJECT TEAM

Unless the organization is a joint venture among highly sophisticated firms with sufficient resources available, or an organization with an internal source of expertise in all fields necessary for operation of an alcohol project, it will be necessary to turn to external sources for expertise. Outside consultants will probably be needed.

It is important to choose the right professionals to handle the organization's needs. Arrangements or contracts should be negotiated and signed with a recognized engineering design firm, a legal firm, a C.P.A. firm, a construction contractor, and, most likely, an investment banker. Because of the rapid development of the alcohol industry, there are a number of young, highly competitive, and aggressive companies in the marketplace offering services in process design, engineering, construction, and financing. Unfortunately, the rapid growth has also enticed a number of marginally qualified firms into this marketplace. Promoters and investors should take great care in selecting qualified firms in order to assure quality work and to protect their investment. It is advisable to obtain personal and business references, and to utilize firms with experience not only in the field of alcohol distilling, but also to choose those who have structured projects of a similar dollar magnitude compared to the project involved.

In evaluating the engineering firm, it is important to recognize that there are no monopolies on alcohol production processes; many choices are available. As each process undoubtedly has strengths and weaknesses, both price and quality should be considered in the selection process. Several specific questions are important in the determination of an engineering firm. The organization should request:

- A detailed economic analysis and description of the process which the firm recommends and an estimated manufacturing cost per gallon

- Specific projects which the firm has engineered in the alcohol industry

- A copy of the process and performance guarantees used by the firm

- A copy of the legal contracts which the firm uses in contracting for services

- The current financial statements and bank references of the engineering firm

The above information should be reviewed by the project's attorneys and C.P.A.'s, and the attorney should finalize all contracts with the engineering firm.

In selecting attorneys, C.P.A.'s, and investment bankers, it is essential to consider the following:

- Experience in the industry

- Location of the professionals, due to the almost daily contact required in the development stage

- The attorney's specific knowledge in areas of taxation, contracts (possibly securities law and underwriting), Federal regulations, patents, etc.

- The C.P.A. firm's special knowledge in project financing, forecasting, taxation, feasibility studies, management analysis, and, possibly, in SEC and underwriting

- The investment banker's experience in successfully placing project financing and, if necessary, the equity portion of a project

The criteria in selecting the contractor are identical to those of the engineering firm, with the addition that the contractor be bondable.

PROJECT FINANCING

Once the team is assembled and the feasibility study is underway, the detailed financing package must be assembled by the C.P.A.'s and investment bankers. This package, in addition to the feasibility study, must include:

- Personel resumes of applicants
- Personal resumes of selected plant managers
- Contracts for supply of raw materials
- Contracts for energy supply
- Contracts for sale of production
- Contracts for sale of coproducts
- Corporate/partnership history and historical financial statements
- Summary of collateral offered
- Summary of projected use of funds
- Summary of projected source of funds

It is important to estimate accurately the time required to accumulate all of the legal and financial documentation required. This is especially true where government forms and applications are necessary. It is also important, when requesting federal financial assistance, to consider aspects which address the government's objectives. For example, government agencies are presently encouraged to give priority assistance to projects which utilize a farm cooperative to supply raw materials and thus contribute to the enrichment of the local economy, or to projects which utilize municipal wastes to fuel the plant. If such options are reasonable, then the loan package should emphasize their expected effects on the local economy. (See Appendix A, Financial Assistance Programs, for financing available from the federal government.)

Identification of federal involvement has another dimension; i.e., a request for financial assistance should give reference to such Federal incentives as the Crude Oil Windfall Profits Tax Act of 1980. The alcohol fuels sections of the Act will help provide long-term market support for gasohol, cut some of the red tape involved in licensing and regulatory procedures, and offer tax incentives for investors in alcohol fuel plants.

Finally, a banking agent for the project should be identified and the bank itself should make some commitment to the project to act as a supplier of funds. This commitment indicates a solid relationship with the bank and gives further creditability to the project. In virtually all instances, with the exception of the costs of documenting the loan package, financing services should be based on results. Investment banking commitments will normally cost 1 to 2 percent of the total loan requested. C.P.A. firms, which professionally handle government applications for financing, will normally charge a negotiable fee for assisting in the feasibility study and preparation of the financing package. This fee, based on the time and billing rates of the individuals involved in the preparation of the financing package, will generally amount to about 1 to 1.5 percent of the loan amount requested.

PLANT CONSTRUCTION AND OPERATION

In the final phase of the business plan, management must become actively involved in the project. The owners will need to hire management and operations personnel to work with the engineering and construction companies to gain in-depth knowledge of the plant operations. It is extremely important that cost controls be effected early in the project development. Thus, the company must implement its internal accounting procedures at the earliest possible date.

Regular meetings among management, engineers, and construction personnel should be maintained throughout the construction phase and into start-up operations. Management should locate necessary skilled and semi-skilled operations personnel, and plan the initial logistics of start-up operations during the project construction period. A timephased approach will prove most effective in overall business planning.

CHAPTER IX
Feasibility Study

The purpose of this chapter is to identify the significant aspects in performing an effective feasibility study and suggest how best to address them. The form and format proposed should not be taken literally; rather, as guidelines for consideration as needed. Knowledge of the audience is mandatory. If the study is to be prepared for a group of investors with varied backgrounds, then more detailed technical explanations of the basic manufacturing process and a historical overview of the industry are probably required. An enunciation of the potential and inherent risks of the proposed venture must be clearly stated, regardless of the audience addressed. If the study is for use by more informed and experienced lending institutions, then too much preamble may dilute the primary points of the proposal.

The basic points to be addressed in the study are how successful the venture will be and how much risk is involved to all parties. The length of the study and the amount of detail required are governed by the size of the project and the amount of control the involved parties (the proposed lender as well as the proposed borrower) have over the decision variables. For example, less information need be presented if the process raw materials are already owned or contracted for, or if the sale of all production of the product and coproducts is guaranteed. Discretion should be used as to the volume of information to be provided.

First, the potential investor must be convinced that the project is worth his time and effort, and a few major issues must be addressed for this prefeasibility phase: namely, the availability of markets, feedstock cost and availability, water, power, and status of legislation.

PREFEASIBILITY PHASE

The initiator of the project must be satisfied that (1) there is a market and it can be reached (conduct market and marketing analysis); (2) resources and means of acquiring them have been identified, e.g., feedstock, water, power, and labor (evaluate factors of production); and (3) all applicable Federal, State, and local requirements can be met (determine environmental, health, and safety factors). With respect to the existence of a market for ethanol and coproducts, the investor must determine how firm the need, at what price, and at what risk to the project. With respect to the necessary feedstocks and chemicals, he must determine their availability, at what price, and with what assurance for each. With respect to the availability of needed water, the same considerations need to be faced. For fuel, the potential investor should check the availability and cost of alternative resources, particularly cogeneration, and coal and natural gas. With respect to laws, regulations, and rules, he should determine which are prohibitive and which are supportive of the project.

FEASIBILITY PHASE

Once the potential investor is convinced that the project deserves an indepth analysis, a feasibility study should be performed in sufficient detail to define the means by which financing, construction, and operation will be accomplished. Each study should address:

- Technical and economic feasibility
- Resource assessment and availability
- Financing alternatives
- Ability to construct and operate a commercial plant in an environmentally acceptable manner on a selected site

The marketing, materials, and legal sections must be reviewed to provide more specificity, and sections on technical and engineering design, site selection and suitability, economic and financial analysis, and management planning should be added. Sufficient detail should be provided to facilitate a lender's decision whether or not to provide the support required.

The following discussion, sample, and outline information elements at the end of this chapter only provide a guide to preparing a feasibility study; the investor must do the convincing. The basic goal of the study should be kept in mind at all times: an answer to the question, "Can this specific project make a profit over the anticipated life of the investment?" The feasibility study should critically examine the proposed project to uncover problem areas and consider all possible risks. Solutions should be offered to reduce risks and unknown economic factors that may cause the venture to fail.

KEY ELEMENTS OF FEASIBILITY STUDY

INTRODUCTION

_____ ✔ Purpose

_____ ✔ Scope

_____ ✔ Limitations

_____ ✔ Summary and opinion

This section sets the foundation for the reader, focusing his attention on the most important conclusions and business factors and their effects on the proposed project. It emphasizes the positive side of the findings, but also indicates where help is needed. The Summary and opinion are very important because many readers want to know the bottom line before reviewing the project in greater detail.

PART 1. MARKET AND MARKETING ANALYSIS

A. ETHANOL

_____ ✔ Market description

_____ ✔ Gasoline demand

_____ ✔ Alcohol demand

_____ ✔ Product analysis

_____ ✔ Competitive factors

_____ ✔ Presecured markets - contracts for sale of product

_____ ✔ Distribution analysis

The objective of this study section is to determine if the total proposed output of the plant can be sold at a profit over the life of the venture. The nature and detail of the analysis will identify and quantify specifics of the market composition, size, and description (e.g., refiners, marketers, consumers, geographic area); competitive factors, both current and future; gasoline and alcohol, current and future pricing and anticipated demand. The intent is to develop a realistic evaluation of immediate market conditions and to prepare realistic market projections for the future.

B. COPRODUCTS

____ ✓ Market description

____ ✓ Market demand

____ ✓ Product analysis

____ ✓ Competitive factors

____ ✓ Promotion analysis - presecured sales

____ ✓ Distribution analysis

The sale of coproducts such as CO_2, yeast, stillage (both wet and as distillers dried grains), and gluten can be a large determinant of the profitability of plant operations. It is likely that for such products as DDGS, the future market will be totally new because of the unprecedented volumes of supply and the untested demand. Price fluctuations of the DDGS have been rapid and irregular with the supply drying up during the summer months. Any long-term contracts for selling the coproducts that can be established in advance are invaluable in establishing the viability of the proposal.

PART 2. TECHNICAL AND ENGINEERING DESIGN

____ ✓ Technological state of development

____ ✓ Process description

____ ✓ Production schedule

____ ✓ Resource requirements

____ ✓ Equipment procurement plan

____ ✓ Construction management plan

____ ✓ Process guarantee/service warranty

____ ✓ Assessment of uncertainties/contingency planning

The objective of this analysis is to determine the technological feasibility of the proposed project. The potential investor must provide evidence that the proposed venture has available a viable process ready for construction and installation according to well-considered timetables. This section provides the rationale for process selection and should analyze the uncertainties surrounding commercial application of the process. It should clearly identify all major areas of project design and operation that represent significant new technological innovation and should delineate what steps (if any) remain to be taken before commercial readiness is achieved.

The process description should include process flow diagrams, energy and mass balances, mechanical specifications, process specifications, major equipment requirements, plant layout sketches, plot plans and off-site requirements, waste-handling procedures, and other general design requirements. Resource requirements include all necessary raw materials such as water and power. All plans should be time-phased and indicate manpower and subcontractor requirements, where appropriate. Ideally, an involved engineering company can provide a process guarantee and subcontractors and suppliers will provide service warranties. An assessment of the uncertainties surrounding the commercial application of the proposed process must be presented, including presentation of your strategy for dealing with unanticipated events.

PART 3. FACTORS OF PRODUCTION

A. FEEDSTOCK

____ ✓ Alternative feedstocks

____ ✓ Selection criteria

____ ✓ Cost and availability of selected feedstock

____ ✓ Contingency plan

The choice of feedstock for ethanol production varies from region to region, and even from site to site. The financial success of the venture is dependent upon the steady procurement of low-cost feedstock. Present availability as well as future agricultural projections should be evaluated for each proposal. The results of current crop development work will influence future feedstock choices and the investor should have full knowledge of the available alternatives.

All available fermentables need to be catalogued and compared to provide as many options as reasonable. Purchase prices at which the use of each alternative feedstock becomes less economical than another feedstock should be constantly updated to provide a basis for purchasing decisions. Factors affecting feedstock supply and availability should be reviewed in this section of the study, including natural factors as well as likely political or competitive events. During your feasibility study, you should identify sources of supply that can be contracted at favorable prices, particularly, those products of substandard grade or classified as waste from some other process.

B. POWER

____ ✔ Alternative sources

____ ✔ Selection criteria

____ ✔ Cost and availability of selected source

____ ✔ Contingency plan

Potential energy sources for process fuel should be analyzed and the best source selected. The criteria usually are availability, cost, and the resulting net energy balance. The better sources of energy, from a national viewpoint, are waste and other renewable resources; and, depending on the site, probably coal and possibly natural gas. Some financing sources, particularly the government, may have explicit guidelines on the use of different fuels. Any likely potential for shortages due to labor shutdowns in other fields should be addressed.

C. WATER

____ ✔ Availability of sources

____ ✔ Requirement

____ ✔ Cost

____ ✔ Contingency plan

Significant amounts of water are used in the ethanol production process (about 16 gallons of water per gallon of ethanol produced). This demand includes requirements for generating steam, cooling, and preparing mashes. The actual amount of water needed will be significantly less due to the recapture and recycling of water used, but in some regions this may be a critical consideration. The source, process requirements, and cost should be analyzed, and a plan developed for acquisition, showing a complete understanding of potential problems.

D. LABOR

____ ✔ Job descriptions/specifications

____ ✔ Availability and rates

____ ✔ Recruiting and training

____ ✔ Labor relations

In this infant industry, production manpower requirements are not well known. Different processes should require specific numbers and types of workers, yet as experience is acquired, the work force required will prob-

ably change. The first estimate of the number of workers necessary can be obtained from the engineering design contractor. The feasibility analysis should deal with the quality and characteristics of the available labor pool; how much they will be paid; how they will be recruited, selected, and trained; and an outline of promotion policies and retention means. Any potential for labor disputes or shutdowns should be addressed for each site considered.

PART 4. SITE SELECTION AND SUITABILITY

____ ✔ Location

____ ✔ Transportation/logistics

____ ✔ Services/utilities

____ ✔ Land acquisition plan

The primary site requirement for location of an ethanol plant is the availability of an abundant water supply. Other site requirements are similar to those for any chemical manufacturing facility, except for the particular material handled and the storage of agricultural and liquid products. In comparing alternative sites, project or anticipate future problems such as the availability/cost of railroad service, regional and local political forces, and distance from the raw materials.

A potential site should be near several sources of different raw materials and several different markets. Multiple modes of transportation are likewise desirable. The site should be located where there will be no unusual circumstances preventing licensing and permits. The costs of these factors should be used in evaluating the best site. The recommended site should be readily available for acquisition.

Any firm agreement to purchase or lease the site should be subject to engineering approval, receipt of required permits, and completion of other legal and environmental requirements.

PART 5. ECONOMIC ANALYSIS

A. COST OF PRODUCTION

____ ✔ Plant size

____ ✔ Fixed costs

____ ✔ Variable costs

____ ✔ Economies of scale

____ ✔ Sensitivity analysis

A detailed economic analysis should be presented for the proposed plant. As described elsewhere in this document, the size chosen for the plant can mean the difference between purchasing raw materials and developing, storing, and treating them; thus incurring a different cost structure for the plant. Detailed assumptions concerning the cost of operation, i.e., fixed and variable costs, need to be identified for the near- and long-term, as a function of output. An accounting model of the plant should be developed and the effects of changing the cost of the key variables should be evaluated. This sensitivity analysis will illustrate the range of possible consequences and help the project developer organize a plan to reduce dependence of the project on uncontrollable variables.

B. INVESTMENT ANALYSIS

_____ ✔ Total capital requirements

_____ ✔ Income and cash flow projections

_____ ✔ Profitability analysis

An analysis of the total capital requirements must be prepared, including appropriate breakouts of the capital components such as the elements of fixed capital requirements, and others, such as working capital. Of course, the final projections must wait until many of the project characteristics are finalized, such as site-specific construction requirements and the size, terms, and interest rates of the necessary loans. However, assumptions should be made based on findings from other sections of this feasibility study in order to determine bottom-line estimates as early as possible.

Working capital requirements should be estimated in conjunction with operating characteristics of the engineering design. Actual amounts would differ, depending on the inventory requirements, the lines of credit opened with a commercial lender, and the amount of product produced. Less than full-capacity operation should be planned, especially during the start-up period, and most probably thereafter. Allow for sufficient funds to cover contingencies, as well.

As the income and cash flow projections are made, the least favorable conditions also should be considered. Profitability analysis should include projections of the classical ratios, including return on investment, payback period, discounted cash flows, net present value, and so on.

The funding options available to the entrepreneur should be a fundamental part of these analyses. The investor should include the likelihood of Federal, State, and local government support as well as taxes, and discuss their effects on the project's financial picture.

C. RISK ANALYSIS

_____ ✔ Government intervention

_____ ✔ Competitive reaction

_____ ✔ Product/process obsolescence

Examination of the potential risks involved with alcohol fuel production is of primary importance to both the entrepreneur and the investor. Not only will it help to identify the possibilities for failure, but if the project is undertaken, it will identify potential problem areas for management. This early identification of economic risk also will aid in determining the optimum project configuration to allow for the most flexibility in dealing with changing economic and technological climates. In conjunction with the economic analysis, an examination of the available margins at various market conditions should be made to determine the effects of changing markets. The purpose of this section is to highlight potential pitfalls to prevent producing a product which costs too much, or having to sell a product at too low a price.

PART 6. MANAGEMENT PLAN

_____ ✔ Background and experience

_____ ✔ Responsibilities

_____ ✔ Management systems

As in any new venture, and especially a pioneering effort such as alcohol fuels production, management capabilities are a most critical component. The management plan should indicate the involvement and responsibilities to be assumed for the entire life span of the project. Responsibilities and authorities must be clearly identified and each position filled with an experienced individual. The technical and business backgrounds are critical, with agribusiness, as well as refinery and plant management experience, preferred. The lending institution must be assured that top-rated managers are being hired.

7. ENVIRONMENTAL HEALTH AND SAFETY AND OTHER REGULATORY REQUIREMENTS

_____ ✔ Statutory and regulatory requirements

_____ ✔ Public concerns

_____ ✔ Management responsibilities

The purpose of this assessment is examination of those issues likely to have the most impact on the permitting and scheduling of the proposed facility. The potential investor should research all applicable local, State, and Federal statutory and regulatory requirements in order to determine whether the environmental, health, and safety requirements can be met, and if so, on what schedule.

The baseline environmental quality of the proposed site should be defined; a determination of applicable environmental standards and constraints should be presented; a technical description and cost evaluation of mandated environmental control measures should be prepared; the quantities of air, water, solid, and radiation residuals of the proposed process assuming applicable control technology should be assessed; and an analysis of the impact of residuals on the baseline environmental quality of the site should be presented, using appropriate predictive models.

Assessments of particular concern to the community or state should be included, quantitative analyses should be presented of worker and community exposures to, and disposal of, known or suspected hazardous materials.

Finally, the investor should describe control options, costs and risks, and control measures that will be instituted to meet requirements. The process of making alcohol fuels should be environmentally safe, and the investor should be certain that no moral, ethical, or legal issue will tie up the project for long periods, or even kill it after it is started.

FEASIBILITY STUDY FORMAT

This outline contains elements upon which the potential investor may expand in his feasibility study.

1. Title Page

   - Name of firm submitting the proposal
   - Mailing addresses
   - Telephone numbers
   - Person to contact if further information is needed

2. Table of Contents

   - All chapters
   - All appendices
   - All tables, charts, & maps

3. The Objective

   - Who/What/When/Where/Why/Cost/Annual Capacity

4. Background

   - History of this particular venture and alcohol fuels

5. Market Description of the Feedstock(s) and Products

   - Summary of the specific feedstocks for the production of alcohol and the market for the products of the particular plant.
   - Who are the sellers and buyers?

6. Organization and Founders

   - Briefly describe all the owners
   - Outline the organization that will direct this project
   - Indicate the distribution of ownership

7. Project Components

   - Outline the major tasks to accomplish the objective
   - Delineate the management responsibility for each task
   - Indicate the integration of appropriate tasks

8. Time Schedule

   - List all tasks in summary form in time sequence
   - Depict the tasks on a chart (See example in Figure IX-1)

9. Project Cost and Sources of Funding

   - Summarize the costs
   - List all sources of funding

THE APPENDICES TO FEASIBILITY STUDY

A. Market and Marketing Analysis

The procurement and retailing of products with all internal and external impacts are estimated. The historical and projected value and cost are determined along with the phases of development and operation. The relationship of the project to regional, State, and local energy policy should be noted. Suggested considerations include:

Feedstock(s)

   - Sources and locations (gathering radius)
   - Cost of collection and transportation
   - Quantity and quality for alcohol production
   - Historical, current, and projected cost
   - Availability over the life of the plant

	ENGINEERING, BUSINESS PLAN, FINANCING, CONSTRUCTION FOR FUEL-GRADE ETHANOL PLANT

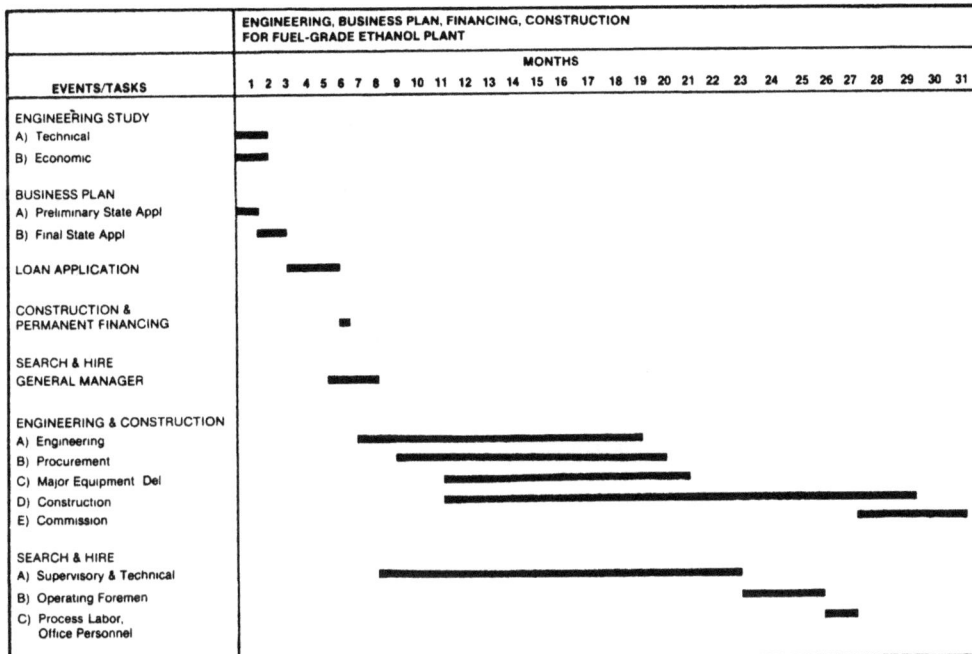

Figure IX-1. Likely Schedule of Events/Tasks

Alcohol

- Market price for expected proof
- Transportation cost
- Current and projected cost

Coproducts

- Value on-site
- Market price

Commitments from sellers, customers, and brokers

Location of the plant

- Include map(s)

- Description of adjacent industries, utilities, general population and the potential complimentary or supplementary relationship with the project

Commitment from utilities, energy suppliers, fire, and police

All written support which has been elicited from the community should be provided. This may take the form of: letters; endorsements; resolutions and ordinances from businesses and non-profit organizations; town, city, county, and State governments; e.g., Jaycees and other community-based organizations; planners; mayors; councils; governors; legislators; or congressmen.

B. Technical and Engineering Design and Factors of Production

The production cycle design will come from the selection of one particular process. The following items need to be taken into account to calculate economic and energy costs of production. Each should be described in the process with particular attention to the production capacity through quantitative descriptions in each step.

Feedstock(s)

- Primary/Alternate

Collection (Harvest and Transportation)

- Includes methods and distances

  Storage of feedstock
  Preparation (cleaning & grinding)
  Preliminary conversion (cooking)
  Saccharification

Fermentation
Distillation
Dehydration
Separation of solids
Denaturing
Drying of solids
Storage of stillage
Use of stillage
Water recycling
Heat source - Primary/Secondary

- Integrated energy generation on-site

- Heat recovery

Use of other coproducts (CO_2, etc.) and wastewater

Detailed energy use calculation on a Btu-per-gallon basis in the plant

- Inputs are all nonrenewable sources for cooking, distillation, & grinding, augering, site transportation

- Outputs include alcohol plus the coproducts

Equipment (instruments/controls/boiler/engines/ pumps), lead time, availability, and costs

- Laboratory facility and equipment on-site for testing and analyzing parts of the process

- Preliminary process flow diagrams

C. Economic Analysis and Financial Plan

A pro forma, or projected, sources and uses of funds statement can be constructed to show how the project plans to acquire and employ funds. This is the basic summary document to show the flow of resources. A projected balance sheet and income statement is required for the financial plan.

An itemized estimate of all project costs including the basis for the estimates is needed for the development of the project.

Cost estimates for the project

- Engineering design
- Site preparation, construction, and installation
- Labor
- Production

    -Feedstocks
    -Utilities
    -Chemicals
    -Bonding, insurance, and taxes

- Maintenance
- Depreciation
- Other overhead

All private and public financing options that will be explored should be listed in as much detail as possible.

The complete financial plan includes the following time periods:

- Development and construction of plant (with a subtotal for each task in the work plan)

- First year of operation, month by month

- Five-year quarterly projection

- Debt service on a quarterly basis

- Previous three years' profit and loss statement and financial statement of existing businesses

Documentation should be complete for deeds, leases, contracts, agreements, options, appraisals, and insurance.

D. Management and Staffing Plan

The organization of the entire project should be detailed with job titles and descriptions. An organization chart provides clarity and brevity for project management structure. List everyone by name, if possible, from the Board of Directors through the supervisory and technical personnel.

The education, technical training, employment, and related business experience of each participant in the project should be detailed on one page in a resume format for each person. Salaries, fees, and overhead cost for labor are included here.

Engineering and construction companies and subcontractors are included with scheduling and reporting requirements. Cost and project management responsibilities need to be clearly indicated.

Provisions for a permanent, trained labor force must be quite specific. This training should be accomplished before the beginning of the operation of the plant.

E. Sample Information Elements

Sample information elements are shown in Figure IX-2, IX-3, and IX-4. The following information should be provided.

1) With respect to each partner or shareholder (e.g., over 5 percent of total shares outstanding):

- Name
- Address
- Citizenship
- Principal occupation
- Percentage of ownership
- Personal financial statement
- Credit reliability/analysis
- Personal interest in any concern which shall have some relationship to project

2) With respect to any corporate organization:

- Parent companies, subsidiaries, affiliates
- Charter of organization
- Financial statement of organization
- SEC 10K report for companies listed on stock exchange
- Tax status

F. Regulatory Compliance Requirements

Obtaining building permits and other necessary certifications, and meeting compliance requirements of the appropriate regulatory agencies having jurisdiction over the project, should indicate the following:

Environmental (Federal and State)
Fire and Safety
Building Codes
Zoning
Utilities
BATF (Federal and State)
Equal Employment Opportunity
National Historic Preservation Act
Occupational Safety and Health Administration
Other permits, ordinances, and regulations

Assuming productive use of the alcohol, stillage, and CO, produced, there still may be a number of environmental factors to be considered. These might include:

- Odors released during the process

- Disposal of wet thin stillage and wastewater. How much backset (recycling of spent thin stillage into fresh fermentation) is built into the plant design? Is there any concentration of thin stillage into syrup? Is there any digestion of thin stillage? What is the projected discharge of wastewater in volume or quality? Does the plant site encompass additional buffer area that could be used for waste disposal lagooning or irrigation? Will a discharge permit be required for the plant?

- Air pollution in boiler operations. Compliance with local and Federal standards. Are fuel alternatives available? Cogeneration prospects. What is the quality of the stack gases from the boiler? Can they be used for stillage drying by direct contact or indirectly?

- Removal of crop residuals causing erosion. Does crop residual removal for fuel feedstock or alcohol feedstock exceed 50 percent of total crop biomass? Will the production of alcohol result in farm and soil management shifts which are very different from current practices?

- Ethanol use of fuel for vehicles and resulting emissions. Seasonal air flow patterns and local air quality history.

In conjunction with the considerations developed under the sections which contain marketing and economic analyses, and the analysis of the production process, alternatives to the final decision and their consequences should be considered. In addition, the environment to be affected should be described (e.g., quality of soil, air, water) prior to undertaking this project.

Completing sample forms similar to those which follow will provide the basic information elements necessary for the preparation of the feasibility analysis.

SALES FORECASTS — GOALS

	81	82	83	84	85

A. ETHANOL

JOBBER: _____

JOBBER: _____

JOBBER: _____

OTHER: _____

    TOTAL _____

SALES FORECAST — TONS

	81	82	83	84	85

B. COPRODUCTS

BUYER: _____

BUYER: _____

BUYER: _____

    TOTAL _____

COPRODUCT CONTENT: PROTEIN   _____ %

                            FAT   _____ %

                            FIBER   _____ %

                            ASH   _____ %

                            MOISTURE   _____ %

Figure IX-2. Sales Forecasts

COMMERCIAL SCALE ETHANOL PRODUCTION AND FINANCING

```
A  • PLANT CAPACITY                                    _____ GAL/HOUR
   • PRODUCTION ETHANOL                                _____ GAL/YR
   • PRODUCTION COPRODUCT                              _____ TON/YR
   • ETHANOL PRICE, F O B  PLANT                       _____ $/GAL
   • COPRODUCT PRICE, F O B  PLANT                     _____ $/TON
   • PLOT SIZE _____FT  X _____ FT
   • PLANT SIZE _____FT  X _____ FT
   • SOIL STRENGTH _____PSF @ _____ FT
   • PLASTICITY INDEX _____
   • _____ pH RATING _____RESISTIVITY LEVEL
   • RAILROAD SIDING _____FEET
   • ACCESS ROAD QUALITY _____
B  • FEEDSTOCK TYPE _____
     FEEDSTOCK CONTENT  STARCH                         _____ % WEIGHT
                        PROTEIN                        _____ % WEIGHT
                        OTHER SOLIDS                   _____ % WEIGHT
                        MOISTURE                       _____ % WEIGHT
     AMOUNT REQUIRED _____ BUSHELS/YR
     JOBBER _____
     NUMBER DAYS INVENTORY _____
   • FEEDSTOCK TYPE _____
     FEEDSTOCK CONTENT  STARCH                         _____ % WEIGHT
                        PROTEIN                        _____ % WEIGHT
                        OTHER SOLIDS                   _____ % WEIGHT
                        MOISTURE                       _____ % WEIGHT
     AMOUNT REQUIRED _____ BUSHELS/YR
     JOBBER _____
     NUMBER DAYS INVENTORY _____
   • ENZYME TYPE _____
     AMOUNT REQUIRED _____ GAL/DAY
     JOBBER _____
     NUMBER DAYS INVENTORY _____
   • ENZYME TYPE _____
     AMOUNT REQUIRED _____ GAL/DAY
     JOBBER _____
     NUMBER DAYS INVENTORY _____
C  • POWER SOURCE _____COAL_____
     AMOUNT REQUIRED _____ TONS/HR
     JOBBER _____
     NUMBER DAYS INVENTORY _____
   • POWER SOURCE _____ELECTRIC_____
     AMOUNT REQUIRED_____ kW/HR
     JOBBER _____
D  • WATER SOURCE _____
     AMOUNT REQUIRED _____ GALS/HP
     JOBBER _____
     NUMBER DAYS INVENTORY _____
E  • LABOR _____
     _____ NUMBER PLANT OPERATION AND MAINTENANCE
     _____ NUMBER ADMINISTRATION AND SUPERVISION
     _____ NUMBER GENERAL ADMINISTRATION AND SALES
```

Figure IX-3. Plant Characteristics

A   COST OF PRODUCTION

    a   CAPITAL COSTS $/YR $/GAL
- EQUIPMENT
- LAND
- INVENTORY
- TAXES
- INSURANCE
- DEPRECIATION
- INTEREST

    b   OPERATING COSTS
- FEED MATERIAL
- SUPPLIES
- FUEL
- WASTE DISPOSAL
- OPERATING LABOR

    c   MAINTENANCE COSTS
- LABOR
- SUPPLIES
- EQUIPMENT

    UNSCHEDULED MAINTENANCE
- LABOR
- SUPPLIES
- EQUIPMENT

B   INVESTMENT ANALYSIS

- REVENUES $_____
  ETHANOL
  STILLAGE
- LESS COST OF SALES
  FEEDSTOCK
  DEPRECIATION
- GROSS PROFIT $_____
- LESS G&A AND O H $_____
- NET PROFIT BEFORE TAXES $_____
- BREAK EVEN QUANTITY   _____ GALS
- PAYBACK PERIOD   _____ YEARS
- RETURN ON INVESTMENT   _____ %

C   RISK ANALYSIS

    PROCESS GUARANTEE   _____ PROOF
                                  _____ FUSEL OIL CONTENT
    PERFORMANCE GUARANTEE   _____ PRODUCTION RATE
                                  _____ QUANTITY ENERGY
                                  _____ QUANTITY PER INPUT

Figure IX-4. Economic Analysis

APPENDIX A

Financial Assistance Programs

INTRODUCTION

The Federal government affects the distribution of national resources by direct loan and loan guarantee programs aimed, directly or indirectly, at private industry. Tax credits, tax exemptions, and price support programs are other forms, but are not discussed here. The purpose of loan programs is to channel funds to those sectors of the economy which could not otherwise obtain credit or could obtain it only at very high costs. Sometimes the channel is indirect, going through state or local governments or other non-profit organizations. The specific goal, in this case, is to enable private ventures to make more alcohol production capability available than if left to normal business dealings.

In addition to the DOE, the Federal agencies with such authority are the Small Business Administration (SBA); the Farmers Home Administration (FmHA) of USDA; Urban Development Action Grants, the Department of Housing and Urban Development (HUD); and Economic Development Administration (EDA), the Department of Commerce. The existing legislation in all cases has already been interpreted to apply to alcohol fuel production. Several legislative efforts are underway that will more specifically address alcohol fuels, and will probably modify some of the characteristics identified in the following discussions.

SUMMARY

This section addresses: (1) characteristics of loan and loan guarantee instruments available to prospective investors; and (2) the process that applicants must follow to obtain the desired assistance. Since there are several instruments available and the applicant can and does need guidance as to which best fits his or her needs, Figure A-1 presents a simplified view of the content of these programs.

The figure covers the two types of assistance considered here, i.e., loan guarantees and direct loans. With the exception of DOE (75 percent loan guarantee), the guarantee and direct loan is usually set at 90 percent of the project cost but can be lower or as high as 100 percent (not shown here). The direct loan amount may be as low as 65 percent of cost. The ceiling on funds available per project varies from under $200,000 to over $25 million. The maturity of the assistance vehicle is defined to be the maximum time available to pay back principal, interest, and other costs. The maturity period varies from 4 to 40 years.

The Farmers Home Administration Business and Industrial Loan Guarantee Program (FmHA-B&I) provides a guarantee up to 90 percent of the loan, with the applicant responsible for furnishing a minimum of 10 percent of equity (Figure A-1, Item 11). FmHA has imposed an administrative ceiling of $20 million for certain ventures in the past; during 1978, the range was from $11,000 to $33 million, and the average was $824,000. Although premature, a ceiling of $25 million is shown on this chart for alcohol fuel plants. Maximum maturity terms are 30 years for land, buildings, and permanent fixtures; 15 years for machinery or equipment; and 7 years for working capital.

The Small Business Administration Local Development Companies Program (SBA-LDC) is identified as Item 10 in Figure A-1. The bank loan is guaranteed by SBA to 90 percent of the loan or $500,000, whichever is the lesser. Although maximum maturity is 25 years plus the estimated time required to complete construction, conversion, or expansion of the facility, the usual period is 15 to 20 years.

Items 2 and 9 identified under "Source" are the Small Business Administration Business Loans Program (SBA-BL). SBA is obligated to purchase not more than 90 percent of the outstanding balance of the authorized loan in the event of borrower default. The ceiling for guaranteed loans normally is $350,000 but may be up to $500,000 for exceptional circumstances. For direct and guaranteed loans, the maximum maturity date is normally 10 years, with working capital loans limited to 6 years, and portions of loans for construction and acquisition of real estate may have a maximum of 20 years.

Items 3 and 5 identified under "Source" are the Farmers Home Administration Farm Loan Program (FmHA-FL). The limit on a direct farm operating loan is $300,000 and the maximum term is 7 years. The limit on a direct farm ownership loan is $200,000 and the maximum term is 40 years.

The program identified as Item 8 under "Source" is the EDA Program. EDA will guarantee up to 90 percent of the unpaid balance of loans for the acquisition of fixed assets or for working capital. The expected EDA participation for a small-scale fuel plant should generally not exceed $5 million and the maximum period for fixed asset loans is 30 years.

Item 7 of Figure A-1 is the DOE loan guarantee program represented by Public Law 96-126. The amount of a loan guarantee shall not exceed 75 percent of the estimated costs, the maximum is to be sufficient to carry out the project, and full repayment is to be made within 20 years. The program identified as Item 6 under "Source" is the Department of Housing and Urban Development Action Grant Program (HUD-UDAG). Ideally, HUD is seeking projects that generate substantially more private commitments than the Action Grant money requested. However, HUD will take into account the various types of projects in considering the degree of

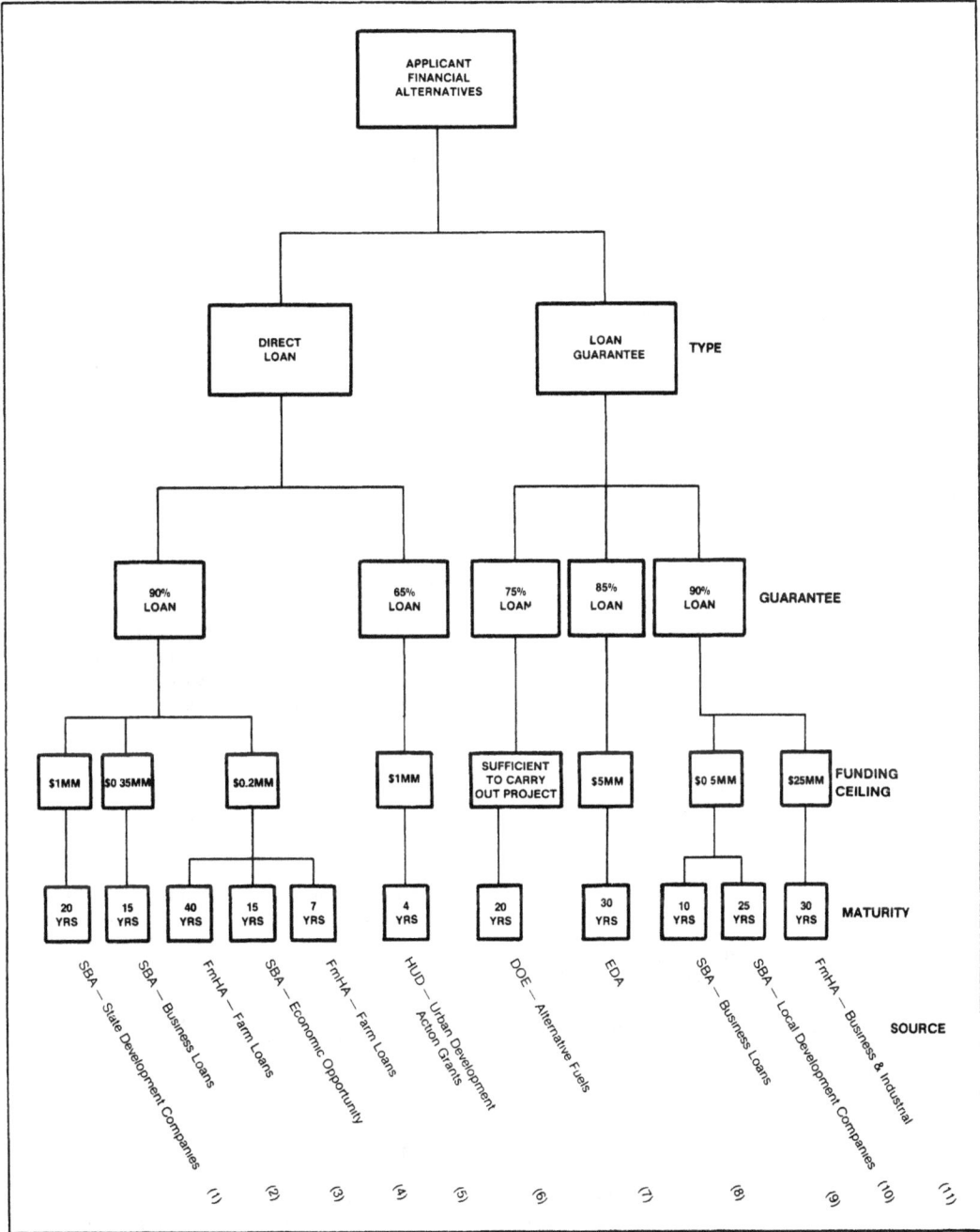

Figure A-1. Financial Assistance Program Tree

private leverage. This program is shown on the figure at the 65-percent guarantee point of the direct loan portion, for illustrative purposes. The FY 1978 average for small cities was almost $1 million, but $6 million was contributed in Utah. Assistance is provided for discrete projects that can be completed in approximately four years. The Small Business Administration Economic Opportunity Program (SBA-EO), Item 4, is a direct loan program, 90 percent guarantee, $200,000 to any one borrower for a maximum of 15 years fixed assets, and 10 years maximum for working capital.

Finally, the Small Business Administration State Development Companies Program (SBA-SDC), Item 1, covers direct loans, $500,000 to $1 million, for a maximum of 20 years.

SOURCES OF FINANCIAL SUPPORT

Small Business Administration

The objective of the SBA is to aid, counsel, assist, and protect the interests of small business concerns and to help victims of disasters. The basic goal is to provide procurement and technology, management, and financial assistance to those needy firms classified as small businesses.

SBA, in support of alcohol fuels, makes available direct energy loans and energy loan guarantees to begin, continue, or expand small businesses that are developing, manufacturing, selling, installing, or servicing specific energy conservation measures.

Program Area(s)

SBA provides a relatively broad range of support to small business firms involved in producing alcohol fuels, including support to local development companies, direct and guaranteed loans to private business, supporting state development companies, purchasing, and/or guaranteeing Small Business Investment Company debentures, and financing economic opportunities to the disadvantaged.

Local Development Companies. SBA is authorized to make loans to local development companies to finance plant acquisition, construction, conversion, or expansion, including the acquisition of land, provided that such loans will assist an identifiable small business concern in accomplishing a sound business purpose. A local development company is a corporation chartered under any applicable State corporation law to operate in a specified area within a State. These loans are otherwise known as Section 502 loans.

Business Loans. SBA is authorized to make both direct loans and participation loans in cooperation with banks and other lending institutions. The loans are commonly referred to as stemming from Section 7(a) and/or 7(l) of the Act. The Act provides that all energy loans shall be of such sound value as reasonably to assure repayment, recognizing that greater risk may be associated with energy loans under Section 7(l) than for loans under Section 7(a).

State Development Companies. SBA is authorized to make loans to state development companies organized under or pursuant to a special legislative act to operate on a State-wide basis. The companies must be formed for the purpose of furthering the economic development of their communities and environs, and with authority to promote and assist the growth and development of small business concerns in the areas covered by their operations. The proceeds of loans made to state development companies shall be used only to provide equity capital or make long-term loans, or both, to small business concerns. These loans are otherwise known as Section 501 loans.

Small Business Investment Companies. Congress established the Small Business Investment Company (SBIC) program in 1958. SBIC's are privately owned, SBA-licensed and SBA-regulated companies whose primary purpose is to provide equity capital and long-term financing to small firms for the sound financing of their business operations and for their growth, expansion, and modernization. The SBA may purchase debentures of a licensee to augment investment funds available to him, and may also guarantee debentures of the licensee issued to others.

Economic Opportunity Loans. The principal purpose of the Economic Opportunity Loan Program is to make funds available on reasonable terms and maturities to small business concerns located in areas with high proportions of unemployment, low income individuals, or small business concerns owned or to be established by persons with low incomes; and to provide management assistance to such persons. Particular emphasis is placed on preservation or establishment of small business concerns located in urban and rural areas.

Organizational Structure

The central and principal office of the SBA is at 1441 L Street, N.W., Washington, DC 20416. Field offices include 10 regional offices and 99 district and branch offices and "posts-of-duty."

Headquarters Office. The Administrator is responsible to the President and Congress for exercising direction, authority, and control over the SBA. Assisting the Administrator are the heads of five offices and one council, two Associate Deputy Administrators, five Associate Administrators, and four Assistant Administrators.

The five offices are: Office of Equal Employment Opportunity and Compliance; Office of Hearings and Appeals; General Counsel; the Inspector General; and the Chief Counsel for Advocacy. Staff support is provided to the National Advisory Councils.

Regional Offices. Regional offices, headed by a Regional Administrator, are the principal field offices of the agency and are responsible and responsive to the central office. District offices are headed by a District Director and are located in a city within a defined, limited, and contiguous geographical area within a region. Branch offices are headed by a Branch Manager and are located in a city within a defined, limited, and contiguous geographical area within a district. Post-of-duty stations are headed by an Officer-in-Charge and are also located in a city within a defined, limited, and contiguous geographical area within a district. Locations of SBA field offices are listed in the References section of this document.

Farmers Home Administration, Department of Agriculture

The objective of the FmHA is to provide credit for those in rural America who are unable to get credit from other sources at reasonable rates and terms. The basic goal is to create and maintain a healthy economic climate in rural communities.

FmHA makes loans under the farm loan program and guarantees those made by private lenders under the business and industrial loan guarantee program for plants for the production of fuel alcohol. Farm loans generally are intended to finance production for on-farm use; commercial production may be financed through the business and industrial loan guarantees.

Program Area(s)

FmHA makes loans with funds borrowed from the U.S. Treasury and those derived from sales to the Federal Financing Bank of Certificates of Beneficial Ownership, which represent actual loans made by the agency.

Business and Industry Loans. Loans made by private lenders but guaranteed by FmHA may be provided to any legal entity, including individuals, public and private organizations, and Federally recognized Indian tribal groups. Priority is given to projects in the country and in communities of 25,000 population and smaller. Loans may be made in cities and towns of less than 50,000 population and in immediately adjacent urbanized areas with a population density of fewer than 100 persons per square mile. Non-farm distilleries can qualify under this program.

Farm Loans. Direct loans at cost of borrowing and loan guarantees for farm ownership and farm operations are authorized to farmers and farm cooperatives to improve farm technology, administration, and productivity of farms. The development of alternative energy sources such as biomass conversion and gasohol production has been determined by the USDA to be projects within this program's jurisdiction. Acceptable uses of farm ownership loans include construction, improvement, or repair of farm homes and service buildings; installation of pollution control or energy conservation measures; and establishing non-agricultural enterprises that help farmers supplement their farm income. Operating loans can be used for pollution abatement and operation of non-agricultural income-earning enterprises.

Organizational Structure

The FmHA is one of two such groups (the other is the Rural Electrification Administration) reporting to the Assistant Secretary for Rural Development, USDA. The Headquarters Office of FmHA is located at Fourteenth Street and Independence Avenue, S.W., Washington, DC 20250. Field support is provided by 46 state offices and numerous county offices.

Headquarters Office. The Administrator directs and supervises the activities of the Deputy Administrator for Financial and Administrative Operations, the Deputy Administrator for Farm and Family, the Deputy Administrator for Rural Development, and the Associate Administrator for Policy Management. The Business and Industry Loan Program reports to the Deputy Administrator for Rural Development, and the Farm Loan Program reports to the Deputy Administrator for Farm and Family.

Regional Offices. Applications for loans are made at the agency's 2,200 local county and district offices, generally located in county-seat towns.

Locations of the FmHA state offices are listed in the References section of this document.

Department of Housing and Urban Development

The objective of the Department of HUD is to administer the principal programs that provide assistance for housing and development of the nation's communities. The basic goal is to administer mortgage insurance programs; a rental subsidy program; antidiscrimination policies; and other programs that aid neighborhood rehabilitation and the preservation of urban centers from blight and decay.

HUD provides grants to cities and urban counties for energy conservation and alternative energy projects. These projects are of particular interest, not only because they are in the national interest with respect to conservation of scarce fuels, but because they can be of importance to distressed urban areas in stabilizing energy costs, making such areas more attractive to commercial and industrial facilities, and helping to alleviate the hardship of escalating energy costs upon low- and moderate-income groups.

Program Area(s)

The Urban Development Action Grant (UDAG) Program is a highly flexible economic development tool which seeks to create partnerships among government, community, and private industry to overcome problems of development. Action grant funds are available to carry out projects in support of a wide variety of economic revitalization of neighborhood reclamation activities that involve partnerships with the private sector. These activities may include a broad range of development actions like land clearance; site improvements; providing infrastructure; rehabilitation; and building public, commercial, industrial, and residential structures.

The program can be applied to the following situations:

- To provide "front-end" funding that allows communities to capture and leverage significant private investments

- To respond to unique, perhaps one-time opportunities while they are appropriate

- To make substantial resources available when needed to join other Federal departments in meeting distressed cities' reinvestment needs

Organizational Structure

The department is administered under the supervision and direction of the Secretary, who is responsible for the administration of all programs, functions, and authorities of the department; regulation of the Federal National Mortgage Association; administration of the Government National Mortgage Association; and advising the President on Federal policy, programs, and activities relating to housing and community development. Headquarters is located at 451 Seventh Street, S.W., Washington, DC 20410. Field installations include 10 regional, 39 area, and 38 insuring offices.

Headquarters Office. The Assistant Secretary for Community Planning and Development directs and supervises the UDAG program, among others, along with his Directors of Community Planning and Pro-

gram Coordination, Management, Evaluation, Legislative and Urban Policy, Policy Planning, Block Grant Assistance, and Field Operations and Monitoring.

Regional Offices. The field operations are carried out through a series of regional, area, and services offices. The regional offices have boundaries and headquarters locations prescribed by the secretary. Each regional office is headed by a regional administrator, who is responsible to the Secretary and Under Secretary for the overall satisfaction of the department's goals and objectives and for the management of the offices within the region.

Locations of the field offices are listed in the References section of this document.

Economic Development Administration, Department of Commerce

The objective of the EDA is to provide assistance to economically distressed areas and regions in order to alleviate conditions of substantial and persistent unemployment and underemployment and to establish stable and diversified economies. The basic goal is the creation of permanent jobs in areas of high unemployment, job loss, or low incomes.

EDA is authorized to provide grants, direct loans and loan guarantees, technical assistance, and other Title II development finance authorities. Specifically, these will be public works grants to rural communities and non-profit economic development organizations, and loan guarantees to private enterprises desiring to establish new alcohol fuel production facilities. Financial assistance is limited by the capacity of the Project to create long-term employment opportunities.

Program Area(s)

EDA is authorized to provide grants, direct loans and loan guarantees, technical assistance, and other special economic development and adjustment assistance.

Public Works and Development Facility Grants. Direct grants are authorized for: (a) public works, public service, and development facility projects which directly or indirectly contribute to long-range economic growth or benefit long-term unemployed and members of low-income facilities in redevelopment areas, and parts of economic development districts; and (b) public works, public service, and development facility projects which provide immediate useful work to the unemployed and underemployed of the project area.

Supplementary grants to augment the direct grants received under the Act or to augment the basic grants under other Federal grant-in-aid programs may be provided to public works, public service growth, or benefit long-term unemployed and members of low-income facilities in redevelopment areas and parts of economic development districts.

Loans and Guarantees. Loan guarantees are authorized for loans made to private enterprises by private lending institutions only in designated redevelopment areas and parts of economic development districts.

Technical Assistance. Technical assistance in the form of direct assistance by EDA personnel, payment to other Federal agencies, contracts, and grants may be extended to redevelopment areas and and other areas that have substantial need. Planning and administrative grants are also available to eligible applicants. EDA conducts a continuing program of study, training, and research in the problems of economic development through members of its staff, payments to other Federal agencies, contracts, and grants. EDA also provides technical assistance for national projects to national or public associations or public bodies. Finally, EDA aids redevelopment and other areas by furnishing interested individuals, communities, industries, and enterprises within such areas any assistance, technical information, market research, or other forms of assistance, information, or advice which could be useful in alleviating or preventing conditions of excessive unemployment or underemployment within such areas.

Special Economic Development and Adjustment Assistance. Grants may be made to States, political subdivisions of States, redevelopment areas, economic development districts, and Indian tribes to meet special needs related to existing or threatened long-term unemployment or low family income levels and arising from economic dislocation. Such grants may be used for public facilities, public services, business development, planning, unemployment compensation, rent supplements, mortgage payment assistance, research, technical assistance, training, relocation of individuals, and other appropriate assistance.

Organizational Structure

The central and principal office of the EDA is in the Department of Commerce, 14th Street at Constitution Avenue, N.W., Washington, DC 20230. Six regional offices cooperate with and assist local areas in organizing and carrying out economic development.

Headquarters Office. The Assistant Secretary for Economic Development directs the programs and is responsible for the conduct of all activities of the EDA subject to the policies and directions prescribed by the Secretary of Commerce. The Deputy Assistant Secretary for Economic Development directs and supervises the Investigations and Inspections Staff, the Special Assistant for Environmental Affairs, the Special Assist-

ant for Field Operations, and the Special Assistant for Indian Affairs. The Deputy Assistant for Economic Development Operations directs and supervises the Office of Business Development, Office of Public Works, and Office of Technical Assistance. The Deputy Assistant Secretary for Economic Development Planning directs and supervises the Office of Development Organizations, Office of Economic Research, and Office of Planning and Program Support.

Regional Offices. The Regional Director, who reports to and is under the supervision and direction of the Assistant Secretary, directs the program and is responsible for the conduct of all activities of the Regional Office. Each office includes a Civil Rights Division, Special Programs Division, Regional Counsel, Business Development Division, Public Works Division, Technical Assistance Division, Technical Support Division, Planning Division, and Economic Development Representatives.

Locations of the field offices are listed in the References section of this document.

Department of Energy

The objective of the Office of Alcohol Fuels is to encourage the development and utilization of alcohol fuels. The Office is responsible for providing research, engineering testing, evaluation, and demonstration skills needed to develop programs to support the objectives of the President's Alcohol Fuels Program. The President has set a domestic production capacity target of 500 million gallons of alcohol fuel during 1981.

The Office provides financial incentives, marketing analysis, technical support, educational programs, and procurement guidance for projects involving alcohol fuels. It also serves as a point-of-contact on alcohol fuels for the DOE, acting as a liaison between the Department and other Federal, regional, state, and local agencies, and consumers.

Program Area(s)

$2.208 billion is available to the Secretary of Energy to stimulate domestic commercial production of alternative fuels.* This appropriation is divided into the following areas:

---

*This appropriation was established through Public Law 96-126, the Department of Interior and Related Agencies Appropriations Act for FY 1980, which created an energy security reserve for alternative fuels production of $19 billion Biomass is an acceptable domestic resource and ethanol is an acceptable derivative

- $100 million for project development feasibility studies, not to exceed $4 million each, was set aside for applications from eligible applicants during the spring of 1980. Individuals, businesses, institutions, and communities were considered eligible.

- $100 million for cooperative agreements, not to exceed $25 million each, to support commercial-scale development was available to applicants during the spring of 1980. Eligibility was open to any individual, business, institution, or community. Federal agencies were excluded from this program.

- $1.5 billion for purchase commitments or price guarantees. DOE is currently drafting solicitations for this incentive.

- $500 million for a reserve to cover any defaults from loan guarantees issued to finance construction of alternative fuels production facilities, provided that the indebtedness guaranteed or committed to be guaranteed does not exceed $1.5 billion. DOE is currently drafting solicitations for this incentive.

- $8 million for program management.

Organizational Structure

The Office of Alcohol Fuels reports to the Secretary of Energy and is currently headed by an Acting Director. Headquarters are in the Forrestal Building, 1000 Independence Ave., S.W., Washington, DC 20585. Field installations for DOE include 10 regional offices.

Headquarters Office. The Director, Office of Alcohol Fuels, directs and supervises the Financial Incentives Division; the Technology Development and Utilization Division; the Information, Liaison, and Public Response Division; and the Program Control and Evaluation Division.

Regional Offices. Locations of the field offices are listed in the References section of this document.

LOAN APPLICATION PROCESSES

Small Business Administration Loan Application Process (See Figure A-2)

1. A small business concern is independently owned and operated and is not dominant in the field of operation in which it is bidding on government contracts. As an example, in the petroleum industry, a concern is considered small if its employees do not exceed 1,000 persons and its crude oil capacity does not exceed 30,000 barrels per day.

2. Applications for financial assistance may be considered only when there is evidence that the desired credit is not otherwise available on reasonable terms. Proof of refusal must contain the date, amount and terms requested, and the reasons for not granting the desired credit. (S.120.2)

3. If unable to obtain the entire loan from a bank or other source, the applicant should ascertain whether a financial institution will make the loan if SBA agrees to purchase an immediate participation or if SBA agrees to guarantee a portion of the bank loan. (S.22.15(b)).

4. The applicant should apply to the SBA office serving the territory in which the applicant's business is located. If he desires to obtain counseling or assistance in filing an application from a regional office which is geographically closer to his business, he may do so. (S.122.15(d))

5. The applicant should file SBA Form 4, Schedule A, Summary of Collateral; and back this up with the following: SBA 641, Request for Counseling; GPO 942, History or Description of Business and Personal Data Management Experience; SBA 035-100, Format for a Business to be Purchased/A New Business to be Established/or An Existing Business to be Expanded or Improved; SBA-413 Personal Finance Statement; SBA 912 Statement of Personal History; SBA 1100 Monthly Cash Flow Projection; SBA 1099 Operating Plan Forecast. (SBA-0353-101, Literature, Brochures, and Forms).

6. Regional Directors are responsible for the proper evaluation of collateral offered. A technical evaluation by SBA appraisers/engineers is also required. (S.122.16(d) and (e)).

7. When SBA approves a loan application, a formal loan authorization is issued by SBA. This authorization is neither a contract to lend nor a loan agreement. Instead, it states the conditions which the borrower must meet before financial assistance will be extended. When the borrower is prepared to meet these conditions, SBA or the financial institution will arrange a date, time, and place for closing the loan. (S.122.19).

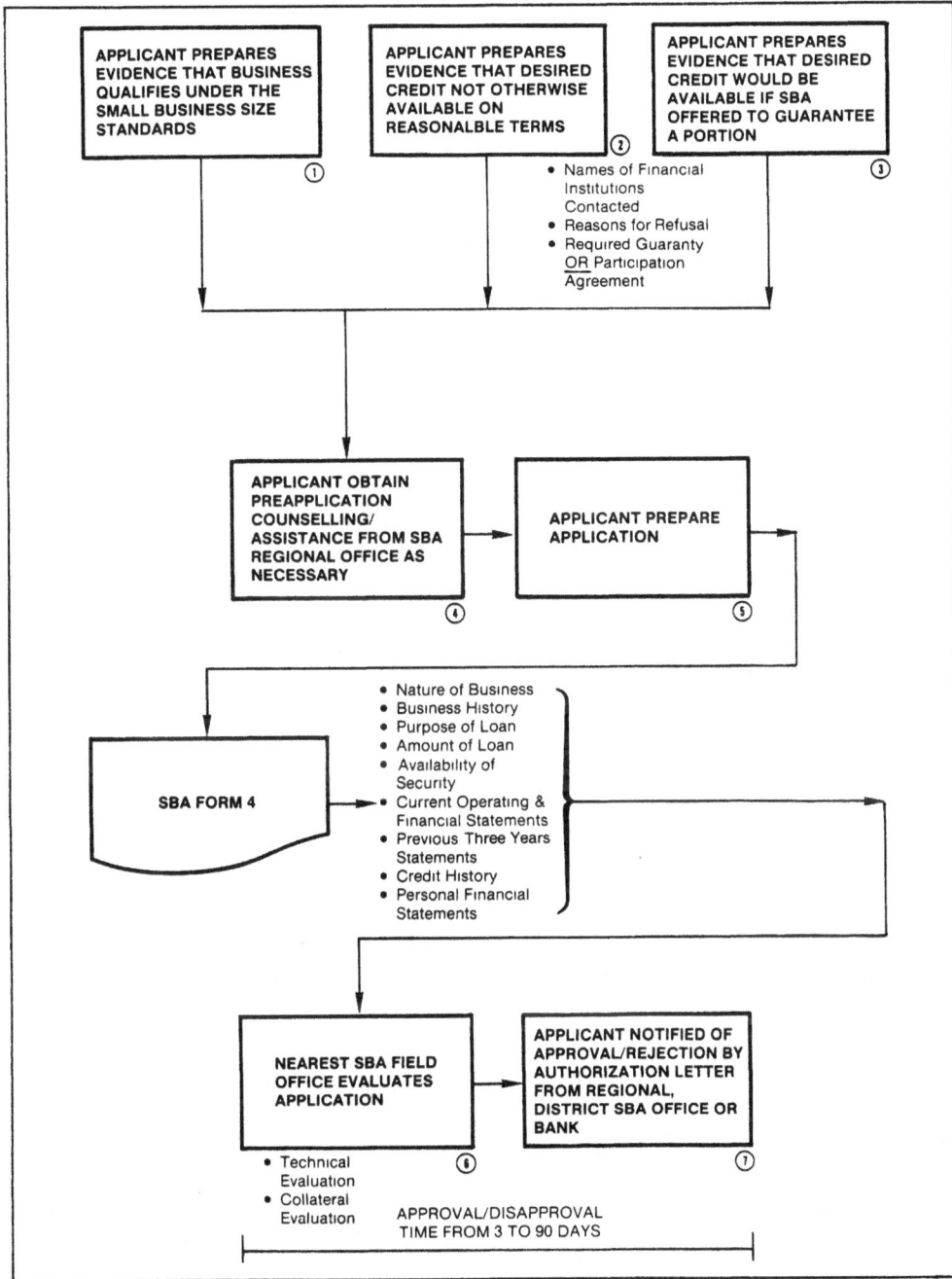

Figure A-2. SBA Loan Application Procedure

The flowchart contains the following elements:

Box 1: APPLICANT PREPARES EVIDENCE THAT BUSINESS QUALIFIES UNDER THE SMALL BUSINESS SIZE STANDARDS

Box 2: APPLICANT PREPARES EVIDENCE THAT DESIRED CREDIT NOT OTHERWISE AVAILABLE ON REASONALBLE TERMS

- Names of Financial Institutions Contacted
- Reasons for Refusal
- Required Guaranty OR Participation Agreement

Box 3: APPLICANT PREPARES EVIDENCE THAT DESIRED CREDIT WOULD BE AVAILABLE IF SBA OFFERED TO GUARANTEE A PORTION

Box 4: APPLICANT OBTAIN PREAPPLICATION COUNSELLING/ASSISTANCE FROM SBA REGIONAL OFFICE AS NECESSARY

Box 5: APPLICANT PREPARE APPLICATION

SBA FORM 4:
- Nature of Business
- Business History
- Purpose of Loan
- Amount of Loan
- Availability of Security
- Current Operating & Financial Statements
- Previous Three Years Statements
- Credit History
- Personal Financial Statements

Box 6: NEAREST SBA FIELD OFFICE EVALUATES APPLICATION
- Technical Evaluation
- Collateral Evaluation

Box 7: APPLICANT NOTIFIED OF APPROVAL/REJECTION BY AUTHORIZATION LETTER FROM REGIONAL, DISTRICT SBA OFFICE OR BANK

APPROVAL/DISAPPROVAL TIME FROM 3 TO 90 DAYS

Farmers Home Administration Loan Application Process (See Figure A-3)

1. Applicant files 449-1 with the nearest A-95 clearinghouse as notice of intent to file an application. Check with the appropriate Federal Regional Council to obtain name and address of appropriate clearinghouse.

2. Applicant may file a preapplication with the County Office including a description of the project, market information, financial data, A-95 comments, and a personal history. (1980.451(f))

3. Alcohol production must meet certain BATF requirements of the U.S. Treasury Department. It is the applicant's responsibility to satisfy those requirements. (FmHA Alcohol Fuels Fact Sheet.)

4. A-95 clearinghouse reviews completed application and notifies applicant of acceptability of project.

5. BATF reviews application and certifies acceptability of information as provided by the applicant.

6. An eligible lender is any Federal- or State-chartered bank, Federal land bank, production credit association, bank for cooperatives, savings and loan association, building and loan association, or SBIC. Others must file 449-18, "Lenders or Holders Request for Approval", and proof of financial capability, charter or license, credit history, etc. (1980.13(b)).

7. FmHA will submit 449-22 to the U.S. Department of Labor for the necessary certification that the proposal will not be in conflict with specified employment/unemployment regulations. (1980.451(h)).

8. If it appears that the project is eligible, has sufficient priority, is economically feasible, and loan guarantee authority is available, FmHA will inform the lender and applicant in writing and request that they complete the application. (1980.451(g)).

9. Application will include: (1) 449-1, "Application for Loans and Guarantees"; (2) 449-2, "Statement of Collateral"; (3) 449-10, "Applicant's Environmental Impact Evaluation"; (4) A&E plans; (5) cost estimates and contingency funding requirements; (6) appraisal reports; (7) financial reports for existing businesses; (8) financial forecasts for new business; (9) credit reports; (10) 400-1, for construction costs greater than $10,000; (11) building permits; (12) personal and corporate statements from guarantors; (13) proposed loan agreement; (14) completed feasibility study, when required; and, (15) other such data. (1980.451(i)).

10. The County Supervisor and District Director determine if material and information submitted are complete. They then prepare comments and recommendations and submit them to the State Director. The County Supervisor furnished 410-9 and 410-10 forms dealing with Privacy Act statements to all individuals involved.

11. The State Director: (1) provides assistance to the County Supervisor; (2) prepares additional explanatory forms; (3) forwards all forms to the National Office; (4) checks the applicant's credit; and (5) submits transmittal letter with recommendations. (1980.451(b)).

12. FmHA will make a determination whether the borrower is eligible, the proposed loan is for an eligible purpose, and there is reasonable assurance of repayment ability, sufficient collateral, and sufficient equity. If FmHA is able to guarantee the loan, it will provide the lender and the applicant with 449-14, listing all requirements for such guarantees. (1980.452).

13. As conditions for FmHA making or guaranteeing a loan, the applicant must provide written statements dealing with the National Register of Historic Places (1980.44) and other Federal, State, and local requirements. (1980.45).

14. When FmHA finds that all requirements have been met, the lender and FmHA will execute 449-35, "Lender's Agreement." Upon receipt of a signed 449-35, and after all requirements have been met, FmHA will execute 449-34, "Loan Note Guarantee." In the event the lender assigns the guaranteed portion of the loan to a holder(s), the lender, holder, and FmHA will execute 449-36, "Assignment Guarantee Agreement." (1980.61(a) (b) (c)).

Department of Housing and Urban Development Grant Application Process (See Figure A-4)

1. Only applicants whose eligibility has been determined in accordance with requirements set forth in 24 CFR Part 570.452 of the Regulations may apply for Action Grants. Potential applicants must request a determination of eligibility from the HUD Area Office, using Standard Form 424, as modified. (UDAG Application Contents, p.1).

```
┌─────────────────┐   ┌─────────────────┐   ┌─────────────────┐
│ APPLICANT FILES │   │ APPLICANT FILES │   │ APPLICANT FILES │
│ PREAPPLICATION  │   │ PREAPPLICATION  │   │ PREAPPLICATION  │
│ WITH A-95       │   │ WITH FmHA       │   │ WITH BUREAU OF  │
│ CLEARINGHOUSE   │   │ COUNTY OFFICE   │   │ ALCOHOL, TAX,   │
│                 │   │                 │   │ AND FIREARMS    │
│            ①    │   │            ②    │   │            ③    │
└─────────────────┘   └─────────────────┘   └─────────────────┘

┌─────────────────┐   ┌─────────────────┐   ┌─────────────────┐
│ BUREAU OF       │   │   FmHA          │   │ A-95            │
│ ALCOHOL, TAX,   │ ◄─│   FORM          │─► │ CLEARINGHOUSE   │
│ & FIREARMS      │   │   449-1         │   │ REVIEWS,        │
│ AUTHORIZES      │   │                 │   │ COMMENTS,       │
│ PROJECT         │   │                 │   │ & CERTIFIES     │
│            ⑤    │   │                 │   │ ELIGIBILITY ④   │
└─────────────────┘   └─────────────────┘   └─────────────────┘

                      ┌─────────────────┐   ┌─────────────────┐
                      │ DEPARTMENT OF   │   │ LENDER FILES    │
                      │ LABOR CERTIFIES │   │ REQUEST FOR     │
                      │ NO ADVERSE      │   │ APPROVAL        │
                      │ LABOR EFFECTS   │   │                 │
                      │            ⑦    │   │            ⑥    │
                      └─────────────────┘   └─────────────────┘

                      ┌─────────────────┐   ┌─────────────────┐
                      │ FmHA MAKES      │   │   FmHA          │
                      │ PRELIMINARY     │ ◄─│   FORM          │
                      │ DETERMINATION   │   │   449-18        │
                      │ & NOTIFIES      │   │                 │
                      │ APPLICANT &     │   │                 │
                      │ LENDER      ⑧   │   │                 │
                      └─────────────────┘   └─────────────────┘

                      ┌─────────────────┐   • A&E Plans
                      │ APPLICANT       │   • Cost Estimates
                      │ FILES           │   • Compliance Evidence
                      │ APPLICATION     │   • Personal & Corporate Financial
                      │                 │     Reports
                      │                 │   • Feasibility Study (Possible)
                      │            ⑨    │   • EIA/EIS (Possible)
                      └─────────────────┘
```

```
┌ ─ ─ ─ ─ ─ ─ ─ ─ ─ ─ ─ ─ ─ ─ ─ ─ ─ ─ ─ ─ ─ ─ ─ ─ ─ ─ ─ ─ ─ ─ ─ ┐

 ┌───────────────┐  ┌───────────────┐  ┌───────────────┐  ┌───────────────┐
 │ COUNTY        │  │ STATE         │  │ FmHA          │  │   FmHA        │
 │ SUPERVISOR/   │  │ DIRECTOR      │  │ NATIONAL      │  │   FORM        │
 │ DISTRICT      │─►│ PROVIDES      │─►│ OFF EVALUATES │─►│   449-14      │
 │ DIRECTOR      │  │ ASSISTANCE &  │  │ APPLICATION & │  │               │
 │ REVIEWS &     │  │ ADVICE &      │  │ ISSUES        │  │               │
 │ PREPARES      │  │ FILES FORMS   │  │ CONDITIONAL   │  │               │
 │ RECOMMEND-    │  │ WITH NAT'L.   │  │ GUAR.         │  │               │
 │ ATIONS    ⑩  │  │ OFFICE    ⑪  │  │ COMMITMENT ⑫ │  │               │
 └───────────────┘  └───────────────┘  └───────────────┘  └───────────────┘

 ┌───────────────┐  ┌───────────────┐  ┌───────────────┐  ┌───────────────┐
 │   FmHA        │  │   FmHA        │  │ FmHA ISSUES   │  │ LENDER        │
 │   FORM        │ ◄│   FORM        │ ◄│ LENDER        │ ◄│ CERTIFIES     │
 │   449-34      │  │   449-35      │  │ AGREEMENT,    │  │ THAT ALL      │
 │               │  │               │  │ LOAN NOTE     │  │ COMMITMENTS   │
 │               │  │               │  │ GUARANTEE,    │  │ & REQUIRE-    │
 │               │  │               │  │ AND, POSSIBLY,│  │ MENTS HAVE    │
 │               │  │               │  │ ASSIGNMENT    │  │ BEEN MET      │
 │               │  │               │  │ GUARANTEE ⑭  │  │           ⑬  │
 └───────────────┘  └───────────────┘  └───────────────┘  └───────────────┘

 ┌───────────────┐
 │   FmHA        │
 │   FORM        │                              TOTAL ELAPSED TIME
 │   449-36      │                              60 TO 90 DAYS
 │               │
 └───────────────┘
└ ─ ─ ─ ─ ─ ─ ─ ─ ─ ─ ─ ─ ─ ─ ─ ─ ─ ─ ─ ─ ─ ─ ─ ─ ─ ─ ─ ─ ─ ─ ─ ┘
```

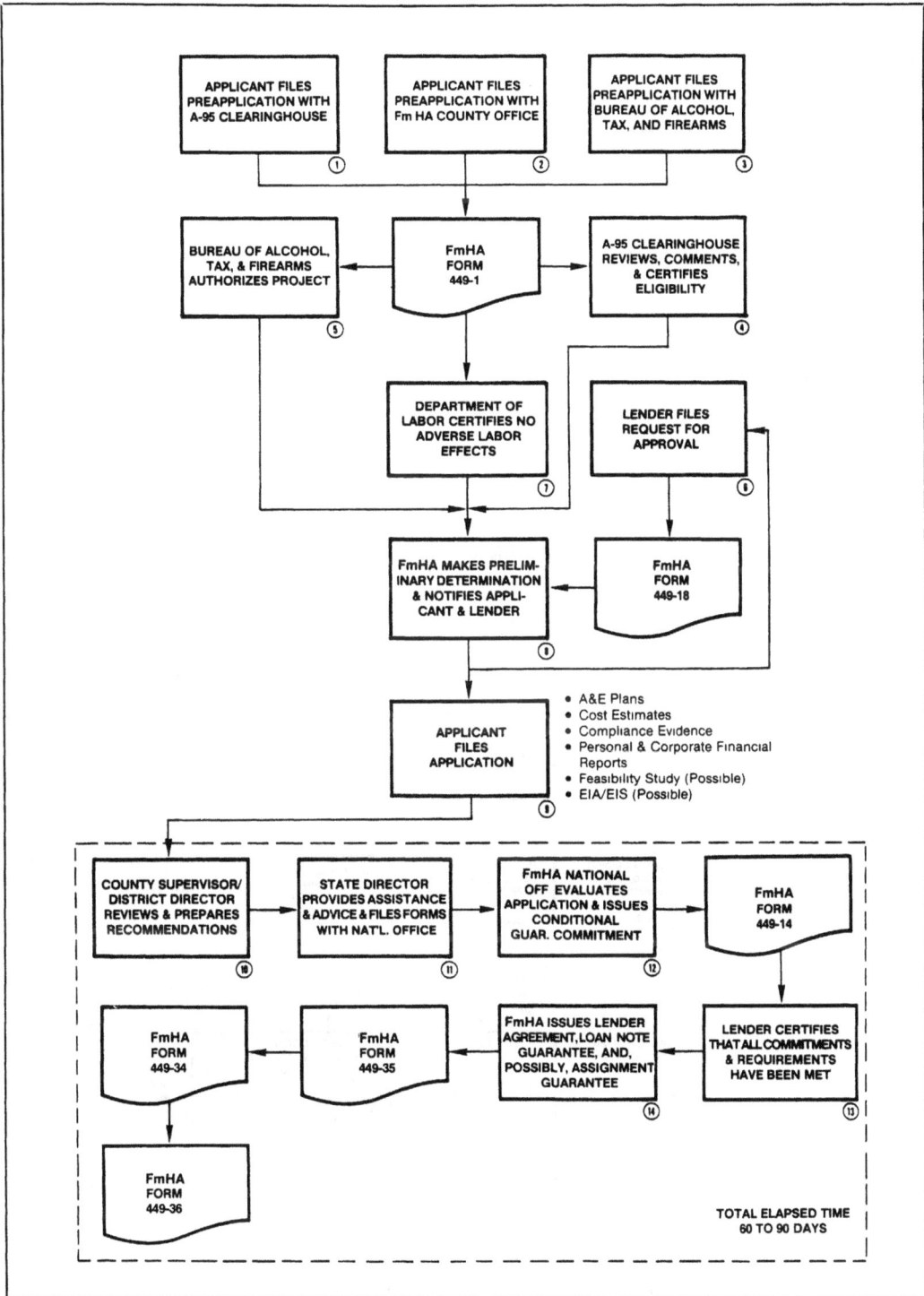

Figure A-3. FmHA Loan Application Procedure

APPLICANT REQUESTS HUD AREA OFFICE FOR DETERMINATION OF ELIGIBILITY ①

- First time applicants allow at least 60 days prior to application submittal

STANDARD FORM 424

APPLICANT NOTIFIES A-95 CLEARING HOUSE OF NOTICE OF INTENT TO FILE APPLICATION ②

- Return to applicant as soon as practical

A-95 CLEARING HOUSE REVIEWS, COMMENTS, & CERT ELIGIBIL. ⑦

HUD AREA OFFICE CERT ELIGIBIL OR ASST SECY COMM PLANNING & DVLPM CERT ELIGIBIL OR REJECT REQUEST ③

- File at least 30 days before intended submittal to HUD area office

APPLICANT PREPARES GRANT APPLICATION PACKAGE ⑥

- Description of project
- Evidence of meeting program objectives
- Project approval information
- Assurances

APPLICANT PREPARES SUPPORTING PLAN FOR CITIZEN PARTICIPATION ④

APPLICANT FILES ORIGINALS TO DHUD OFFICE OF ACTION GRANTS & TWO COPIES TO HUD AREA OFF ⑧

- Metropolitan cities & urban counties by end of Jan Apr Jul Oct
- Small cities by end of Feb May Aug Nov

- Acknowledge receipt & preliminary approval or necessary additions revisions (notes time available for consideration this qtr)

APPLICANT PREPARES SUPPORTING ENVIRONMENT IMPACT ASSMT ⑤

HUD AREA OFFICE SCREENS APPLICATIONS FOR COMPLETENESS ⑨

- Capacity of applicant to conduct project
- Adequacy of proposed resources
- Relevancy of project to selection criteria
- A-95 comments from clearing house
- Negotiating role might be necessary

HUD AREA OFF REVIEWS FINAL APPLICATION PACKAGE ⑩

DHUD OFFICE OF ACTION GRANTS REVIEW FINAL APPLICATION PACKAGE ⑫

HUD REGIONAL OFF REVIEWS FINAL APPLICATION PACKAGE ⑪

- Relevancy of project to selection criteria
- Assessing of field recommend /comments
- Negotiating role might be necessary

- General comments and recommendations possible
- Negotiating role might be necessary

- 60 to 90 days after submittal of grants application pkg

- Funding held until completion of environmental assessment private commitments other contractural commitments

APPLICANT RECEIVES AWARD CONTRACT FOR COMPLETION & RETURN TO HUD AREA OFFICE ⑭

ASST SECY COMMUN PLAN & DEVELOPMENT NOTIFIES APPLICANT OF FINDINGS NOTICE OF AWARDS MADE TO DESIGNATED STATE CENTRAL INFORMATION RECEPTION AGENCY ⑬

- Amount approved or
- Reasons for rejection

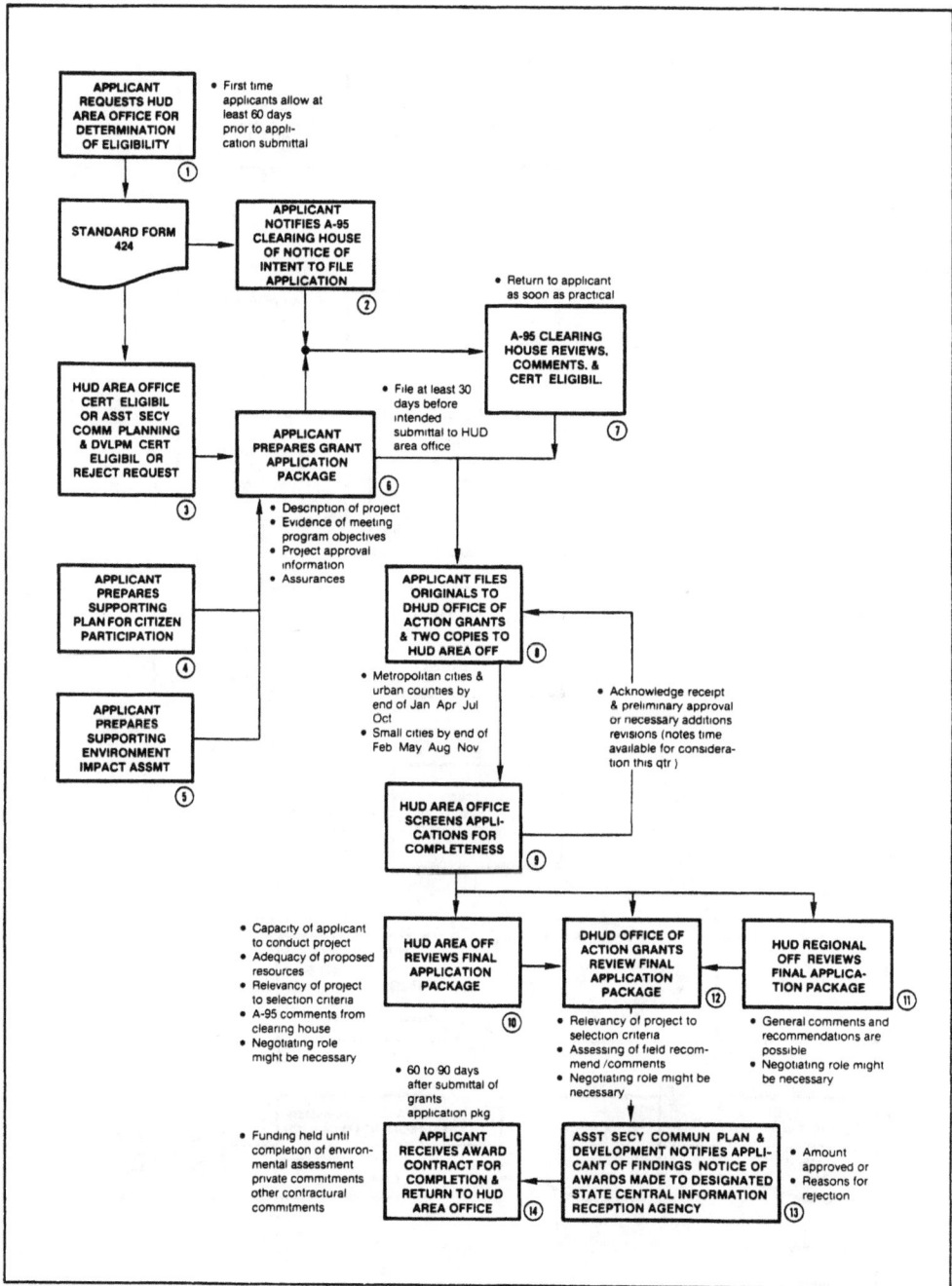

Figure A-4. HUD Loan Application Procedure

COMMERCIAL SCALE ETHANOL PRODUCTION AND FINANCING

2. At the same time as the Standard Form 424 is filed with the HUD Area Office, submit it or a summary to the A-95 clearinghouse as notice of the intent to file an application (UDAG Application Contents, p.1.)

3. To certify eligibility, determination must show that the community: (1) meets the minimum level of physical and economic distress; (2) has demonstrated results in providing housing for low- and moderate-income persons; and (3) has demonstrated results in providing equal opportunity in housing and employment for low- and moderate-income persons and members of minority groups. If the HUD Area Office is unable to conclude that a community is eligible, the case will be referred to the Assistant Secretary for Community Planning and Development for a final determination. (The Action Grant Information Book, p.5).

4. Public hearings and other citizen participation activities must take place before a community can submit an application. (The Action Grant Information Book, p.4.).

5. HUD will not accept applications for funding consideration unless a level of environmental clearance has been completed for the project, as covered in 24 CFR Part 58.15. (UDAG Application Contents, p.15).

6. Application package consists of five parts: (1) Standard Form 424; (2) Description of Proposed Project: A. brief summary of project and its participants; B. detailed project description; (3) Evidence of Meeting Program Objectives: A. alleviation of physical distress; B. alleviation of economic distress; C. fiscal improvement; D. impact on the special problems of low- and moderate-income persons and minorities; E. employment plan to ensure that the private sector jobs established are available to the unemployed; F. record of applicant; (4) Project Approval Information: A. technical requirements; B. citizen participation, Civil Rights, Equal Employment, and Housing Authority; (5) Assurances. (UDAG application contents).

7. Submit a full application to the A-95 clearinghouse for review and comment at least 30 days prior to submission of all Action Grant applications. Submit A-95 comments to the Office of Action Grants with the application. (UDAG Application Contents, p.1).

8. Applications must be received by HUD from metropolitan cities during the first month of each quarter and from small cities in the second month if they are to be considered for funding in the quarter. (The Action Grant Information Book, p.5).

9. All applications will be screened to assure they are complete before being officially accepted for review. Copies of complete applications will then be forwarded to the regional office and Washington. (The Action Grant Information Book, p.6).

10. The review period will take up a large part of the quarter. The area office will review and comment on the capacity of the applicant to carry out the project and complete it in a timely manner, the adequacy of the proposed resources, the relevance of the proposed project to selection criteria in 570.457 (C) through (K), and the A-95 comments. Area office comments and recommendations will be forwarded to Washington for review. (The Action Grant Information Book, p.6).

11. Regional offices may also submit comments and recommendations on the proposed project. Regional Administrators, along with area office staff, will participate in the negotiations with applicants and in the final recommendations for funding. (The Action Grant Information Book, p.6).

12. Washington will have the primary responsibility for (1) reviewing all applications against the selection criteria listed in the regulations -570.457 (C) (K); (2) assessing the recommendations and comments from the field; (3) negotiations between the applicant and the department; and (4) making recommendations to the Assistant Secretary for Community Planning and Development. (The Action Grant Information Book, p.6)

13. During the review process, negotiations with the applicant may be necessary. HUD will make the final decisions on approvals no later than the last day of each quarter.

14. An approved applicant will be notified in writing and will receive an award contract which must be completed and returned to the HUD area office.

Economic Development Administration Loan Application Process (See Figure A-5)

1. All prospective applicants should forward project assistance requests to the Economic Development Representative (EDR) in their State or area. The request should indicate the type and amount of EDA assistance desired, the project location, and

the applicant's assessment of the local economic development benefits to be derived from the project.

2. The EDR will set up an appointment with each prospective applicant and local community officials, if appropriate. The preapplication conference enables a quick check on the economic development benefits, project feasibility, the matching with EDA goals and strategy, and the available funding.

3. Feasibility studies are required prior to filing an application for assistance. EDA's primary aim in setting this requirement is to assure that: (1) the project can be economically successful; (2) the production technology can meet established cost and production targets; (3) a viable local or regional market exists; and (4) the particular applicant's operation is so designed as not to pose a threat to either the environment or the safety of its employees.

4. Once the feasibility study is completed, the EDR will forward it through the appropriate regional office to DOE headquarters for a technical evaluation.

5. All project applications must include an acceptable employment plan. Preference will be given to projects that exceed the established agency goal for hiring the long-term unemployed. The Department of Labor is to be asked for a critique early in this process.

6. After the feasibility study is reviewed at headquarters DOE, EDA will select projects for funding not only on the basis of DOE recommendations, but review of the local economic development benefits commensurate with its existing budgetary capabilities.

7. No EDA funds will be disbursed until the applicant has met the bonding requirements for and subsequently obtained a commercial distilled spirits permit from BATF, or sufficient evidence is provided that a perfected application with BATF is in the process of adjudication.

8. Once initial EDA project considerations are made, project application materials will be presented by the RegionalOffice Development Finance Staff at an initial application conference.

9. Most plants larger than one million gallons of alcohol per year will be subject to final review by the National Office Development Financing Staff. Applicants and their consultants may be

required to furnish additional information at this time.

10. EDA is required by statute to make a finding that reasonable assurance of repayment exists for each loan guarantee extended.

11. Alcohol fuel plants require an additional approval by the Office of Private Investment at headquarters EDA since they are new to EDA.

12. The letter of notice of award is signed by the Assistant Secretary for Economic Development (EDA).

13. The Office of Congressional Affairs, EDA, notifies the State representatives in Congress of the loan guarantee offer.

14. Finally, the applicant receives written notification and must contact regional development counsel and arrange for closing.

Department of Energy Loan Guarantee Application Process (See Figure A-6)

1. DOE will publish an Announcement in the Federal Register soliciting applications for loan guarantees. A synopsis of the Announcement will be published simultaneously in the Commerce Business Daily. In addition, copies of the Announcement will be mailed to any person known to be interested in responding. The solicitation announcement will contain: (1) a brief description of the size and type of projects covered; (2) the time period during which responses may be filed; and (3) the name and address of the DOE representative who may be contacted.

2. The applicant must provide sufficient detail to establish credibility that: (1) the project will be in compliance with established environmental regulations; (2) the project will conform to provisions of Executive Orders dealing with Floodlands Management, and Protection of Wetlands; (3) the project is environmentally acceptable; (4) any affected Indian Tribe(s) has been involved in the planning; (5) there is sufficient evidence that the applicant will initiate and complete the project in a timely, efficient, and acceptable manner; and (6) the project will be built and operated in the United States.

3. Upon receiving an application for a loan guarantee for a demonstration facility, DOE determines whether: (1) the application is complete and in compliance with the solicitation request; (2) the type of technology, scope, size,

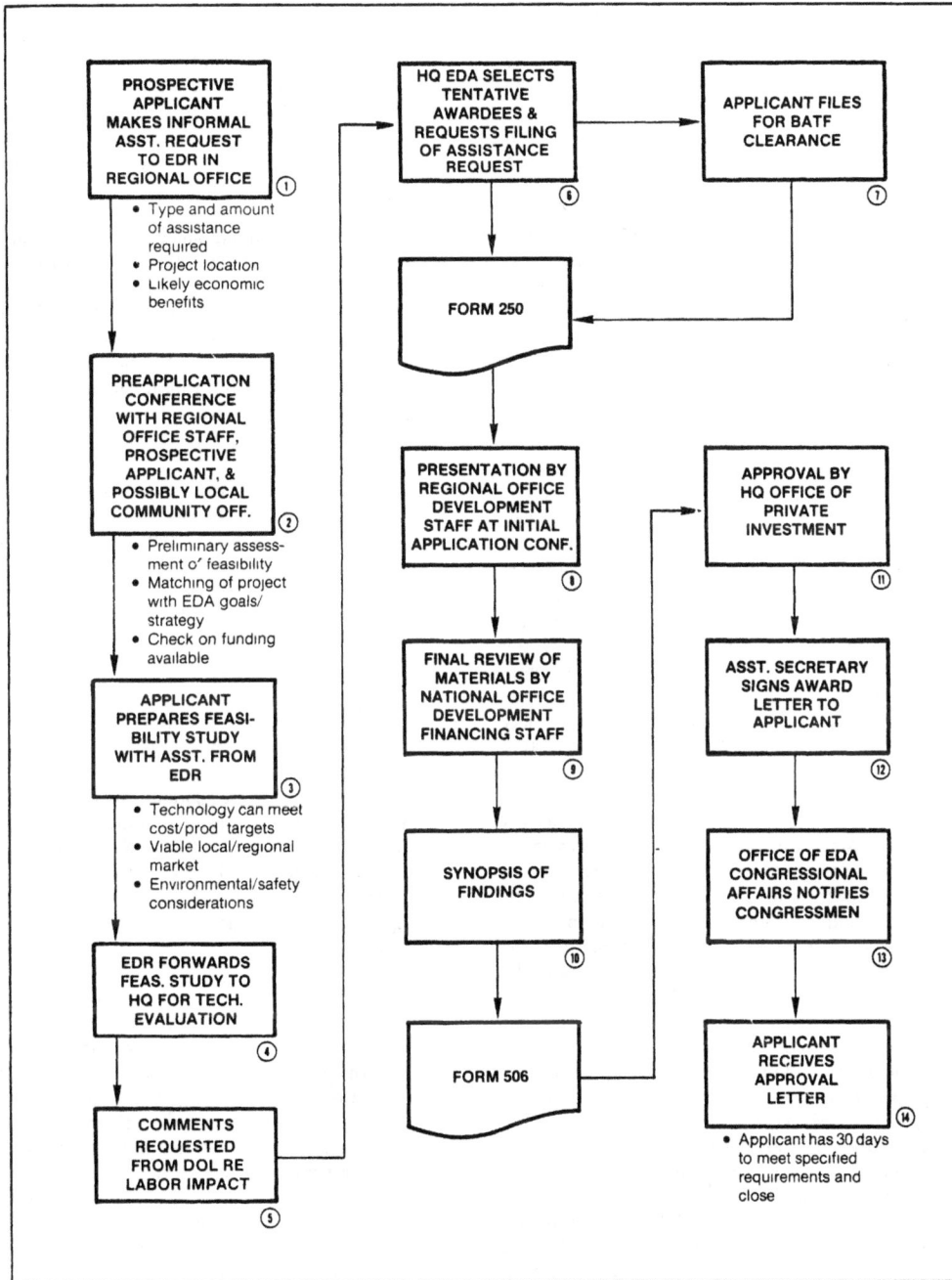

Figure A-5. EDA Loan Application Procedure

The flowchart contains the following boxes:

① PROSPECTIVE APPLICANT MAKES INFORMAL ASST. REQUEST TO EDR IN REGIONAL OFFICE
- Type and amount of assistance required
- Project location
- Likely economic benefits

② PREAPPLICATION CONFERENCE WITH REGIONAL OFFICE STAFF, PROSPECTIVE APPLICANT, & POSSIBLY LOCAL COMMUNITY OFF.
- Preliminary assessment of feasibility
- Matching of project with EDA goals/strategy
- Check on funding available

③ APPLICANT PREPARES FEASIBILITY STUDY WITH ASST. FROM EDR
- Technology can meet cost/prod targets
- Viable local/regional market
- Environmental/safety considerations

④ EDR FORWARDS FEAS. STUDY TO HQ FOR TECH. EVALUATION

⑤ COMMENTS REQUESTED FROM DOL RE LABOR IMPACT

⑥ HQ EDA SELECTS TENTATIVE AWARDEES & REQUESTS FILING OF ASSISTANCE REQUEST

⑦ APPLICANT FILES FOR BATF CLEARANCE

FORM 250

⑧ PRESENTATION BY REGIONAL OFFICE DEVELOPMENT STAFF AT INITIAL APPLICATION CONF.

⑨ FINAL REVIEW OF MATERIALS BY NATIONAL OFFICE DEVELOPMENT FINANCING STAFF

⑩ SYNOPSIS OF FINDINGS

FORM 506

⑪ APPROVAL BY HQ OFFICE OF PRIVATE INVESTMENT

⑫ ASST. SECRETARY SIGNS AWARD LETTER TO APPLICANT

⑬ OFFICE OF EDA CONGRESSIONAL AFFAIRS NOTIFIES CONGRESSMEN

⑭ APPLICANT RECEIVES APPROVAL LETTER
- Applicant has 30 days to meet specified requirements and close

financial assistance sought, and geographic location of the project fall within the solicitation objectives; (3) sufficient loan guarantee authority and appropriations are available to conduct the project; (4) the adverse community impacts resulting from the proposed facility have been adequately evaluated by the applicant; and (5) the terms and conditions of the proposed loan as set forth in the application are generally acceptable to DOE; and (6) whether there is satisfactory evidence that the lender is willing, competent, and capable of performing the terms and conditions of the loan guarantee agreement. DOE will notify the applicant whether its application meets the initial requirements. It may be necessary to disapprove the application or request additional information.

4. When the application meets the requirements, DOE will inform: (1) the governor of the state, officials of each political subdivision, and Indian tribe(s), as appropriate; (2) the general public by notice in the Federal Register and in local newspapers to the extent appropriate; (3) the Secretary of Interior, if the facility is to be located on Indian land or if a substantial portion of the feedstock is to be obtained pursuant to an agreement with any Indian tribe; and (4) the Advisory Council on Historic Preservation for any project which affects property listed or eligible for listing in the National Register of Historic Places.

5. The governor of the state in which the proposed facility would be located shall be requested to submit a recommendation for or against the location of such facility in his state within a reasonable period after the initial notification. DOE shall not guarantee or make a commitment to guarantee a loan if the governor recommends, within a reasonable period, that such action not be taken unless the Secretary, DOE, finds that there is an overriding national interest in taking such action. DOE may provide the following forms of community impact assistance to aggrieved parties: (1) a planning assistance grant to a governor of a state and Indian tribe, as appropriate; (2) a management grant to a political subdivision or Indian tribe, as appropriate; (3) a loan guarantee to a political subdivision or Indian tribe, as appropriate; (4) a requirement that the applicant obtaining financial assistance advance funds for management and essential community development; and (5) direct loans by DOE for financing essential community development.

6. If the total cost of each individual project from which the selection is to be made is $50 million or less, then the selection official shall be the ap-

propriate Assistant Secretary of Energy, or designated representative, and the evaluation will be conducted by an appointed panel. If the total cost exceeds $50 million, then the selecting official shall be the Secretary, or designated representative, and the application evaluation will be conducted by a specifically constituted Source Evaluation Board.

7. If DOE decides to guarantee or make a commitment to guarantee despite the affected governor's recommendation that such action not be taken, DOE presents its reasons for such determination to the governor in writing. The decision is final unless otherwise determined upon judicial review by the U.S. Court of Appeals upon application made within 90 days from the date of the Secretary's decision.

8. DOE forwards the selected applications to the U.S. Attorney General and to the Chairman of the FTC. Written views, comments, and recommendations are requested for each application selected and forwarded from both of these officials concerning the impact of a loan guarantee on competition and concentration in the production of energy.

9. The Secretary, after considering a negative recommendation from the Attorney General or Chairman of the FTC, may nevertheless proceed with the award of the guarantee to such applicant if the Secretary determines that the energy project meets the overall goals of DOE and outweighs the negative recommendations. Alternatively, the Secretary may notify the applicant of the disapproval of its application with reasons for such disapproval.

10. If the Secretary determines to award the guarantee to an applicant in spite of the negative recommendation by the Attorney General or the Chairman of the FTC, the Secretary shall forward the decision and negative recommendation papers to the President for a written decision on whether it is in the national interest to have the Secretary award a loan guarantee to the applicant. If the President upholds the negative recommendation, the Secretary will notify the applicant that the application has been disapproved and the reasons for the disapproval.

11. If the President upholds the Secretary's determination to proceed, or the Secretary has selected an applicant after receiving favorable recommendations from both the Attorney General and the Chairman of the FTC, DOE forwards the relevant documents to the Secretary of the Treasury

COMMERCIAL SCALE ETHANOL PRODUCTION AND FINANCING

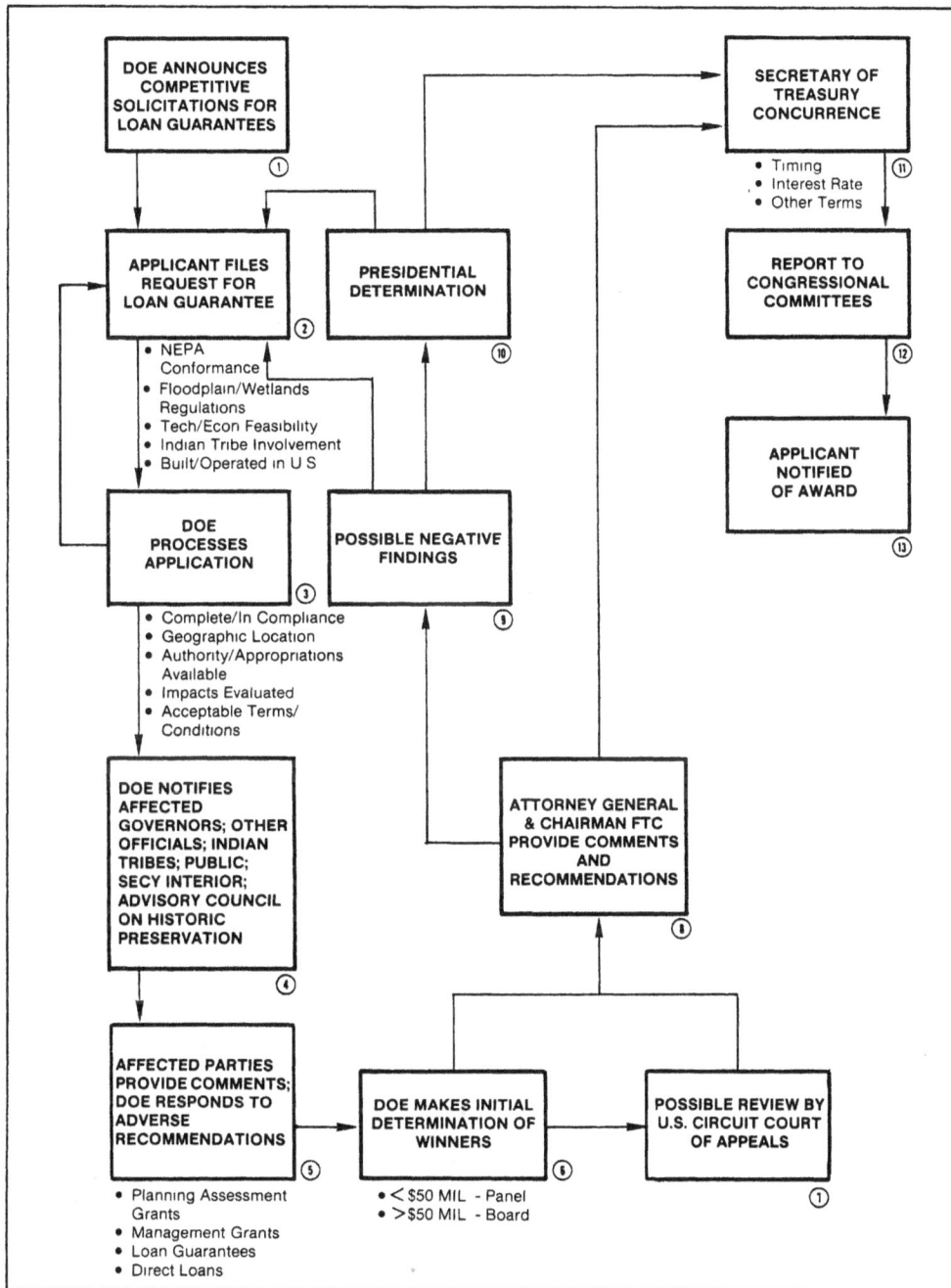

Figure A-6. DOE Loan Application Procedure

for concurrence, with respect to the timing, interest rate, and other such terms. The Secretary of Treasury shall ensure to the maximum extent feasible that such terms will have the minimum possible impact on the capital markets of the United States.

12. DOE will forward a full and complete report on each project to the Chairman of the Committee on Science and Technology of the House of Representatives, and to the Chairman of the Committee on Energy and Natural Resources of the Senate. DOE cannot finalize the loan guarantee or commitment to guarantee until 90 days after receipt by Congress.

13. Upon obtaining the concurrence of the Secretary of Treasury and the expiration of the time specified for Congressional reaction, the Secretary of DOE may award the loan guarantee or commitment to guarantee to the successful applicant.

The characteristics of Financial Assistance Programs discussed in this Appendix are shown in Figure A-7.

	LOAN OBJECTIVES	INTEREST RATE	ELIGIBILITY

A. SBA PRINCIPAL LENDING PROGRAMS
 	
• **LOCAL DEVELOP-MENT COMPANIES**
 	
Authorization:
 	
Public Law:
  699 SB1A
  15 USC Sec. 661
	To help a development company acquire land and building, construct a new plant, purchase necessary machinery and equipment, expand or convert an existing plant, provided the project will assist a specific small business	(a) Guaranteed loans legal and reasonable rate on entire loan balance, (b) immediate participation legal and reasonable rate on bank share and published rate on SBA share, (c) first mortgage legal and reasonalbe rate on bank loan, published rate on SBA loan, (d) direct SBA loan published rate	Any Corporation which (a) is formed by public-spirited citizens interested in planned economic growth of a community with at least 75 percent ownership and control held by persons living or doing business in the community, b) has been incorporated either for profit or non-profit under laws of the state in which it expects to do business, (c) is authorized to promote and assist the growth and development of small business in its area of operations, and (d) has a minimum of 25 stockholders or members

• **BUSINESS LOANS**
 	
Authorization:
 	
Public Law:
  85-536 SBA 1
  15 USC 636
	Business construction, conversion or expansion, purchase of equipment, facilities, machinery, supplies or materials and working capital	6-⅝ per annum on direct loan and SBA share of an immediate participation loan On the bank's share of an immediate participation loan, the lending institution may set reasonable and legal rate with a maximum ceiling set by SBA from time to time On a quaranty loan, bank may set legal and reasonable rate, with a maximum ceiling set by SBA from time to time	Most businesses including farms that are (1) independently owned and operated and not dominant in their fields, (2) unable to obtain private financing on reasonable terms (3) qualified as "small" under SBA's size standards, based on dollar volume of business or number of employees

• **STATE DEVELOP-MENT COMPANIES**
 	
Authorization:
 	
Public Law:
 	
  699 SB1 A
  15 USC SEC 661
	To help state development company provide equity capital and long-term loans to small business involved in the production of alcohol fuels	Published annually, but not to exceed 8 percent	Any corporation organized under or pursuant to a special act of the State Legislature, with authority to operate statewide and to assist the growth and development of business concerns, including small businesses, in its area

• **SMALL BUSINESS INVESTMENT**
 	
Authorization:
 	
Public Law:
 	
  699 SB 1A
  15 USC Sec. 661
  Title 3
	To provide an SBIC or Section 301 (d) SBIC with funds for financing eligible small business for their growth, modernization, and/or expansion, in the area of alcohol fuels	Debentures sold to the Federal Financing bank bear interest at comparable Agency rates for paper of a similar maturity Section 301(d) SBIC debentures sold to SBA bear interest at a rate not less than a rate determined by the Secretary of Treasury plus any additional charge toward covering other costs of the program which SBA may determine consistent with its purpose Section 301(d) SBICs are eligible for a subsidized interest rate for the first five years on their debentures sold to SBA, subject to repayment of that subsidy before any distribution is made to stockholders other than SBA	A regular small business investment company (SBIC) or small business investment company organized under Section 301 (d) of the SBIC Act, solely to assist disadvantaged entrepreneurs Although statutory minimum capital is less, SBA, by policy, requires a minimum capital of at least $500,000 An SBIC must show that such capital is adequate to operate actively and profitably A Section 301 (d) Licensee with limited capital must show that funds will be provided to cover its operating expenses without depleting its capital

Figure A-7. Financial Assistance Programs Characteristics

COMMERCIAL SCALE ETHANOL PRODUCTION AND FINANCING

SOURCE OF FUNDS	MATURITY	CEILING	COLLATERAL/SECURITY
(a) Bank loan guaranteed by SBA to 90 percent of the loan or $500,000 whichever is the lesser, (b) bank loan with immediate participation by SBA, (c) bank first mortgage loan and SBA direct second mortgage loan, or (d) direct from SBA	Maximum maturity of 25 years plus estimated time required to complete construction, conversion, or expansion Usually 15 to 20 years	$500,000 for each identifiable small business to be assisted as a prerequisite to obtaining SBA financing, a development company must provide a reasonable share of cost of project in funds raised by sale of stock, debentures, memberships, or cash equivalent (e g , land) Minimum amount to be provided by development company will generally be 20 percent of cost of project SBA will take a second lien position when the local lending institutions will participate in the SBA's first mortgage plan	A lien on the fined assests acquired with loan proceeds so as to reasonably assure repayment of the loan
Banks and other lending institutions, excluding small business investment companies licensed by SBA	Maximum of 10 years as a rule However, working capital loans generally are limited to 6 years, while portions of loans for construction and acquisition of real estate may have maximum of 20 years	$350,000 to any one borrower is the maximum SBA share of an immediate participation loan, where SBA and private lending institution each put up part of loan funds immediately, and the maximum SBA direct loan, one made by the agency For guaranteed loans, made by a bank and partially guaranteed by SBA, the maximum is also $350,000 normally but may be up to $500,000 for exceptional circumstances	Real estate or chatel mortgage, assignment of warehouse receipts for marketable merchandise, assignment of certain types of contracts, guarantees or personnel endorsements, in some instances assignment of current receivables
Direct from SBA	Maximum of 20 years, in actual practice these loans are usually requested from a 5 to 10 year term	As much as state development company's total outstanding borrowings from all other sources Based on experience, the average loan request from a State development company is $500,000 to $1,000,000 (Total available limited by annual budget allocations)	Security for SBA loan on an equal basis with funds borrowed by development company from any other sources after August 21, 1958 (SBA funds may be secured on a ratable basis with other borrowings of the State development company)
SBA provides the authorized leverage funds to an SBIC through its 100 percent gurantee of the debentures sold to the Federal Financing Bank	Maximum 15 years	An SBIC is eligible to borrow $3 for every $1 of private capital up to a maximum of $35 million An SBIC which has 65 percent or more of its total funds available invested in venture capital is eligible to borrow $4 for every $1 of its private capital, up to a total maximum of $35 million, provided it has private capital of $500,000 or more The same eligibiliy applies to Section 301(d) Licensees, except that the required venture capital investment is 30 percent and there is no limit on funding	Evidence of indebtedness is a debenture of the SBIC or Section 301(d) Licensee subordinated to any other debenture bonds, promissory notes, or other debts and obligations of the SBIC or Section 301(d) SBIC unless the SBA, in the exercise of reasonable investment prudence. determines otherwise Adequately capitalized Section 301(d) Licensees are eligible for SBA purchases of their 3 percent cumulative preferred stock in an amount equivalent to a portion or all of their paid-in capital and paid-in surplus, the proceeds thereof constituting a part of their authorized leverage

Figure A-7. Financial Assistance Programs Characteristics (Continued)

	LOAN OBJECTIVES	INTEREST RATE	ELIGIBILITY
• ECONOMIC OPPORTUNITY Authorization: Public Law: 93-384 SBA 1 15 USC 636 (f)	Any use which will carry out the purposes the same as other business loans	On direct loans and SBA share of immediate participation plans, the rate is set periodically, based on a statutory formula Bank rate same as on other business loans	Low income or disadvantaged persons who have lacked the opportunity to start or strengthen a small busines and cannot obtain the necessary financing from other sources on reasonable terms
B. FmHA LOANS AND GUARANTEES FOR ALCOHOL FUELS Authorization: Rural Development Act of 1972 SEC 3108 USE 1932 PL92-419 TITLE 1	The Farmers Home Administration (FmHA) makes loans under the Farm Loan Program and guarantees loans by private lenders under the Business and Industrial (B&I) Loan Guarantee Program for plants for the production of fuel alcohol	Guaranteed loan rates will be negotiated between lender and borrower, They may be fixed or variable Insured loan rates to public bodies, nonprofit associations, and Indian tribes are at the rate of 5% per annum Other insured loans will be at the rate presented by FmHA at the time	Loans made by private lenders but guaranteed by FmHA may be made to any legal entity, including individuals, public and private organizations and Federally recognized Indian tribal groups Priority is given to projects in the country and in communities of 25,000 population and smaller, although loans may be made in cities and towns of less than 50,000 population and its immediately adjacent urbanized areas with a population density or less than 100 persons per square mile
Authorization: Agricultural and Credit Act of 1978	Intended to finance production of alcohol for on-farm use, or for partial off-farm use and sale, only if it is necessary to make the farm operation commercially viable	Farm operating loans Interest rates currently at 10 5 percent for operating loans Farm ownership loans. Interest rates are 10 percent annually	Operators of family farms who are U S citizens, are of legal age, who cannot obtain sufficient credit elsewhere at reasonable rates and terms, and who otherwise meet the requirements for an FmHA farm loan may be eligible The county or area committee of FmHA — consisting of three persons who know local farming and credit conditions — determines eligibility of applicants
C. HOUSING & URBAN DEVELOPMENT, URBAN DEVELOP. ACTION GRANTS Authorization: Title 1 of the Housing and Community Devel-opment Act of 1974 Public Law 93-383, 42 U.S.C. 5301-5317 as amended by Title 1 of the Housing and Community Develop. Act of 1977, Section 110, Public Law 93-128, HL U.S.C. 5304	To assist severely distressed cities and severely distressed urban counties in alleviating physical and economic deterioration through economic development and neighborhood revitalization Alcohol fuel production is an example of such efforts	Not applicable — grant	Applicant Eligibility Eligible applicants are distressed cities and distressed urban counties which meet the following criteria specified in Section 470 452 of the regulations (a) Minimum standards of physical and economic distress, (b) demonstrated results in providing housing for persons of low and moderate income, and (c) demonstrated results in providing equal opportunity in housing and employment for low and moderate income persons and members of minority groups Beneficiary Eligibility Same as Applicant Eligibility

Figure A-7. Financial Assistance Programs Characteristics (Continued)

SOURCE OF FUNDS	MATURITY	CEILING	COLLATERAL/SECURITY
Direct from SBA and/or banks and other lending institutions, excluding small business investment companies licensed by SBA	Maximum of 15 years Working capital loans generally limited to a 10-year maximum	$100,000 to any one borrower, as SBA share of loan	Any worthwhile collateral which is available or will be acquired with the proceeds of the loan
FmHA may guarantee up to 90 percent of a loan, with the applicant responsible for furnishing a minimum of 10 percent of equity The maximum guarantee for alcohol fuel production is expected to be lower	Maximim terms are 30 years for land, buildings and permanent fixtures, 15 years for machinery or equipment, or the life of the machinery or equipment, whichever is shorter, 7 years for working capital The interest rate will be determined between the lender and borrower	No ceilings, pre-screening for $5,000,000	Whatever the bank requires for loan guarantees, such as a lien on fixed assets
Guaranteed loans are made and serviced by legally organized private lending institutions, such as commercial banks, Federal Land Banks, Production Credit Associations, insurance companies, and savings and loan associations FmHA provides the lender with a guarantee to reimburse up to 90 percent of any loss the lender takes on a loan	Farm operating loans Operating loans may be repaid in from 1 to 7 years In some cases, borrower may be given an additional 7 years to repay

Farm ownership loans Farm ownership loans have a maximum term of 40 years | Farm operating loans The limit on an operating loan made directly by FmHA is $100,000, a private loan guaranteed by FmHA has a limit of $200,000

Farm ownership loans The limit on a farm ownership loan made directly by FmHA is $200,000 on a guaranteed loan $300,000 | Farm operating loans. Fixtures

Farm ownership loans Real estate, security, or liens on fixed assets |
| Projects which include financial assistance from the State or other public entities will receive more favorable consideration Other public resource may be provided by matching other Federal grants, or by firm commitments or other Federal or local resources No activities will be funded unless there is a firm commitment of private resource to the proposed project | Assistance is for a discrete project which can be completed in approximately 4 years No additional funding will be available in subsequent years to complete a project approved in a prior year Funds are made available through a Letter of Credit | FY 78 small cities range $77,700-$5,700,000, average. $934,300 FY 78 metro cities range $85,000-$13,500,000, average $2,950,000 | Not applicable — grant |

Figure A-7. Financial Assistance Programs Characteristics (Continued)

	LOAN OBJECTIVES	INTEREST RATE	ELIGIBILITY
D. ECONOMIC DEVELOPMENT ADMIN. PUBLIC WORKS GRANTS TO RURAL COMMUNITIES & NON-PROFIT RURAL-BASED ORGANIZATIONS	To sustain industrial and commercial viability in designated areas by providing financial assistance to business that create or retain permanent jobs expand or establish plants in redevelopment areas for projects where financial assistance is not available from other sources, on terms and conditions that would permit accomplishment of the project and further economic development in the area	Interest rates are negotiable — may be fixed for variable	Any individual, private or public corporation, or Indian tribe provided that the project to be funded is physically situated in an area designated as eligible under the Act at the time the application is filed. Neither business development loans nor guarantees of any kind will be extended to applicants who (1) have, within the previous 3 years, relocated any or all of their facilities to another city, or State, (2) contemplate relocating part or all of their existing facilities with a resultant loss of employment at such facilities, and (3) produce a product or service for which there is a sustained and prolonged excess of supply over demand
E. DEPARTMENT OF ENERGY, ALTERNATIVE FUELS PRODUCTION **Authorization:** **Public Law 96-126**	To expedite the domestic development and production of alternative fuels and to reduce dependence on foreign supplies of energy resource by establishing such domestic production at maximum levels at the earliest time practicable	A rate, determined by the Secretary in consultation with the Secretary for Treasury, to be reasonable taking into account the range of interest rate prevailing in the private sector for similar government guaranteed obligations of comparable risk	Organizations interested in pursuing commercial production of alternative fuels who are capable of providing development and production facilities at the earliest time practicable
F. ENERGY SECURITY ACT OF 1980 **Authorization:** **Public Law 96-294** **Title II**	To reduce the dependence of the United States on imported petroleum and natural gas by all economically and environmentally feasible means, including the use of biomass energy resources, and to formulate and implement a national program for increased production and use of biomass energy that does not impare the nation's ability to produce food and fiber on a sustainable basis for domestic and export use	(a) Insured loans made by the Secretary of Agriculture shall bear interest at rates determined by the Secretary of Agriculture, taking into consideration the current average market yield on outstanding marketable obligations of the United States with remaining periods to maturity comparable to the maturities of such loans, plus not to exceed one per centum, as determined by the Secretary of Agriculture, and adjusted to the nearest one-eighth of one per centum (b) Interest rates are not specified for loan guarantees provided by the Secretary of Energy and/or the Secretary of Agriculture	All interested persons, e.g., defined as any individual, company, cooperative, partnership, corporation, association, consortium, unincorporated organization, trust, estate, or any entity organized for a common business purpose, and state or local government or any agency or instrumentality thereof, or any Indian tribe or tribal organization

Figure A-7. Financial Assistance Programs Characteristics (Continued)

COMMERCIAL SCALE ETHANOL PRODUCTION AND FINANCING

SOURCE OF FUNDS	MATURITY	CEILING	COLLATERAL/SECURITY
The Federal participation in a direct, fixed asset loan may not exceed 65 percent of project fixed asset costs A local development corporation or State agency usually participates to the extent of five percent Of the remaining 30 percent, 10 percent or 15 percent if no local or State participation must be in the form of applicant's equity and the balance from a conventional commercial lender Applicants are encouraged to increase their equity participation beyond the minimum Comparable 15 percent equity requirements also apply to guarantees	Fixed asset loans, 25 years, maximum, working capital loans 5 years, guarantees, life of loan or lease EDA loan funds will only be disbursed after all other funds have been injected into project	Long-term business development loans up to 65 percent of the cost may be used for the acquisition of fixed assets only (i e land, building, machinery, and equipment, including land preparation and building rehabilitation) Funds may be used for most kinds of new industrial or commercial facilities or to expand one already in existence Loans for working capital needs are not limited by statute, but are available only for short periods In addition, the government will guarantee up to 90 percent of the unpaid balance of loans for the acquisition of fixed assets or for working capital, and up to 90 percent of the rental payments required by guaranteed lease FY 78 range $260,000 to $5,200,000, FY 78 Average $1,500,000	Personal guarantees usually will be required in order to ensure an ongoing management commitment over the life of the credit
Federal guarantee up to 75 percent of the estimated construction, startup, and related costs, lender is any individual, partnership, corporation, federal entity or other legal entity formed for the purpose or engaged in the business of lending money	Full repayment is to be made over the lesser of 20 years or a period equal to 90 percent of the expected average useful economic life of the project's major physical assets	The amount of the loan guaranteed, when combined with other funds available to the applicant, shall be sufficient to carry out the project, including adequate contingency funds	The project assets and other collateral or surety as determined by the Secretary to be necessary
To the extent provided in advance in appropriation Acts for the two year period beginning October 1, 1980, there is authorized to be appropriated and transferred $1 45 billion from the Energy Security Reserve established in the Treasury of the United States under P L 96-126 as follows (1) $600 million to the Secretary of Agriculture, (2) $600 million to the Secretary of Energy of which at least $500 million shall be available to the Office of Alcohol Fuels and (3) $250 million to the Secretary of Energy for municipal waste biomass energy	Not specified	(a) Insured loans made by the Secretary of Agriculture may not exceed 90 per centum of the total estimated cost of construction and may not exceed $1 million per project In the event the construction costs exceed the original estimates, the Secretary may make an insured loan for so much of the additional total costs, not to exceed 10 per centum of the initial estimates (b) Any guarantee of a loan made by either Secretary may not exceed 90 per centum of the cost of construction as estimated by the Secretary on the date of the guarantee or commitment to guarantee In the event the construction costs exceed the original estimates, the Secretary may guarantee a loan for up to 60 per centum of the difference between the original and final estimates	Not specified

Figure A-7. Financial Assistance Programs Characteristics (Continued)

APPENDIX B

Regulations and Permits

This section reviews regulations which affect the ethanol production industry. It includes BATF, FDA, transportation regulations and local building codes.

BATF REGULATIONS

The design, construction, and operation of alcohol fuel plants require attention from Federal, State, and local authorities in order to ensure maintenance of the public good. Specific aspects of alcohol fuel production relating to safety and environmental concerns have been addressed in the main body of this document; other more general requirements, such as meeting building codes and zoning regulations, are site-specific and must be complied with, based on the individual case. Again, specific to alcohol fuels, this appendix addresses BATF regulations and other regulations and permits as they concern the potential investor.

In the Energy Tax Act of 1978 (Public Law 95-618), Congress required the Secretary of the Treasury to propose legislative recommendations to facilitate fuel use of distilled spirits. In compliance with this mandate, the BATF submitted a proposal in 1979. Congress incorporated the proposed legislation, with modifications, in Public Law 96-223.

Public Law 96-223 added a new section to Subchapter B of Chapter 51 of the Internal Revenue Code of 1954. New Section 5181, Distilled Spirits for Fuel Use, provides for the establishment of distilled spirits plants solely for the purpose of producing, processing and storing, and using or distributing distilled spirits to be used exclusively for fuel use. The Secretary of the Treasury is directed to expedite all applications, establish a minimum bond, and generally encourage and promote (through regulation or otherwise) the production of alcohol for fuel purposes. The Law gives the Secretary authority to provide by regulation for the waiver of any provision of Chapter 51 (except Section 5181 or any provision requiring the payment of tax). This waiver authority may be exercised with respect to alcohol fuel plants as necessary to carry out the provisions of Section 5181.

Plants which are established for the production of alcohol for purposes other than fuel, such as beverage or other industrial use, must qualify to operate under previously existing provisions of the law. You may wish to contact BATF for further details (see Appendix E).

The law exempts alcohol fuel plants which produce less than 10,000 proof gallons of distilled spirits annually from furnishing a bond to cover the tax liability on the spirits. Also, persons wishing to establish such plants are assured by law that their applications for permits to operate will be acted upon within 60 days after receipt of a completed application. If this limit is exceeded, the application is automatically approved.

Distilled spirits may be withdrawn from alcohol fuel plants free of tax after having been rendered unfit for beverage use. The term "rendered unfit for beverage use" is used to clearly distinguish this requirement for fuel spirits from "denaturation", which is required for spirits withdrawn from other distilled spirits plants.

A principal reason for encouraging production of distilled spirits for fuel use is that these spirits are substituted for scarce fossil fuels. Therefore, in defining distilled spirits for purposes of the alcohol fuel provisions, the law excludes distilled spirits produced from petroleum, natural gas, or coal.

Finally, Public Law 96-223 amended Section 5601 to impose criminal penalties on any person who withdraws, uses, sells, or otherwise disposes of distilled spirits produced under Section 5181 for other than fuel use.

This section addresses those portions of regulations dealing with large alcohol fuel plants (defined by BATF as more than 500,000 proof gallons of spirits per calendar year). We will cover the following permits; bonds; construction, equipment, and security; supervision; accounting for spirits; transfers and withdrawals; and transfers between plants. Our intent is to be informative, but we suggest that the Bureau be contacted for more specific interpretation and guidance where needed.

The following is a brief summary of the regulations now in force:

DEPARTMENT OF THE TREASURY

**Bureau of Alcohol, Tobacco, and Firearms
(27 CFR Part 19)
T.D. ATF-71**

**FUEL USE OF DISTILLED SPIRITS -
IMPLEMENTING A PORTION OF THE CRUDE OIL
WINDFALL PROFITS TAX ACT OF 1980
(PUBLIC LAW 96-223)**

AGENCY: Bureau of Alcohol, Tobacco, and Firearms, Department of the Treasury.

ACTION: Temporary rule (Treasury decision).

SUMMARY: This temporary rule implements Section 232, Alcohol Fuels, in Part III of Title II of the Crude Oil Windfall Profits Tax Act of 1980 (Public Law 96-223). The temporary rule provides for the establishment and operation of distilled spirits plants solely for producing, processing and storing, and using or distributing distilled spirits (ethyl alcohol) exclusively for fuel use. Public Law 96-223 vests the Secretary of the Treasury with authority to simplify and liberalize statutory requirements for such alcohol fuel plants. The BATF will issue final regulations only after careful consideration of the comments received on these temporary regulations.

EFFECTIVE DATE OF TEMPORARY REGULA-TIONS: July 1, 1980.

FOR FURTHER INFORMATION CONTACT:

John V. Jarowski, Alcohol Fuels Coordinator
Bureau of Alcohol, Tobacco, and Firearms
Washington, DC 20226
Telephone 202-566-7626

(Persons interested in applying for a permit should contact their BATF regional office. A list of addresses and toll-free telephone numbers is provided in Appendix E.)

SUPPLEMENTARY INFORMATION: This document contains temporary regulations implementing a portion of the Crude Oil Windfall Profits Tax Act of 1980 (Public Law 96-223). The temporary regulations provided by this document will remain in effect until superseded by final regulations on this subject.

REGULATIONS

1. New Subpart Y of 27 CFR Part 19.

Regulations applicable to alcohol fuel plants have been included in a new subpart, Subpart Y, of Part 19, Distilled Spirits Plants. Except for those portions of Subparts A through X of Part 19 which have been incorporated in the new subpart by reference, alcohol fuel plants are not subject to the provisions applicable to other distilled spirits plants. While this document provides simplified BATF regulations for alcohol fuel plants, persons wishing to establish a plant should check with other Federal (EPA), state, and local authorities as to their specific requirements.

2. Definitions.

Where necessary, terms have been defined as used in the new subpart. In some instances, BATF needed to redefine a term used elsewhere in Part 19. For instance, the term "distilled spirits", as used in Subpart Y, does not include distilled spirits produced from petroleum, natural gas, or coal. In addition, BATF introduces some new terms. For example, the term "alcohol fuel plant" means a distilled spirits plant established under Subpart Y solely for producing, processing and storing, and using or distributing distilled spirits to be used exclusively for fuel use. The term "fuel alcohol" means spirits which have been rendered unfit for beverage use as provided in Subpart Y and are eligible to be withdrawn free of tax exclusively for fuel use.

3. Permits.

Any person wishing to establish an alcohol fuel plant must first make application for and obtain an Alcohol Fuel Producer's Permit. For purposes of the type of application required (as well as for bonding and other requirements), BATF classifies alcohol fuel plants into three categories:

(a) Small plants - produce (including receipts of) not more than 10,000 proof gallons per calendar year;

(b) Medium plants - produce (including receipts of) more than 10,000 but not more than 500,000 proof gallons per calendar year; and,

(c) Large plants - produce (including receipts of) more than 500,000 proof gallons per calendar year.

These classes are based on the tax liabilities incident to the various levels of activity, as well as differences in capital expenditures and attendant technical sophistication required to set up plants of different sizes. By law, small plants are exempt from bonding and special rules apply to their applications.

In determining the level of activity at a plant for purposes of bonding and qualification, BATF considers spirits received from another plant as additional production. With respect to small plants, this approach has no effect on a proprietor who will receive no spirits from other plants. Proprietors of other small plants, however, may wish to distill less than 10,000 proof gallons of spirits per year and receive additional spirits from other plants. Those proprietors may produce and receive up to a total of 10,000 proof gallons per year and still be exempt from bonding. Counting receipts as additional production simplifies the regulations for bonds and qualification of plants.

4. Application Requirements.

Applications are filed on BATF Form 5110.74 furnished by the Bureau (See Figure B-1). For small plants, only the form needs to be filed. For medium and large plants, a bond must be placed and some additional information (to be identified in Section 6) must be provided. Also, the law requires bonds for the medium and large plants. Applications are to be submitted to the regional regulatory administrator of the BATF region in which the plant is located.

Having an application form should avoid confusion as to the information required to be submitted and serve to expedite processing of applications.

Although BATF expects numerous applications to be filed in response to the temporary regulations, BATF anticipates that permits will be issued expeditiously. With respect to the small plants in particular, action well within the 60-day statutory limitation should be possible.

In general, a permit will be issued to any person who files a completed application (together with a bond for medium and large plants). Criteria for denial of applications, and for subsequent suspension or revocation of permits in the case of false or incomplete applications, are set forth in Sections 19.936 and 19.953 of the regulations. Time permitting, BATF may conduct an investigation and on-site inspection prior to issuance of the permit.

5. Changes Affecting Permits.

Permits may not be sold or transferred. If a different person purchases or acquires control of the plant, that person must file a new application. Also, BATF issues permits to cover specified plant premises. Any change in the location or extent of plant premises requires submission of an application to the regional regulatory administrator to amend the permit. Sections 19.945 through 19.953 provide more details on changes affecting permits and administrative and procedural matters relative to permits.

6. Bonds.

Bonds give the government security against possible loss of distilled spirits tax revenue. While tax is not collected on distilled spirits for fuel use lawfully produced and used as fuel, liability for tax, by law, attaches to the spirits on creation. The proprietor generally obtains a bond from a surety or insurance company, by paying a premium based on the penal sum required for the bond.

The law exempts small plants producing (including receipts of) not more than 10,000 proof gallons per year from furnishing bonds. For medium and large plants which are required to file bonds, the penal sum required is based on the tax liability to be covered. There are two sources of this tax liability:

(a) The tax on spirits that will be produced (including spirits rendered unfit for beverage use in the process of production, as provided in Section 19.983); and,

(b) The tax on spirits that will be received by transfer in bond (see Sections 19.996 through 19.999).

In the temporary regulations, the penal sum of the bond is based on the quantity in proof gallons of spirits that will be produced (including receipts) in a calendar year.

The bond for a medium plant producing (including receipts of) between 10,000 and 20,000 proof gallons per year is $2,000. For each increase of 10,000 proof gallons (or fraction thereof) in total annual production (including receipts), the bond penal sum is increased $1,000. The maximum bond for a medium plant is

$50,000 (applicable to plants producing (including receipts) between 490,000 and 500,000 proof gallons per year).

For large plants, the amount of the bond is $50,000 plus an additional $2,000 for each 10,000 proof gallons (or fraction) produced and received in excess of 500,000. The maximum bond for a large plant is $200,000. Any plant producing (including receipts) in excess of 1,240,000 proof gallons of spirits in a calendar year must furnish a bond in the amount of $200,000.

Documentation required is BATF Form 5110.56, as shown in Figure B-2.

7. Premises, Equipment, and Security.

The premises of an alcohol fuel plant include all areas where distilled spirits are produced, processed, stored, used, or held for distribution. BATF requires buildings and equipment used in alcohol fuel plant operations to be constructed and arranged to enable the proprietor to maintain security adequate to deter diversion of the spirits. Regional regulatory administrators can require additional measures if security at a plant is found to be inadequate.

8. Supervision of Operations.

Regional regulatory administrators may assign BATF officers to alcohol fuel plants to supervise operations or to conduct inspections and audits. Authorities of BATF officers to carry out their duties have been incorporated by reference in the temporary regulations. BATF officers have a right of access to plant premises at all times to perform their official duties. The regulations require proprietors to render assistance to BATF officers in performing gauges and examining equipment and containers, as necessary, to determine that all spirits are lawfully accounted for.

9. Accounting for Spirits.

The regulations require proprietors to determine and record the quantities of spirits produced, received, rendered unfit for beverage use, and used or removed from the premises. The records must be adequate to allow BATF officers to verify that the proprietor has lawfully disposed of all spirits. BATF requires no government forms or specific formats for these records. Commercial invoices, books of account, and other proprietary records are sufficient, so long as they show the information required by the regulations.

The regulations also require proprietors to file periodic reports of their operations. Frequency of reporting varies with the size of the plant: annually for small plants, semiannually for medium plants, and quarterly for large

COMMERCIAL SCALE ETHANOL PRODUCTION AND FINANCING

plants. The required form is BATF Form 5110.75, as shown in Figure B-3.

10. Withdrawal of Spirits.

Distilled spirits withdrawn from alcohol fuel plants may be used only for fuel purposes. Before withdrawing spirits from plant premises, the proprietor must render the spirits unfit for beverage use. The only exception to this requirement is for spirits transferred to other plants.

The regulations in Section 19.992 provide that fuel alcohol is considered unfit for beverage use when, for each 100 gallons of spirits, the proprietor has added to the spirits five gallons or more of gasoline, automotive gasoline, kerosene, deodorized kerosene, rubber hydrocarbon solvent, methyl isobutyl ketone, mixed isomers of nitropropane, any combination of the foregoing, or 5 gallons of isopropyl alcohol and 1/8-ounce of denatonium benzoate N.F. (Bitrex). BATF selected these substances based on denatured alcohol and solvent formulations currently provided by regulation or ruling, and on requests and inquiries from persons currently involved in producing alcohol fuels. Proprietors may request authorization to use substitute materials under Section 19.993 of the regulations.

Once rendered unfit for beverage use, fuel alcohol is eligible to be withdrawn free of tax. However, BATF does require proprietors to account for fuel alcohol remaining on plant premises and keep records of all sales and other dispositions of fuel alcohol.

11. Use on Premises.

Proprietors may use spirits for fuel on the premises of an alcohol fuel plant without rendering the spirits unfit for beverage use. The proprietor must record the quantities of spirits so used.

12. Transfers in Bond.

Proprietors may transfer spirits to another alcohol fuel plant. The proprietor need not render the spirits unfit for beverage use prior to transfer. Generally, liability for the tax on the spirits while in transit is the responsibility of the receiving proprietor.

The temporary regulations prescribe procedures to be followed when transferring spirits between alcohol fuel plants. A commercial invoice or shipping document is required to cover each shipment of spirits. These procedures are necessary to guard against diversion of spirits to beverage use.

Spirits also may be transferred between other distilled spirits plants and alcohol fuel plants. For such transfers, procedures and forms are prescribed in Sections 19.506 through 19.510 of the regulations.

The following are the Bureau of Alcohol, Tobacco and Firearms regulations now in force, effective date of temporary regulations is July 1, 1980.

DEPARTMENT OF THE TREASURY

Bureau of Alcohol, Tobacco and Firearms

FUEL USE OF DISTILLED SPIRITS-IMPLEMENTING A PORTION OF THE CRUDE OIL
WINDFALL PROFIT TAX ACT OF 1980
(Public Law 96-223)

AGENCY: Bureau of Alcohol, Tobacco and Firearms, Department of the Treasury

ACTION: Temporary rule (Treasury decision)

SUMMARY: This temporary rule implements Section 232, Alcohol Fuels, in Part III of Title II of the crude Oil Windfall Profit Tax Act of 1980 (P.L. 96-223). The temporary rule provides for the establishment and operation of distilled spirits plants solely for producing, processing and storing, and using or distributing distilled spirits (ethyl alcohol) exclusively for fuel use. P.L. 96-223 vests the Secretary of the Treasury with authority to simplify and liberalize statutory requirements for such alcohol fuel plants. The Bureau of Alcohol, Tobacco and Firearms (ATF) will issue final regulations only after careful consideration of the comments received on these temporary regulations.

FOR FURTHER INFORMATION CONTACT:

John V. Jarowski, Research and Regulations Branch, Bureau of Alcohol, Tobacco and Firearms, Washington, DC 20226, Telephone 202-566-7626.

(Persons interested in applying for a permit should contact their ATF regional office. A list of addresses and toll free telephone numbers is provided in Appendix E.)

SUPPLEMENTARY INFORMATION: This document contains temporary regulations implementing a portion of the Crude Oil Windfall Profit Tax Act of 1980 (Public Law 96-223). The temporary regulations provided by this document will remain in effect until superseded by final regulations on this subject. A notice of proposed rulemaking with respect to final regulations appears elsewhere in this issue of the Federal Register.

Legislative Background

In the Energy Tax Act of 1978 (P.L. 95-618) Congress required the Secretary of the Treasury to propose legislative recommendations to facilitate fuel use of distilled spirits. In compliance with this mandate, the

Bureau submitted a proposal in 1979. Congress incorporated the proposed legislation, with modifications, in P.L. 96-223.

Public Law 96-223 added a new section to Subchapter B of Chapter 51 of the Internal Revenue Code of 1954. New Section 5181, Distilled Spirits for Fuel Use, provides for the establishment of distilled spirits plants solely for the purpose of producing, processing and storing, and using or distributing distilled spirits to be used exclusively for fuel use. The Secretary of the Treasury is directed to expedite all applications, to establish a minimum bond, and to generally encourage and promote (through regulation or otherwise) the production of alcohol for fuel purposes. The law gives the Secretary authority to provide by regulation for the waiver of any provision of Chapter 51 (except section 5181 or any provision requiring the payment of tax). This waiver authority may be exercised with respect to alcohol fuel plants as necessary to carry out the provisions of Section 5181.

The law exempts alcohol fuel plants which produce not more than 10,000 proof gallons of distilled spirits annually from furnishing a bond to cover the tax liability on the spirits. Also, persons wishing to establish such plants are assured by law that their applications for permits to operate will be acted upon within 60 days or less after receipt of a completed application. If this limit is exceeded, the application is automatically approved.

Distilled spirits may be withdrawn from alcohol fuel plants free of tax after having been rendered unfit for beverage use.The term "rendered unfit for beverage use" is used to clearly distinguish this requirement for fuel spirits from "denaturation," which is required for spirits withdrawn from other distilled spirits plants.

A principal reason for encouraging production of distilled spirits for fuel use is that these spirits are substituted for scarce fossil fuels. Therefore, in defining distilled spirits for purposes of the alcohol fuel provisions, the law excludes distilled spirits produced from petroleum, natural gas, or coal.

Finally, P.L. 96-223 amended Section 5601 to impose criminal penalties on any person who withdraws, uses, sells, or otherwise disposes of distilled spirits produced under Section 5181 for other than fuel use.

Regulations

1. **New Subpart Y of 27 CFR Part 19.** Regulations applicable to alcohol fuel plants have been included in a new subpart, Subpart Y, of Part 19, Distilled Spirits Plants. Except for those portions of Subparts A through X of Part 19 which have been incorporated in the new subpart by reference, alcohol fuel plants are not subject to the provisions applicable to other distilled spirits plants. While this document provides simplified ATF regulations for alcohol fuel plants, persons wishing to establish a plant should check with other Federal (EPA), State and local authorities as to their specific requirements.

2. **Definitions.** Where necessary, terms have been defined as used in the new subpart. In some instances, ATF needed to redefine a term used elsewhere in Part 19. For instance, the term "distilled spirits" as used in Subpart Y does not include distilled spirits produced from petroleum, natural gas, or coal. In addition, ATF introduces some new terms. For example, the term "alcohol fuel plant" means a distilled spirits plant established under Subpart Y solely for producing, processing and storing, and using or distributing distilled spirits to be used exclusively for fuel use. The term "fuel alcohol" means spirits which have been rendered unfit for beverage use as provided in Subpart Y and are eligible to be withdrawn free of tax exclusively for fuel use.

3. **Permits.** Any person wishing to establish an alcohol fuel plant must first make application for and obtain an Alcohol Fuel Producer's Permit. For purposes of the type of application required (as well as for bonding and other requirements), ATF classifies alcohol fuel plants into three categories:

(a) Small plants—producing (including receipts of) not more than 10,000 proof gallons per calendar year;

(b) Medium plants—produce (including receipts of) more than 10,000 but not more than 500,000 proof gallons percalendar year; and

(c) Large plants—produce (including receipts of) more than 500,000 proof gallons per calendar year.

These classes are based on the tax liabilities incident to the various levels of activity, as well as differences in capital expenditures and attendant technical sophistication required to set up plants of different sizes. By law, small plants are exempt from bonding and special rules apply to their applications.

In determining the level of activity at a plant for purposes of bonding and qualification, ATF considers spirits received from another plant as additional production. With respect to small plants, this approach has no effect on a proprietor who will receive no spirits from other plants. Proprietors of other small plants, however, may wish to distill less than 10,000 proof gallons of spirits per year and receive additional spirits from other plants. Those proprietors may produce and receive up to a total of 10,000 proof gallons per year and still be exempt from bonding. Counting receipts as additional production simplifies the regulations for bonds and qualification of plants.

4. **Application Requirements.** Applications are filed on a form (ATF Form 5110.74) furnished by the Bureau. For small plants, only the form needs to be filed. For medium and large plants, some additional information must be provided. Also, the law requires bonds for the medium and large plants. Applications are to be submitted to the regional regulatory administrator of the ATF region in which the plant is located.

Having an application form should avoid confusion as to the information required to be submitted and serve to expedite processing of applications. Comments as to the design of the form, the information required, and as to whether the use of form is desirable, are specifically requested. Although ATF expects numerous applications to be filed in response to the temporary regulations, ATF anticipates that permits will be issued expeditiously. With respect to the small plants in particular, action well within the 60 day statutory limitation should be possible.

In general, a permit will be issued to any person who files a completed application (together with a bond for medium and large plants). Criteria for denial of applications, and for subsequent suspension or revocation of permits in the case of false or incomplete applications, are set forth in Sections 19.936 and 19.953 of the regulations. Time permitting, ATF may conduct an investigation and on-site inspection prior to issuance of the permit.

5. **Changes Affecting Permits.** Permits may not be sold or transferred. If a different person purchases or acquires control of the plant, that person must file a new application. Also, ATF issues permits to cover specified plant premises. Any change in the location or extent of plant premises requires submission of an application to the regional regulatory administrator to amend the permit. Sections 19.945 through 19.953 provide more details on changes affecting permits and administrative and procedural matters relative to permits.

6. **Bonds.** Bonds give the Government security against possible loss of distilled spirits tax revenue. While tax is not collected on distilled spirits for fuel use lawfully produced and used as fuel, liability for tax, by law, attaches to the spirits on creation. The proprietor generally obtains a bond from a surety or insurance company, by paying a premium based on the penal sum required for the bond.

The law exempts small plants producing (including receipts of) not more than 10,000 proof gallons per year from furnishing bonds. For medium and large plants which are required to file bonds, the penal sum required is based on the tax liability to be covered. There are two sources of this tax liability:

(a) The tax on spirits that will be produced (including spirits rendered unfit for beverage use in the process of production, as provided in Sec. 19.983); and,

(b) The tax on spirits that will be received by transfer in bond (see Sections 19.996 through 19.999).

In the temporary regulations, the penal sum of the bond is based on the quantity in proof gallons of spirits that will be produced (including receipts) in a calendar year.

The bond for a medium plant producing (including receipts of) between 10,000 and 20,000 proof gallons per year is $2,000. For each increase of 10,000 proof gallons (or fraction thereof) in total annual production (including receipts), the bond penal sum is increased $1,000. The maximum bond for a medium plant is $50,000 (applicable to plants producing (including receipts) between 490,000 and 500,000 proof gallons per year).

For large plants, the amount of the bond is $50,000 plus an additional $2,000 for each 10,000 proof gallons (or fraction) produced and received in excess of 500,000. The maximum bond for a large plant is $200,000. Any plant producing (including receipts) in excess of 1,240,000 proof gallons of spirits in a calendar year must furnish a bond in the amount of $200,000.

7. **Premises, Equipment and Security.** The premises of an alcohol fuel plant includes all areas where distilled spirits are produced, processed, stored, used or held for distribution. ATF requires buildings and equipment used in alcohol fuel plant operations to be constructed and arranged to enable the proprietor to maintain security adequate to deter diversion of the spirits. Regional regulatory administrators can require additional measures if security at a plant is found to be inadequate.

8. **Supervision of Operations.** Regional regulatory administrators may assign ATF officers to alcohol fuel plants to supervise operations or to conduct inspections and audits. Authorities of ATF officers to carry out their duties have been incorporated by reference in the temporary regulations. ATF officers have a right of access to plant premises at all times to perform their official duties. The regulations require proprietors to render assistance to ATF officers in performing gauges and examining equipment and containers, as necessary to determine that all spirits are lawfully accounted for.

9. **Accounting for Spirits.** The regulations require proprietors to determine and record the quantities of spirits produced, received, and rendered unfit for beverage use, and used or removed from the premises. The records must be adequate to allow ATF officers to verify that the proprietor has disposed of all spirits lawfully. ATF requires no Government forms or specific formats for these records. Commercial invoices, books of account, and other proprietary records are sufficient, so long as they show the information required by the regulations.

The regulations also require proprietors to file periodic reports of their operations. Frequency of reporting varies with the size of the plant: annually for small plants, semiannually for medium plants, and quarterly for large plants.

10. **Withdrawal of Spirits.** Distilled spirits withdrawn from alcohol fuel plants may be used only for fuel purposes. Before withdrawing spirits from plant premises, the proprietor must render the spirits unfit for beverage use. The only exception to this requirement is for spirits transferred to other plants.

The regulations in Sec. 19.992 provide that fuel alcohol is considered unfit for beverage use when for each 100 gallons of spirits, the proprietor has added to the spirits five gallons or more of gasoline, automotive gasoline, kerosene, deodorized kerosene, rubber hydrocarbon solvent, methyl isobutyl ketone, mixed isomers of nitropropane, any combination of the foregoing, or 5 gallons of isopropyl alcohol and 1/8 ounce of denatonium benzoate N.F. (Bitrex). ATF selected these substances based on denatured alcohol and solvent formulations currently provided by regulation or ruling, and on requests and inquirites from persons currently involved in producing alcohol fuels. Proprietors may request authorization to use substitute materials under Sec. 19.993 of the regulations.

Once rendered unfit for beverage use, fuel alcohol is eligible to be withdrawn free of tax. However, ATF does require proprietors to account for fuel alcohol remaining on plant premises and to keep records of all sales and other dispositions of fuel alcohol.

11. **Use on Premises.** Proprietors may use spirits for fuel on the premises of an alcohol fuel plant without rendering the spirits unfit for beverage use. The proprietor must record the quantities of spirits so used.

12. **Transfers in Bond.** Proprietors may transfer spirits to another alcohol fuel plant. The proprietor need not render the spirits unfit for beverage use prior to transfer. Generally, liability for the tax on the spirits while in transit is the responsibility of the receiving proprietor.

The temporary regulations prescribe procedures to be followed when transferring spirits between alcohol fuel plants. A commercial invoice or shipping document is required to cover each shipment of spirits. These procedures are necessary to guard against diversion of spirits to beverage use.

Spirits also may be transferred between other distilled spirits plants and alcohol fuel plants. For such transfers, procedures and forms are prescribed in Sections 19.506 through 19.510 of the regulations.

Issues for Comment

While all comments on these regulations are welcome and will receive careful consideration before final regulations are issued, ATF specifically requests comments on the following two issues.

1. **Location of Plants in dwellings.** Under existing regulations in 27 CFR Part 19, plants may not be located in any dwelling house, in any yard, shed, or enclosure connected with a dwelling house, or in certain other areas. Since ATF has not incorporated these restrictions in the temporary regulations, they do not apply to alcohol fuel plants.

As to yards, sheds, or enclosures, ATF feels the existing restrictions would place undue constraints on farm production of alcohol fuel. For urban and suburban areas, ATF feels that local zoning ordinances, fire codes, and similar provisions will preclude establishment of plants at inappropriate locations. The alcohol fuel producer's permit does not relieve the proprietor of the obligations imposed by State or local law.

However, ATF will consider prohibiting the establishment of plants in residential dwelling units. Accordingly, comments are requested as to whether such a prohibition is necessary or desirable.

2. **Specifications for fuel alcohol and materials authorized for use in rendering spirits unfit for beverage use.** In contrast to existing denatured alcohol regulations in 27 CFR Parts 211 and 212, the temperary regulations do not require fuel alcohol to be of any particular proof. In addition, proprietors have the option to choose among a number of materials or combinations thereof to use in rendering spirits unfit. Proprietors will decide both on the proof of the spirits and on what materials to use to render the spirits unfit, based on cost factors and on what works effectively for the particular fuel uses selected by themselves or their customers. The regulations do point out, however, that if gasoline is to be added to the spirits, unleaded gasoline may be required if the fuel alcohol will be used in certain engines.

For some uses higher proof spirits (anhydrous or nearly so) are necessary. For other uses, lower proof spirits may be adequate.

ATF gives the proprietor options in this area because, beyond the blended motor fuels (e.g., gasohol) which are coming into widespread use, there is interest in using alcohol (once rendered unfit) for fuel purposes without blending with other fuels. ATF recognizes, however, that there may be a demand for tighter controls. For this reason, ATF will consider modifying the requirements for spirits rendered unfit, including specifying minimum proof levels for fuel alcohol. Accordingly, ATF specifically requests comments on the following questions:

(a) Are minimum proof standards necessary or desirable for fuel alcohol?

(b) Should the list of materials authorized for rendering spirits unfit for beverage use be broadened to include other substances, or sould some materials or options permitted under the temporary regulations be deleted?

(c) Are more detailed specifications for rendering spirits unfit necessary? If adopted, should these specifications be geared to the end use of the fuel alcohol?

Experimental Plants

1. **General.** Many persons have obtained permits to operate experimental distilled spirits plants (Section 5312) to produce distilled spirits for fuel purposes. The majority of these operations would qualify as small alcohol fuel plants under the temporary regulations. The authorized operations for experimental plants are generally more limited than under these temporary regulations, particularly as to removal or distribution of any spirits produced. The proprietor must furnish a bond. Also, experimental distilled spirits plant permits are issued for a limited time, usually two years. In contrast to experimental distilled spirits plants, alcohol fuel plants:

(a) Hold a permit which generally continues in effect indefinitely with no renewal required:

(b) Are authorized to sell or distribute the fuel alcohol they produce;

(c) Are exempt from bonding of annual production (including receipts) totals not more than 10,000 proof gallons; and,

(d) May transfer spirits between plants. Given the advantages of operating as an alcohol fuel plant, ATF

has developed a simplified procedure to convert outstanding experimental distilled spirits plant permits to alcohol fuel producer's permits, on the assumption that most proprietors will elect conversion. The proprietor of an experimental plant does not have to file an application to have the permit converted. With respect to any bond that was furnished, ATF regional offices will advise proprietors how to cancel surety bonds or secure refunds of cash or securities deposited in lieu of obtaining a surety bond.

2. **Status of Existing Experimental Plants.** Effective July 1, 1980, any person holding a permit to operate an experimental distilled spirits plant for fuel purposes will be considered as authorized to operate under these regulations. ATF regional offices are notifying all affected proprietors. Before expiration of the outstanding permit, a new alcohol fuel producer's permit will be issued to each such proprietor. Applications which are currently pending to establish experimental distilled spirits plants for fuel purposes will be treated as applications for alcohol fuel producer's permits. Proprietors who do not wish to be converted to alcohol fuel plant status should file a written request with the regional regulatory administrator.

Addresses of Regional Regulatory Administrators

Any person who wishes to apply for a permit under the regulations in this document may obtain the required forms and further information from the regional regulatory administrator of the region in which the plant will be located. A list of addresses and toll-free telephone numbers is provided in Appendix E.

Effective Date

The Energy Tax Act of 1978 (Pub. L. 95-618, 92 Stat. 3174) directed the Secretary to expedite, to the maximum extent possible, the applications of persons desiring to produce distilled sirits for fuel use, and to suggest legislative amendments which could reduce the amount of regulation to which such fuel producers would be subject. In furtherance of the first mandate, existing waiver authority under 26 U.S.C. 5312 was used to facilitate establishment of alcohol fuel plants on an experimental basis. The proposed legislative amendments have not been enacted as part of Pub. L. 96-223.

Pub. L. 96-223 gives the Secretary broad authority to waive or reduce existing regulatory requirements for plants which will produce distilled spirits exclusively for fuel use, such as, by allowing simplified application and recordkeeping procedures, and by providing reduced control and bonding requirements. In addition, an expedited application procedure (with no bond) is provided for small producers. For such producers, the law requires action on the issuance of a permit within sixty days of submission of a completed application. Pub. L. 96-223 is effective July 1, 1980.

Immediate guidance is necessary for the affected parties, both potential new applicants to produce distilled spirits for fuel purposes and persons currently operating as proprietors of experimental distilled spirits plants, to reap the benefits of the new law upon its becoming effective.

The issuance of this Treasury decision with notice and public procedure under 5 U.S.C. 553(d) is impracticable and not in the public interest, because absence of these regulations would create a serious delay in the production of alcohol fuel, inconsistent with the intent of the Energy Tax Act of 1978 (Pub. L. 95-618, 92 Stat. 3174) and the Crude Oil Windfall Profit Tax Act of 1980 (Pub. L. 96-223, 94 Stat. 229) to encourage the expeditious production of alcohol fuels. Immediate action is necessary to avoid delay and facilitate production of alcohol fuel, in implementing the Crude Oil Windfall Profit Tax Act of 1980, which is effective July 1, 1980. Accordingly, this Treasury decision becomes effective on July 1, 1980.

Authority and Issuance

These regulations are issued under the authority contained in 26 U.S.C. 5181 (94 Stat. 278) and 26 U.S.C. 7805 (68A Stat. 917, as amended).

Accordingly, Title 27 Code of Federal Regulations is amended as follows:

PART 19—DISTILLED SPIRITS PLANTS

Paragraph 1. The table of sections is amended to reflect the addition of §19.63a immediately following §19.63 and the addition of Subpart Y—Distilled Spirits for Fuel Use. As amended, the table of sections reads as follows:

Subpart D—Administrative and Miscellaneous Provisions

§19.63a Alcohol fuel plants.

Subpart Y—Distilled Spirits for Fuel Use

Permits

Changes Affecting Permits

Permanent Discontinuance of Business

Suspension or Revocation of Permits

Bonds

Construction, Equipment and Security

Supervision

Accounting for Spirits

COMMERCIAL SCALE ETHANOL PRODUCTION AND FINANCING

Paragraph 2. A new §19.63a is added immediately following §19.63. The new section incorporates authority to waive provisions of law and regulations with respect to alcohol fuel plants under the authoritites of the Director in Part 19. As added, §19.63a reads as follows:

§19.63a Alcohol fuel plants

Under the provisions of Subpart Y of this part, distilled spirits plants may be established solely for producing, processing and storing, and using distributing distilled spirits to be used exclusively for fuel use. To the extent the Director finds it necessary to carry out the provisions of 26 U.S.C. 5181, the Director may waive any provision of 26 U.S.C. Chapter 51 or this part (other than 26 U.S.C. 5181, this section, Subpart Y, or any provisions requiring the payment of tax).

(§232, Pub. L. 96-223, 94 Stat. 278 (26 U S C 5181))

Paragraph 3. §19.505 is amended to provide for the transfer of spirits between plants qualified under 26 U.S.C. 5171 and alcohol fuellants. The statutes at large citation is also amended. As amended, §19.505 reads as follows:

§19.505 Authorized transfers.

(a) *Spirits.* Pursuant to approval of an application as provided for in §19.506, bulk spirits (including denatured spirits) may be transferred in bond between bonded premises in bulk conveyances, or by pipeline, or in bulk containers into which spirits may be filled on bonded premises. However, spirits (including denatured spirits) produced from petroleum, natural gas, or coal, may not be transferred

to alcohol fuel plants. Spirits transferred in bond from alcohol fuel plants to plants qualified under 26 U.S.C. 5171 shall be accounted for separately by the consignee proprietor, and shall not subsequently be withdrawn, used, sold or otherwise disposed of for other than fuel use.

(b) * * *

(§201, Pub L 85-859, 72 Stat 1362, as amended, 1380, as amended (26 U S C 5212, 5362); §232, Pub L 96-223, 94 Stat 278 (26 U S C 5181))

Paragraph 4. A new Subpart Y—Distilled Spirits for Fuel Use, is added to provide regulations for alcohol fuel plants. As added, Subpart Y reads as follows:

Subpart Y—Distilled Spirits for Fuel Use

§19.931 Scope of subpart.

This subpart relates to the qualification and operation of distilled spirits plants established solely for producing, processing and storing, and using or distributing distilled spirits to be used exclusively for fuel use. Except where incorporated in this subpart by reference, the provisions of Subparts A through X of this part do not apply to alcohol fuel plants (see §19.63a).

(§232, Pub L 96-223, 94 Stat. 278 (26 U.S C 5181))

§19.932 Taxes

(a) *Distilled spirits tax.* Distilled spirits may be withdrawn free of tax from the premises of an alcohol fuel plant exclusively for fuel use in accordance with this subpart. Payment of tax will be required in the case of diversion of spirits to beverage use or other unauthorized dispositions. The provisions of Subpart C of this part are applicable to distilled spirits for fuel use as follows:

(1) Imposition of tax liability (§§19.21 through 19.25);

(2) Assessment of tax (§§19.31 and 19.32); and,

(3) Claims for tax (§§19.41 and 19.44).

(b) *Still tax.* A commodity tax is imposed by 26 U.S.C. 5101 on the manufacturer for each still or condenser for distilling made by him. Manufacturers of stills are subject to a special occupational tax. However, a proprietor manufacturing stills or condensers exclusively for use in his plant or plants is exempt from these taxes. In addition, proprietors of alcohol fuel plants are exempt from the requirement of 26 U.S.C. 5105 to file an application and obtain a permit before setting up distilling apparatus. Provisions relating to stills are contained in 27 CFR Part 196.

(§201, Pub. L 85-859, 72 Stat. 1314, as amended, 1339 (26 U S C 5001, 5101, 5103); §232, Pub L 96-223, 94 Stat 278 (26 U S C 5181))

§19.933 Status of existing experimental distilled spirits plants

Notwithstanding any other provisions of this subpart, effective July 1, 1980, the prior application and permit of existing experimental plants for alcohol fuel production under Section 5312 will be considered as approved applications to operate under the provisions of 26 U.S.C. 5181 as alcohol fuel plants. Such existing plants may continue to operate and the operations shall be conducted pursuant to the provisions of this subpart. A new permanent permit as an alcohol fuel plant will be issued in lieu of, and prior to the expiration date of, the existing permit as an experimental plant. However, persons who wish to retain their permits under 26 U.S.C. 5312 as experimental distilled spirits plants, instead of converting to alcohol fuel plant status, may do so by filing a written request with the regional regulatory administrator.

§19.934 Meaning of terms

When used in this subpart, and in forms prescribed under this subpart, terms shall have the meaning given in this section. Words in the plural form include the singular and vice versa, and words indicating the masculine gender include the feminine. The terms "includes" and "including" do not exclude things not enumerated which are in the same general class.

Alcohol fuel plant or plant. An establishment qualified under this subpart solely for producing, processing and storing, and using or distributing distilled spirits to be used exclusively for fuel use.

Alcohol fuel producer's permit. The document issued pursuant to 26 U.S.C. 5181 authorizing the person named therein to engage in business as an alcohol fuel plant.

ATF officer. An officer or employee of the Bureau of Alcohol, Tobacco and Firearms (ATF) authorized to perform any function relating to the administration or enforcement of this subpart.

Bonded Premises. The premises of an alcohol fuel plant where distilled spirits are produced, processed and stored, and used or distributed. Premises of small alcohol fuel plants, which are exempt from bonding under §19.938, shall be treated as bonded premises for purposes of this subpart.

CFR. The Code of Federal Regulations.

Director. The Director, Bureau of Alcohol, Tobacco and Firearms, the Department of the Treasury, Washington, D.C.

Fuel alcohol. Distilled spirits which have been rendered unfit for beverage use at an alcohol fuel plant as provided in this subpart.

Gallon or wine gallon. The liquid measure equivalent to the volume of 231 cubic inches.

Person. An individual, trust, estate, partnership, association, company or corporation.

Proof. The ethyl alcohol content of a liquid at 60 degrees Fahrenheit, stated as twice the percent of ethyl alcohol by volume.

Proof gallon. A gallon of liquid at 60 degrees Fahrenheit which contains 50 percent by volume of ethyl alcohol having a specific gravity of 0.7939 at 60 degrees Fahrenheit referred to water at 60 degrees Fahrenheit as unity, or the alcoholic equivalent thereof.

Proprietor. The person qualified under this subpart to operate the alcohol fuel plant.

Region. A Bureau of Alcohol, Tobacco and Firearms region.

Regional regulatory administrator. The principal regional official responsible for administering regulations in this subpart.

Secretary. The Secretary of the Treasury or his delegate.

Spirits or distilled spirits. That substance known as ethyl alcohol, ethanol, or spirits of wine in any form (including all dilutions and mixtures thereof by whatever process produced), but not fuel alcohol unless specifically stated. For purposes of this subpart, the term does not include spirits produced from petroleum, natural gas, or coal.

This chapter. Title 27, Code of Federal Regulations, Chapter I (27 CFR Chapter I).

Transfer on bond. The transfer of spirits between alcohol fuel plants or between a distilled spirits plant qualified under 26 U.S.C. 5171 and an alcohol fuel plant.

Type of plant. The following three types of alcohol fuel plants are recognized in this subpart:

(a) *Small plant.* An alcohol fuel plant which produces –including receipts) not more than 10,000 proof gallons of spirits per calendar year.

(b) *Medium plant.* An alcohol fuel plant which produces (including receipts) more than 10,000 and not more than 500,000 proof gallons of spirits per calendar year.

(c) *Large plant.* An alcohol fuel plant which produces (including receipts) more than 500,000 proof gallons of spirits per calendar year.

U.S.C. The United States Code

Permits

§19.935 Application for permit required.

Any person wishing to establish an alcohol fuel plant shall first make application for and obtain an alcohol fuel producer's permit. Alcohol fuel producers permits are continuing. The permit continues in effect unless automatically terminated under §19.945, suspended or revoked as provided in §19.953, or voluntarily surrendered.

(§232, Pub. L. 96-223, 94 Stat. 278 (26 U.S.C. 5181))

§19.936 Criteria for issuance of permit.

In general, an alcohol fuel producer's permit will be issued to any person who completes the required application for permit and who furnishes the required bond (if any). However, the regional regulatory administrator may institute proceedings for the denial of the application, if the regional regulatory administrator determines that:

Persons wishing to establish a small plant shall apply for a permit as provided in this section. Except as provided in paragraph (d) of this section, operations may not be commenced until the permit has been issued.

(a) *The applicant (including, in the case of a corporation, any officer, director, or principal stockholder, and in the case of a partnership, a partner) is, by reason of business experience, financial standing, or trade connections, not likely to maintain operations in compliance with 26 U.S.C. Chapter 51, or regulations issued thereunder; or*

(b) *The applicant has failed to disclose any material information required, or has made any false statement, as to any material fact, in connection with the application; or*

(c) *The premises on which the applicant proposes to conduct the operations are not adequate to protect the revenue. The procedures applicable to denial of applications are set forth in 27 CFR Part 200.*

(§201, Pub. L. 85-859, 72 Stat. 1370, as amended (26 U.S.C. 5271); §232, Pub. L. 96-223, 94 Stat. 278 (26 U.S.C. 5181))

§19.937 Small plants.

Persons wishing to establish a small plant shall apply for a permit as provided in this section. Except as provided in paragraph (d) of this section, operations may not be commenced until the permit has been issued.

(a) *Application for permit.* The application (ATF Form 5110.74) shall be submitted to the regional regulatory administrator and shall set forth the following information:

(1) Name and mailing address of the applicant, and the location of the alcohol fuel plant if not apparent from the mailing address;

(2) A diagram of the plant premises and a statement as to the ownership of the premises (if the premises are not owned by the proprietor, the owner's consent to access by ATF officers must be furnished);

(3) A description of all stills and a statement of their maximum capacity;

(4) The materials from which spirits will be produced;

(5) A description of the security measures to be used to protect premises, buildings, and equipment where spirits are produced, processed, and stored; and,

(6) A statement as to the environmental impact of the proposed operation.

(b) *Receipt by the regional regulatory administrator*

(1) Notice of receipt—Within 15 days of receipt of the application, the regional regulatory administrator shall send a written notice of receipt to the applicant. The notice will include a statement as to whether the application meets the requirements of paragraph (a). If the application does not meet those requirements, the application will be returned and a new 15-day period will commence upon receipt by the regional regulatory administrator of the amended or corrected application.

(2) Failure to give notice—If the required notice of receipt is not sent, and the applicant has a receipt indicating that the regional regulatory administrator has received the application, the 45-day period provided for in paragraphs (c) and (d) will commence on the fifteenth day after the date the regional regulatory administrator received the application.

(3) Limitation—The provisions of subparagraphs (1) and (2) will apply only to the first application submitted with respect to any one small plant in any calendar quarter. However, an amended or corrected first application will not be treated as a separate application.

(c) *Determination by the regional regulatory administrator.* Within 45 days from the date the regional regulatory administrator sent the applicant a notice of receipt of a completed application, the regional regulatory administrator shall either (1) issue the permit, or (2) give notice in writing to the applicant, stating in detail the reason that a permit will not be issued. Denial of an application will not prejudice any further application for a permit made by the same applicant.

(d) *Presumption of approval.* If, within 45 days from the date of the notice to the applicant of receipt of a completed application, the regional regulatory administrator has not notified the applicant of issuance of the permit or denial of the application, the application shall be deemed to have been approved and the applicant may proceed as if a permit had been issued.

(§232, Pub. L. 96-223, 94 Stat. 278 (26 U.S.C. 5181))

§19.938 Waiver of bond requirement for small plants.

No bond is required for small plants.

(§232, Pub. L. 96-223, 94 Stat. 278 (26 U.S.C. 5181))

§19.939 Medium plants.

Any person wishing to establish a medium plant shall make application for and obtain an alcohol fuel producer's permit and furnish a bond as provided in this section. Operations may not be commenced until the application has been approved and the permit issued.

(a) Application for permit. The application (ATF Form 5110.74) shall be submitted to the regional regulatory administrator and shall set forth the following information:

(1) The information required by §19.937(a);

(2) Statement of maximum total proof gallons of spirits that will be produced and received during a calendar year;

(3) Information identifying the principal persons involved in the business and a statement as to whether the applicant or any such person has ever been convicted of a felony or misdemeanor under Federal or State law; and,

(4) Statement of the amount of funds invested in the business and the source of those funds.

(b) *Bond required.* A bond of sufficient penal sum, as prescribed in 19.958, must be submitted and approved before a permit may be issued.

(§232, Pub. L. 96-223, 94 Stat. 278 (26 U.S.C. 5181))

§19.940 Large Plants.

Any person wishing to establish a large plant shall make application for and obtain an alcohol fuel producer's permit and furnish a bond as provided in this section. Operations may not be commenced until the application has been approved and the permit issued.

(a) *Application for permit.* The application (ATF Form 5110.74) shall be submitted to the regional regulatory administrator and shall set forth the following information:

(1) The information required by §19.937(a);

(2) Statement of the maximum proof gallons of spirits that will be produced and received during a calendar year (not required if the bond is in the maximum sum);

(3) Information identifying the principal persons involved in the business and a statement as to whether the applicant or any such person has ever been convicted of a felony or misdemeanor under Federal or State law;

(4) Statement of the amount of funds invested in the business and the source of those funds; and,

(5) Statement of the type of business organization and of the persons interested in the business, supported by the items of information listed in §19.941.

(b) *Bond required.* A bond of sufficient penal sum, as prescribed in §19.958, must be submitted and approved before a permit may be issued.

(§232, Pub. L. 96-223, 94 Stat. 278 (26 U.S.C. 5181))

§19.941 Organizational documents.

The supporting information required by paragraph (a)(5) of §19.940, includes, as applicable, copies of—

(a) *Corporate documents.*

(1) Corporate charter or certificate of corporate existence or incorporation.

(2) List of directors and officers, showing their names and addresses. However, do not list officers who have no responsibilities in connection with the operation of the alcohol fuel plant.

(3) Certified extracts or digests of minutes of meetings of board of directors, authorizing certain individuals to sign for the corporation.

(4) Statement showing the number of shares of each class of stock or other evidence of ownership, authorized and outstanding, and the voting rights of the respective owners or holders.

(b) *Statement of interest.*

(1) Names and addresses of the 10 persons having the largest ownership or other interest in each of the classes of stock in the corporation, or other legal entity, and the nature and amount of the stockholding or other interest of each, whether the interest appears in the name of the interested party or in the name of another for him. If a corporation is wholly owned or controlled by another corporation, those persons of the parent corporation who meet the above standards are considered to be the persons interested in the business of the subsidiary, and the names thereof need be furnished only upon request of the regional regulatory administrator.

(2) In the case of an individual owner or partnership, the name and address of each person interested in the plant, whether the interest appears in the name of the interested party or in the name of another for that person.

(c) *Availability of additional documents.* The originals of documents required to be submitted under this section and additional items required under §19.942 such as the articles of incorporation, bylaws, State certificate authorizing operations, or articles of partnership or association (in the case of a partnership where required by State law) shall be made available to any ATF officer upon request.

(§201, Pub. L. 85-859, 72 Stat. 1349, as amended, 1370, as amended (26 U.S.C. 5172, 5271); §232, Pub. L. 96-223, 94 Stat. 278 (26 U.S.C. 5181))

§19.942 Information already on file and supplemental information

If any of the information required by §§19.937 through 19.941 is on file with the regional regulatory administrator, that information, if accurate and complete, may be incorporated by reference and made a part of the application. When required by the regional regulatory administrator, the applicant shall furnish as a part of the application for permit, additional information as may be necessary to determine whether the application should be approved.

(§232, Pub. L. 96-223, 94 Stat. 278, (26 U.S.C. 5181))

Changes Affecting Permits

§19.945 Automatic termination of permits.

(a) *Permits not transferable.* Permits issues under this subpart shall not be transferred. In the event of the lease, sale, or other transfer of such a permit, or of the authorized operations, the permit automatically terminates.

(b) *Corporations.* In the case of a corporation holding a permit under this subpart, if actual or legal control of the permittee corporation changes, directly or indirectly, whether by reason of change in stock ownership or control (in the permittee corporation or in any other corporation), by operation of law, or in any other manner, the permittee shall, within 10 days of such change, give written notice, executed under the penalties of perjury, to the regional regulatory administrator; the permit may remain in effect until the expiration of 30 days after the change, whereupon the permit will automatically terminate. However, if operations are to be continued after the change in control, and an application for a new permit is filed within 30 days of the change, then the outstanding permit may remain in effect until final action is taken on the new application. When final action is taken on the application, the outstanding permit automatically terminates.

(§201, Pub. L. 85-859, 72 Stat. 1370, as amended (26 U.S.C. 5271))

§19.946 Change in type of alcohol fuel plant.

(a) *Small plants.* If the proprietor of a small plant wishes to increase production (including receipts) to a level in excess of 10,000 proof gallons of spirits per calendar year, the proprietor shall first furnish a bond and obtain an amended permit by filing application under §§19.939 or 19.940, as applicable. Information filed with the original application for permit need not be resubmitted, but may be incorporated by reference in the new application.

(b) *Medium plants.* Where the proprietor of a medium plant intends to increase production (including receipts) above 500,000 proof gallons of spirits per calendar year, the proprietor shall first obtain an amended permit by filing an application under §19.940. A new or strengthening bond may be required (see §19.957(a)). Information already on file may be incorporated by reference in the new application.

(c) *Curtailment of activities.* Proprietors of large or medium plants who have curtailed operations to a level where they are eligible to be requalified as medium or small plants may, on approval of a letter of application by the regional regulatory administrator, be relieved from the additional requirements incident to their original qualification. In the case of a change to small plant status, termination of the bond and relief of the surety from further liability shall be as provided in Subpart H of this part.

(§201, Pub. L. 85-859, 72 Stat. 1370, as amended (26 U.S.C. 5271); §232, Pub. L. 96-223, 94 Stat. 278 (26 U.S.C. 5181))

§19.947 Change in name of proprietor.

Where there is to be a change in the individual, firm, or corporate name, the proprietor shall, within 30 days of the change, file an application to amend the permit; a new bond or consent of surety is not required.

(§101, Pub. L. 85-859, Stat. 1349, as amended, 1370, as amended (26 U.S.C. 5172, 5271); §232, Pub. L. 96-223, 94 Stat. 278 (26 U.S.C. 5181))

§19.948 Changes in officers, directors, or principal persons

Where there is any change in the list of officers, directors, or principal persons, furnished under the provisions of §§19.939, 19.940, or 19.941, the proprietor shall submit, within 30 days of any such change, a notice in letter form including the new list of officers and a statement of the changes reflected in such list.

(§232, Pub. L. 96-233, 94 Stat. 278 (26 U.S.C. 5181))

§19.949 Change in proprietorship.

(a) *General.* If there is a change in the proprietorship of a plant qualified under this part, the outgoing proprietor shall comply with the requirements of §19.952 and the successor shall, before commencing operations, apply for and obtain a permit and file the required bond (if any) in the same manner as a person qualifying as the proprietor of a new plant.

(b) *Fiduciary.* A successor to the proprietorship of a plant who is an administrator, executor, receiver, trustee, assignee or other fiduciary, shall comply with the applicable provisions of §19.186(b).

(§201. Pub. L. 85-859, 72 Stat. 1349, as amended (26 U.S.C. 5172))

§19.950 Continuing partnerships.

If under the laws of the particular State, the partnership is not terminated on death or insolvency of a partner, but continues until the winding up of the partnership affairs is completed, and the surviving partner has the exclusive right to the control and possession of the partnership assets for the purpose of liquidation and settlement, the surviving partner may continue to operate the plant under the prior qualification of the partnership. However, in the case of a large or medium plant, a consent of surety must be filed, wherein the surety and the surviving partner agree to remain liable on the bond. If the surviving partner acquires the business on completion of the settlement of the partnership, he shall qualify in his own name from the date of acquisition, as provided in §19.949(a). The rule set forth in this section shall also apply where there is more than one surviving partner.

(§201, Pub. L. 85-859, 72 Stat. 1349, as amended (26 U.S.C. 5172); §232, Pub. L. 96-223, 94 Stat. 278 (26 U.S.C. 5181))

§19.951 Change in location.

Where there is a change in the location of the plant or of the area included within the plant premises, the proprietor shall file an application to amend the permit and, if a bond is required, either a new bond or a consent of surety on ATF Form 1533 (5000.18). Operation of the plant may not be commenced at the new location prior to issuance of the amended permit.

(§201, Pub. L. 85-859, 72 Stat. 1349, as amended, 1370, as amended (26 U.S.C. 5172, 5271); §805(c), Pub. L. 96-39, 93 Stat. 276 (26 U.S.C. 5173); §232, Pub. L. 96-223, 94 Stat. 278 (26 U.S.C. 5181))

Permanent Discontinuance of Business

§19.952 Notice of permanent discontinuance.

A proprietor who permanently discontinues operations as an alcohol fuel plant shall, after completion of the operations, file a letter head notice with the regional regulatory administrator. The notice shall be accompanied (a) by the alcohol fuel producer's permit, and by the proprietor's request that such permit be canceled; (b) by a written statement disclosing, as applicable, whether (1) all spirits (including fuel alcohol) have been lawfully disposed of, and (2) any spirits are in transit to the premises; and (c) by a report covering the discontinued operations (the report shall be marked "Final Report").

(§201, Pub. L. 85-859, 72 Stat. 1349, as amended, 1370, as amended (26 U.S.C. 5172, 5271); §232, Pub. L. 96-223, 94 Stat. 278 (26 U.S.C. 5181))

Suspension or Revocation of Permits

§19.953 Suspension or revocation

Whenever the regional regulatory administrator has reason to believe that any person holding an alcohol fuel producer's permit—

(a) *Has not in good faith complied with the applicable provisions of 26 U.S.C. Chapter 51, or regulations issued thereunder, or*

(b) *Has violated conditions of the permit; or*

(c) *Has made any false statement as to any material fact in the application therefore; or*

COMMERCIAL SCALE ETHANOL PRODUCTION AND FINANCING

(d) *Has failed to disclose any material information required to be furnished; or*

(e) *Has violated or conspired to violate any law of the United States relating to intoxicating liquor or has been convicted of any offense under Title 16, U.S.C. punishable as a felony or of any conspiracy to commit such offense: or*

(f) *Has not engaged in any of the operations authorized by the permit for a period of more than 2 years; the regional regulatory administrator may institute proceedings for the revocation or suspension of the permit in accordance with the procedures set forth in 27 CFR Part 200.*

(§201, Pub. L. 85-859, 72 Stat. 1370, as amended (26 U.S.C. 5271))

Bonds

§19.956 Bonds.

An operations bond is required for medium and large plants. Surety bonds may be given only with corporate sureties holding certificates of authority from, and subject to the limitations prescribed by, the Secretary as set forth in the current revision of Treasury Department Circular 570. However, in lieu of corporate surety the proprietor may pledge and deposit as surety for his bond, securities which are transferable and are guaranteed as to both interest and principal by the United States, in accordance with the provisions of 31 CFR Part 225. The regional regulatory administrator will not release such securities until liability under the bond for which they were pledged has been terminated.

(§805(c), Pub. L. 96-39,93 Stat. 276 (26 U.S.C. 5173); CH. 390, Pub. L. 80-280, 61 Stat. 648, 650 (6 U.S.C. 6, 7, 15); §232, Pub. L. 96-223, 94 Stat. 278 (26 U.S.C. 5181))

§19.957 Amount of bond.

The penal sum of the bond is based on the total quantity of distilled spirits to be produced (including receipts) during a calendar year. If the level of production and/or receipts at the plant is to be increased, and the bond shall be obtained.

(a) *Medium plants.* A medium plant which will produce (including receipts of) between 10,000 and 20,000 proof gallons of spirits per year requires a bond in the amount of $2,000. For each additional 10,000 proof gallons (or fraction thereof), the bond amount is increased $1,000. The maximum bond for a medium plant is $50,000.

(b) *Large plants.* The minimum bond for a large plant is $52,000 more than 500,000, but not more than 510,000 proof gallons annual production (including receipts)). For each additional 10,000 (or fraction) proof gallons, the amount of the bond is increased

$2,000. The maximum bond for a large plant is $200,000 (more than 1,240,000 proof gallons).

(§805(c, Pub. L. 96-39, 93 Stat. 276 (26 U.S.C. 5173); §232, Pub. L. 96-233, 94 Stat. 278 (26 U.S.C. 5181))

§19.958 Instructions to compute bond penal sum.

(a) *Medium plants.* To find the required amount of your bond, estimate the total proof gallons of spirits to be produced and received in a calendar year. The amount of the bond is $1,000 for each 10,000 proof gallons (or fraction), subject to a minimum of $2,000 and a maximum of $50,000. The following table provides some examples:

Annual production and receipts in proof gallons

More than	but not over	Amount of bond
10,000 to	10,000	$2,000
20,000 to	30,000	3,000
90,000 to	100,000	10,000
190,000 to	200,000	20,000
490,000 to	500,000	50,000

(b) *Large plants.* To find the required amount of your bond, estimate the total proof gallons of spirits to be produced and received in a calendar year. The amount of the bond is $50,000 plus $2,000 for each 10,000 proof gallons (or fraction) over 500,000. The following table provides some examples:

Annual production and receipts in proof gallons

More than	but not over	Amount of bond
500,000 to	510,000	$52,000
510,000 to	520,000	54,000
740,000 to	750,000	100,000
990,000 to	1,000,000	150,000
1,240,000		200,000

(§805(c), Pub. L. 96-39, 93 Stat. 276 (26 U.S.C. 5173); §232, Pub. L. 96-223, 94 Stat. 278 (26 U.S.C. 5181))

§19.959 Conditions of bond.

The bond shall be conditioned on payment of all taxes (including any penalties and interest) imposed by 26 U.S.C. Chapter 51, on compliance with all requirements of law and regulations, and on payment of all penalties incurred or fines imposed for violation of any such provisions.

(§805(c), Pub. L. 96-39, 93 Stat. 276 (26 U.S.C. 5173); §232, Pub. L. 96-223, 94 Stat. 278 (26 U.S.C. 5181))

§19.960 Additional provisions with respect to bonds.

Subpart H of this part contains further provisions applicable to bonds which, where not inconsistent with this subpart, are applicable to bonds of alcohol fuel plants.

Construction, Equipment and Security

§19.972 Construction and equipment.

Buildings and enclosures where distilled spirits will be produced, processed, or stored shall be constructed and arranged to enable the proprietor to maintain security adequate to deter diversion of the spirits. Distilling equipment shall be constructed to prevent unauthorized removal of spirits, from the point where distilled spirits come into existence until production is complete and the quantity of spirits has been determined. Tanks and other vessels for containing spirits shall be equipped for locking and be constructed to allow for determining the quantities of spirits therein.
(§201, Pub. L. 85-859, 72 Stat. 1353, as amended (26 U.S.C. 5178))

§19.973 Security.

Proprietors shall provide security adequate to deter the unauthorized removal of spirits. The proprietor shall store spirits either in a building, a storage tank, or within an enclosure, which the proprietor will keep locked when operations are not being conducted.
(§201, Pub. L. 85-859, 72 Stat. 1353, as amended (26 U.S.C. 5178); §806, Pub. L. 96-39, 93 Stat. 279 (26 U.S.C. 5202))

§19.974 Additional security.

If the regional regulatory administrator finds that security is inadequate to deter diversion of the spirits, as may be evidenced by the occurrence of break-ins or by diversion of spirits to unauthorized purposes, additional security measures may be required. Such additional measures may include, but are not limited to, the following:

(a) *The erection of a fence around the plant or the alcohol storage facility;*

(b) *Flood lights;*

(c) *Alarm systems;*

(d) *Watchman services; or,*

(e) *Locked or barred windows.*

The exact additional security requirements would depend on the extent of the security problems, the volume of alcohol produced, the risk to tax revenue, and safety requirements.
(§201, Pub. L. 85-859, 72 Stat. 1353, as amended (26 U.S.C. 5178); §806, Pub. L. 96-39, 93 Stat. 279 (26 U.S.C. 5202))

Supervision

§19.975 Supervision of operations.

The regional regulatory administrator may assign ATF officers to premises of plants qualified under this subpart. The authorities of ATF officers, provided in §§19.80 through 19.84, and the requirement that proprietors keep premises accessible to and furnish facilities and assistance to ATF officers, provided in §§19.85 and 19.86, apply to plants qualified under this subpart.
(§201, Pub. L. 85-859, 72 Stat. 1320, as amended, 1357, as amended, 1358, as amended 1375, as amended (26 U.S.C. 5006, 5203, 5204, 5213))

Accounting for Spirits

§19.980 Gauging.

(a) *Equipment and method.* Proprietors shall gauge spirits by accurately determining the proof and quantity of spirits. The proof of the spirits shall be determined using a glass cylinder, hydrometer, and thermometer. However, fuel alcohol may be accounted for in wine gallons, so it is not necessary to determine the proof of fuel alcohol manufactured, on-hand, or removed. The proprietor may determine quantity either by volume or weight. A tank or receptacle with a calibrated glass scale installed, a calibrated dipstick, conversion charts, or (subject to approval by the Director) meters, may be used to determine quantity by volume. Detailed procedures for gauging spirits are provided in 27 CFR Part 13.

(b) *When Required.* Proprietors shall gauge spirits and record the results in their records at the following times:

(1) On completion of production of distilled spirits;

(2) On receipt of spirits at the plant;

(3) On addition of materials to render the spirits unfit for beverage use;

(4) Before withdrawal of spirits (including fuel alcohol) from plant premises or other disposition thereof; and,

(5) When spirits are to be inventoried.

§19.981 Inventories.

Proprietors shall take actual physical inventory of all spirits (including fuel alcohol) on bonded premises at least once during each period for which a report is required by §19.988. The results of the inventory shall be

COMMERCIAL SCALE ETHANOL PRODUCTION AND FINANCING

posted in the applicable reocrds required by §19.982.
(§201, Pub. L. 85-859, 72 Stat. 1356, as amended (26 U.S.C. 5201))

§19.982 Records.

(a) *All plants.* All proprietors shall maintain records with respect to:

    (1) The quantity and proof of spirits produced;

    (2) The proof gallons of spirits on-hand and received;

    (3) The quantities and types of materials added to render the spirits unfit for beverage use;

    (4) The quantity of fuel alcohol manufactured; and,

    (5) All dispositions of spirits (including fuel alcohol). Fuel alcohol may be recorded in wine gallons.

(b) *Medium and large plants.* Proprietors of medium and large plants shall also record the kind and quantity of materials used to produce spirits.

(c) *General requirements.* The records must contain sufficient information to allow ATF officers to determine the quantities of spirits produced, received, stored, or processed and to verify that all spirits have been lawfully disposed of or used. However, the proprietor need not prepare records specifically to meet the requirements of this subpart. Records which the proprietor prepares for other purposes (i.e., invoices or other commercial records) are sufficient, so long as they show all the required information.
(§807, Pub. L. 96-39, 93 Stat. 284 (26 U.S.C. 5207))

§19.983 Spirits rendered unfit in the production process.

If the proprietor renders spirits unfit for beverage use before removal from the production system, the production records shall also include the kind and quantity of materials added to each lot of spirits. In such a case, a separate record under §19.985 is not required. This paragraph applies to in-line addition of materials and to systems in which, before any spirits come off the production equipment, the proprietor adds materials for rendering the spirits unfit for beverage use to the first receptacle where spirits are to be deposited.
(§807, Pub. L. 96-39, 93 Stat. 284 (26 U.S.C. 5207); §232, Pub. L. 96-223, 94 Stat. 278 (26 U.S.C. 5181))

§19.984 Record of spirits received.

The proprietor's copy of the consignor's invoice or other document received with the shipment, on which the proprietor has noted the date of receipt and quantity received, constitutes the required record.
(§807, Pub. L. 96-39, 93 Stat. 284 (26 U.S.C. 5207))

§19.985 Record of spirits rendered unfit for beverage use.

The proprietor shall record the kind and quantity of materials added to render each lot of spirits unfit for beverage use and the quantity of fuel alcohol manufactured (which may be given in wine gallons).
(§807, Pub.L. 96-39, 93 Stat. 284 (26 U.S.C. 5207)).

§19.986 Record of dispositions.

(a) *Fuel alcohol.* For fuel alcohol distributed, used, or otherwise disposed of the proprietor shall record the—

    (1) Quantity of fuel alcohol;

    (2) Date of disposition; and

    (3) Name and address of the person to whom distributed or, if used or otherwise disposed of by the proprietor, the purpose for which used or the nature of the other disposition (e.g., destruction or redistillation). Commercial invoices, sales slips, or similar documents are acceptable if they clearly show the required information.

(b) *Spirits.*

    (1) For spirits transferred in bond to another alcohol fuel plant, the commercial invoice or other document required by §19.997 constitutes the required record. For transfers to other distilled spirits plants, the form required by §19.508 is the required record.

    (2) For spirits used or otherwise disposed of (e.g., lost, destroyed, redistilled) on the premises of the alcohol fuel plant, the proprietor shall maintain a record as follows:

        (i) the quantity of spirits (in proof gallons) and the date of disposition;

        (ii) the purpose for which used or the nature of the other disposition.
(§807, Pub. L. 96-39, 93 Stat. 284 (26 U.S.C. 5207))

§19.987 Maintenance and retention of records.

The proprietor shall retain the records required by this subpart for a period of not less than three years from the date thereof or from the data of the last entry made

thereon, whichever is later. The records shall be kept at the plant where the operation or transaction occurs and shall be available for inspection by any ATF officer during business hours. For records maintained on data processing equipment, the provisions of §19.743 apply.
(§807, Pub. L. 96-39, 93 Stat. 284 (26 U.S.C. 5207)).

§19.988 Reports.

Proprietors shall file reports of their operations, depending on the type of plant, as follows:

Type of Plant and Reporting Periods

Small plant—Annually (December 31)
Medium plant—Semiannually (June 30 and December 31)
Large plant—Quarterly (Close of each calendar quarter)

The proprietor shall submit each required report to the regional regulatory administrator within 30 days after the close of the applicable reporting period. The report shall be submitted on a form provided for that purpose.
(§807, Pub. L. 96-39, 93 Stat. 284 (26 U.S.C. 5207))

Transfers and Withdrawals

§19.989 Withdrawal free of tax.

Fuel alcohol produced under this subpart, may be withdrawn free of tax from plant premises exclusively for fuel use.
(§201, Pub. L. 85-859, 72 Stat. 1362, as amended (26 U.S.C. 5214); §232, Pub. L. 96-223, 94 Stat. 278 (26 U.S.C. 5181))

§19.990 Other uses prohibited.

The law imposes criminal penalties on any person who withdraws, uses, sells or otherwise disposes of distilled spirits (including fuel alcohol) produced under this subpart for other than fuel use.
(§201, Pub. L. 85-859, 72 Stat. 1398, as amended (26 U.S.C. 5601); §232, Pub. L. 96-223, 94 Stat. 278 (26 U.S.C. 5181))

§19.991 Requirement for rendering spirits unfit for beverage use.

Before spirits may be withdrawn from plant premises, the spirits must contain, or the proprietor shall add, substances to render the spirits unfit for beverage use as provided in this subpart. However, spirits used for fuel on the premises of the alcohol fuel plant and spirits transferred to other plants need not be rendered unfit for beverage use.
(§232, Pub. L. 96-223, 94 Stat. 278 (26 U.S.C. 5181))

§19.992 Authorized formulas for fuel alcohol.

Spirits will be considered rendered unfit for beverage use and eligible for tax-free withdrawal as fuel alcohol, when for every 100 gallons of spirits, there has been added:

(a) *5 gallons or more of—*

    (1) Gasoline or automotive gasoline (for use in engines which require unleaded gasoline. Environmental Protection Agency regulations and manufacturer's specifications may require that unleaded gasoline be used to render the spirits unfit).

    (2) Kerosene

    (3) Deodorized kerosene

    (4) Rubber hydrocarbon solvent,

    (5) Methyl isobutyl ketone,

    (6) Mixed isomers of nitropropane, or

    (7) Any combination of (1) through (6); or,

(b) *1/8 ounce of denatonium benzoate N.F. (Bitrex) and 5 gallons of isopropyl alcohol.*
(§232, Pub. L. 96-233, 94 Stat. 278 (26 U.S.C. 5181))

§19.993 Substitute Materials.

Other materials may be used to render spirits unfit for beverage use subject to approval by the Director. A proprietor who wishes to use substitute materials to render spirits unfit for beverage use may submit a letter requesting authorization to the Director through the regional regulatory administrator. The letter should state the materials, and the quantity of each, which the proprietor proposes to add to each 100 gallons of spirits. The Director may require the proprietor to submit a sample of the proposed substitute material. The proprietor shall not use any proposed substitute material prior to its approval.
(§232, Pub. L. 96-233, 94 Stat. (26 U.S.C. 5181))

§19.994 Marks.

The proprietor shall conspicuously and permanently mark or securely label each container of fuel alcohol containing 55 gallons or less, as follows:

WARNING—FUEL ALCOHOL—MAY BE HARMFUL OR FATAL IF SWALLOWED

The mark or label shall be placed on the head or side of the container, and shall be in plain legible letters. Proprietors may place other marks or labels on containers

so long as they do not obscure the required mark.
(§232, Pub. L. 96-233, 94 Stat. 278 (26 U.S.C. 5181); §201, Pub. L. 85-859, 72 Stat. 1360, as amended (26 U.S.C. 5206))

§19.995 Container size.

Spirits, including fuel alcohol, shall not be filled at alcohol fuel plants into containers holding less than five gallons. However, smaller containers may be used for reasonable quantities of samples held or removed to a bonafide laboratory for testing or analysis, so long as the containers are marked as samples.
(§232, Pub. L. 96-223, 94 Stat. 278 (26 U.S.C. 5181))

Transfers Between Plants

§19.996 Transfer in bond.

(a) *Transfers between alcohol fuel plants.* Proprietors may remove spirits from the bonded premises of an alcohol fuel plant (including the premises of a small plant) for transfer to another alcohol fuel plant. Bulk conveyances in which spirits are transferred shall be secured with locks or seals. The spirits need not be rendered unfit for beverage use prior to transfer. Spirits so transferred may not be withdrawn, used, sold, or otherwise disposed of from the consignee plant for other than fuel use.

(b) *Transfers to or from other distilled spirits plants.* Spirits (not including spirits produced from petroleum, natural gas, or coal) may be transferred in bond from distilled spirits plants qualified under Subpart G of this part to alcohol fuel plants. Alcohol fuel plants may transfer spirits in bond to distilled spirits plants qualified under Subpart G. Spirits so transferred may not be withdrawn, used, sold, or otherwise disposed of for other than fuel use.

(c) *Transfer procedures.* The procedures in §§19.997 through 19.999 apply only to the transfers between two alcohol fuel plants. See §§19.506 through 19.510 for requirements where one plant is a distilled spirits plant qualified under subpart G of this part.
(§201, Pub. L. 85-859, 72 Stat. 1362, as amended (26 U.S.C. 5212); §232, Pub. L. 96-223, 94 Stat. 278 (26 U.S.C. 5181))

§19.997 Consignor premises.

The consignor shall prepare, in triplicate, a commercial invoice or shipping document to cover each shipment of spirits. The consignor shall enter on the document the quantity of spirits transferred, a description of the shipment (for example, number and size of drums or barrels, tank truck, etc.), the name, address, and permit number of the consignee, and the serial numbers of any seals, locks, or other devices used to secure the con-

veyance. The consignor shall forward the original and one copy of the document to the consignee with the shipment, and retain a copy as a record.
(§201, Pub. L. 85-859, 72 Stat. 1362, as amended (26 U.S.C. 5212))

§19.998 Reconsignment in transit.

When, prior to or on arrival at the premises of a consignee, spirits transferred in bond are found to be unsuitable for the intended purpose, were shipped in error, or, for any other bona fide reason, are not accepted by such consignee, or are not accepted by a carrier, they may be reconsigned, by the consignor, to himself, or to another qualified consignee. In such case, the bond, if any, of the proprietor to whom the spirits are reconsigned shall cover such spirits while in transit after reconsignment on his copy of the document covering the original shipment. Where the reconsignment is to another proprietor, a new document shall be prepared and prominently marked with the word "Reconsignment."
(§201, Pub. L. 85-859, 72 Stat. 1362, as amended (26 U.S.C. 5212); §232, Pub. L. 96-223, 94 Stat. 278 (26 U.S.C. 5181))

§19.999 Consignee premises.

(a) *General.* When spirits are received by transfer in bond, the proprietor shall examine each conveyance to determine whether the locks or seals, if any, are intact upon arrival at this premises. If the locks or seals are not intact, he shall immediately notify the area supervisor, before removal of any spirits from the conveyance. The consignee shall determine the quantity of spirits received and record the quantity and the date received on both copies of the document covering the shipment. The consignee shall return one receipted copy to the consignor and retain one copy as the record of receipt required by §19.984.

(b) *Portable containers.* When spirits are received in barrels, drums, or similar portable containers, the proprietor shall examine each container and unless the transfer was made in a sealed conveyance and the seals or other devices are intact on arrival, verify the contents of container. The proprietor shall record the quantity received for each container on a list, and attach a copy of the list to each copy of the invoice or other document required by §19.997 covering the shipment.

(c) *Bulk conveyances and pipelines.* When spirits are received in bulk conveyances or by pipeline, the consignee shall gauge the spirits received and record the quantity so determined on each copy of the invoice or other document covering the shipment. However, the regional regulatory administrator may waive the requirement for gauging spirits on receipt by pipeline if, because of the location of the premises, there will be no jeopardy to the revenue.

(§201, Pub. L. 85-859, 72 Stat. 1358, as amended 1362, as amended (26 U.S.C. 5204, 5212); §232, Pub. L. 96-223, 94 Stat. 278 (26 U.S.C. 5181))

Signed: May 21, 1980
G. R. Dickerson, Director
Approved: June 9, 1980
Richard J. Davis, Assistant Secretary
(Enforcement Operations)

DEPARTMENT OF TREASURY

Bureau of Alcohol, Tobacco and Firearms
27 CFT Part 19
(Notice No. 345)
Fuel Use of Distilled Spirits—Implemeting a Portion of the Crude Oil Windfall Profit Tax Act of 1980
(Public Law 96-223)

AGENCY: Bureau of Alcohol, Tobacco and Firearms, Department of the Treasury.

ACTION: Proposed rulemaking cross-reference to temporary regulations.

SUMMARY: In the Rules and Regulations portion of this Federal Register, the Bureau of Alcohol, Tobacco and Firearms (ATF) is issuing temporary regulations regarding implementation of Section 232, Alcohol Fuels, in Part III of Title II of the crude Oil Windfall Profit Tax Act of 1980 (Pub. L. 96-223). The temporary regulations also serve as a notice of proposed rulemaking for final regulations.

DATES: The effective date of the temporary regulations is July 1, 1980. Written comments must be delivered or mailed by October 20, 1980.

ADDRESS: Send comments to Chief, Regulations and Procedures Division, Bureau of Alcohol, Tobacco and Firearms, P.O. Box 385, Washington, DC 20044

Disclosure of comments: Any person may inspect the written comments or suggestions during normal business hours at the ATF Reading Room, Office of Public Affairs, Room 4407, Federal Building, 12th and Pennsylvania Avenue, NW, Washington, DC 20226.

FOR FURTHER INFORMATION CONTACT:

John V. Jarowski, Research and Regulations Branch, Bureau of Alcohol, Tobacco and Firearms, Washington, DC 20226, Telephone: 202-566-7626.

SUPPLEMENTARY INFORMATION:

Public Participation: Interested persorns may submit written comments and suggestions regarding the temperary regulations. All communications received within the comment period will be considered before final regulations are issued. Any person who desires an opportunity to comment orally at a public hearing on the temperary regulations should submit a written request to the Director within the comment period. However, the Director reserves the right to determine whether public hearing should be held.

ATF plans to contact State regulatory officials to obtain their comments on these regulations. ATF seeks to coordinate Federal and State regulation of alcohol fuels in an effort to minimize conflicting or duplicative regulations.

The temporary regulations in the Rules and Regulations portion of this issue of the Federal Register revise and add new regulations in 27 CFR Part 19. For the text of the temporary regulations, see 45 FR (T.D. ATF-71) published in the Rules and Regulations portion of this issue of the Federal Register.

Signed: May 21, 1980
G. R. Dickerson, Director
Approved: June 9, 1980
Richard J. Davis, Assistant Secretary (Enforcement and Operations)

FDA AND OTHER ANIMAL FEED REGULATIONS

Most plants designed for the production of fuel alcohol from grain will be designed to produce coproducts in the form of grain residues suitable for use as animal feeds. Therefore, a substantial portion of the alcohol plant's activities will be concerned with producing and marketing these products. Most states have adopted commercial feed laws governing the production and marketing of animal feeds. In addition, the sale of feeds on an interstate basis is regulated at the federal level by various agencies, primarily the Food and Drug Administration (FDA). Thus, regulatory compliance will be an important consideration in developing the plant's animal feed operations.

The regulations in each State should be checked for particular requirements; however, the regulations in the majority of States are very similar. In many States, the State legislators have adopted all or part of the Uniform State Feed Bill.[1] These statutes incorporate by reference provisions of the Food, Drug, and Cosmetic Act and the of-

[1]Officially adopted by the Association of American Feed Control Officials and endorsed by the American Feed Manufacturers Association and the National Feed Ingredients Association

ficial definitions of feed ingredients as defined by the Association of American Feed Control Officials (AAFCO).

For the alcohol plant, as a manufacturer of animal feeds, ther are three primary areas of concern with regard to feed regulation-registration, adulteration, and misbranding. Registration (licensing) is in most instances quite simple and involves registering the products to be manufactured and paying the fees required. The statutory language in some states may require fee tags or stamps to be purchased and attached to the labeling used on packaged feeds or, in the case of feeds sold in bulk, to provide these tags or stamps to the buyer along with the invoice or bill of lading. The fees charged are, for the most part, nominal. They are either set at a flat rate for each product produced and sold or are based on tonnage. Some states have registrations involving a combination of a flat rate plus a tonnage fee.[2] A number of states may also require the filing of affidavits of tonnage on a regular basis in order to advise the state on the level of production at a given plant. In each state the actual mechanism of registration is usually a simple matter of following the statutory guidelines and associated regulations.[3]

After the initial registration has been completed, the primary concerns of management, to ensure continued compliance, will be in the area of quality control. From a licensing standpoint, lapses in quality control can produce charges of adulteration and misbranding. To ensure continuing compliance, most States have (by statute) provided for inspecting, sampling, and analyzing production. Discovery of a violation of the provisions of the Feed Law can result in various penalties (fines) or in detention of the feed (i.e. withholding the feed from sale or distribution).

Contamination of the feed can result in a charge of adulteration. The definition of adulteration is usually specifically provided by statute. The Uniform Feed Bill provides the following definition:

Section 7. Adulteration.

A commercial feed shall be deemed to be adulterated:

(a)(1) If it bears or contains any poisonous or deleterious substance which may render it injurious to health; but in case the substance is not an added substance, such commercial feed shall not be considered adulterated under this subsection if the quantity of such substance in such commercial feed does not ordinarily render it injurious to health; or

(2) If it bears or contains any added poisonous, added deleterious, or added nonnutritive substance which is unsafe within the meaning of Section 406 of the Federal, Food, and Cosmetic Act (other than one which is (i) a pesticide chemical in or on a raw agricultural commodity; or (ii) a food additive); or

(3) If it is, or it bears or contains any food additive which is unsafe within the meaning of Section 409 of the Federal Food, Drug, and Cosmetic Act; or

(4) If it is a raw agricultural commodity and it bears or contains a pesticide chemical which is unsafe within the meaning of Section 408(a) of the Federal Food, Drug, and Cosmetic Act: Provided, that where a pesticide chemical has been used in or on a raw agricultural commodity in conformity with an exemption granted or a tolerance prescribed under Section 408 of the Federal Food, Drug, and Cosmetic Act and such raw agricultural commodity has been subjected to processing such as canning, cooking, freezing, dehydrating, or milling, the residue of such pesticide chemical remaining in or on such processed feed shall not be deemed unsafe if such residue in or on the raw agricultural commodity has been removed to the extent possible in good manufacturing practice and the concentration of such residue in the processed feed is not greater than the tolerance prescribed for the raw agricultural commodity unless the feeding of such processed feed will result or is likely to result in a pesticide residue in the edible product of the animal, which is unsafe within the meaning of Section 408(a) of the Federal Food, Drug, and Cosmetic Act.

(5) If it is, or it bears or contains any color additive which is unsafe within the meaning of Section 706 of the Federal Food, Drug, and Cosmetic Act.

(b) If any valuable constituent has been in whole or in part omitted or abstracted therefrom or any less valuable substance substituted therefor.

---

[2]For example, Wisconsin assesses $10 00 per plant per year plus $0.10 per ton inspection fee for all feed produced annually By contrast, Georgia simply charges a flat fee of $2 00 annually per product produced

[3]Most States have provided authority under the respective Administrative Procedures Act for the agencies to pass regulations further defining or implementing legislation and these regulations should be checked

(c) If its composition or quality falls below or differs from that which it is purported or is represented in possess by its labeling.

(d) If it contains a drug and the methods used in or the facilities or controls used for its manufacture, processing, or packaging do not conform to current good manufacturing practice regulations promulgated by the (authorized State agency) to ensure that the drug meets the requirement of this Act as to safety and has the identity and strength and meets the quality and purity characteristics which it purports or is represented to possess. In promulgating such regulations, the (authorized State agency) shall adopt the current good manufacturing practice regulations for medicated feed premixes and for medicated feeds established under authority of the Federal Food, Drug, and Cosmetic Act, unless he determines that they are not appropriate to the conditions which exist in this State.

(e) If it contains visable weed seeds in amounts exceeding the limits which the (authorized state agency) shall establish by rule or regulation.

The above statute, in addition to the enumerated provisions, incorporates by reference two important areas of federal regualtion.

First, the feed could be adulterated by containing an added substance not on the list of substances Generally Recognized as Safe by the FDA (The GRAS List) or if it contains a substance on the GRAS List in excess of the limits prescribed for that substance in the Code of Federal Regulations (CFR). Thus, a violation could occur if a plant were to utilize a process that added a substance to the feed during the process of manufacturing which is not in compliance with the above regulations. For example, a process for drying feed residue which utilizes waste gases might produce unacceptable levels of contamination or contamination with some substance not on the GRAS List.

Second, a problem could also arise if the grain used as a substrate contained excessive levels of pesticides. The processing might not eliminate enough of the residue to bring the quantity in the final product within the acceptable limits.[4]

In addition to avoiding the charge of adulteration, quality control is essential to ensure regulatory compliance in other areas. To ensure compliance, it is necessary that a feed that purports to be a certain product must meet the definition established for that product.

Failure to meet the definition or standard could result in the charge of misbranding. The Uniform State Feed Bill provides the following definition for misbranding:

Section 6. Misbranding.

A commercial feed shall be deemed to be misbranded:

(a) If its labeling is false or misleading in any particular.

(b) If it is distributed under the name of another commercial feed.

(c) If it is not labeled as required in Section 5 of this Act.

(d) If it purports to be or is represented as a commercial feed, or if it purports to contain or is represented as containing a commercial feed ingredient, unless such commercial feed or feed ingredient conforms to the definition, if any, prescribed by regulation by the (authorized state agency).

(e) If any word, statement, or other information required by or under authority of this Act to appear on the label or labeling is not prominently placed theron with such conspicuousness (as comapred with other words, statements, designs, or devices in the labeling) and in such terms as to render it likely to be read and understood by the ordinary individual under customary conditions of purchase and use.

The definitions referred to in the above statutory provisions for feed products have been established by AAFCO and are incorporated by reference.[5]

This discussion of feed regulations briefly outlines the areas of interest to investors in alcohol fuel plants. Compliance with these regulations can be achieved by properly designing and operating the plant. The grain and feed portions of the operation can be designed to minimize problems with contamination and provide adequate controls to monitor and manage product quality in order to eliminate or minimize regulatory complaince problems. By recognizing and considering these regulatory requirements during the planning and design phases, new plants can do a great deal to minimize potential problems in this area.

[4]Limits on pesticides under Section 408 of FDCA are determined by the EPA and are set forth in the CFR's

[5]Official and tentative definitions of feed ingredients as established by the Association of American Feed Control Officials (AAFCO) contains definitions for a wide variety of products These include Maize (corn products) 48 1 to 48 30; Fermentation Products 36 1 to 36 12, Distiller's Products 27 1 to 27 7

TRANSPORTATION REGULATIONS

The surface transportation of raw materials and finished product may be regulated by State and Federal agencies, depending upon the location of the plant and its proximity to the source of materials and its markets. If the plant, material source, and the market are located in the same State, it is highly probable that only the individualistic State law will apply to regulate surface transportation, at least insofar as for-hire motor truck transportation is concerned. Some States, such as Texas, Nebraska, and South Dakota, choose to regulate truck transportation rather extensively, while others have little or no concern for regulation of the basic types of commodities flowing to and from a plant. Generally, grain transportation is not regulated, while coal and alcohol fuel are. Several States, including Florida, have recently deregulated the transporting of all commodities within their borders, while several others are actively studying deregulation proposals.

Assuming interstate commerce is involved in either obtaining materials or in reaching the market place, the nature of regulation is determined by the Interstate Commerce Commission. Currently, the truck transportation of grain is exempt (or not regulated), while, under the Motor Carrier Reform Act of 1980 (H.R. 6418), the movement of DDG will likely also be exempt. Truck movement of coal and alcohol, on the other hand, is currently regulated. In this context, lawful interstate movements of such commodities can be performed only by a carrier holding Federal ICC authority.*

Primary transportation considerations will necessarily include plant site selection vis a vis existing, visible rail facilities. This is particularly critical where a plant is to be located any distace (perhaps more than 200 or 250 miles) from the primary source of any major raw material (grain or coal).

LOCAL BUILDING CODES

Like zoning matters, the State's jurisdictional power is the basic authority under which building codes are enacted. Some State legislatures enact statewide building codes while others delegate the authority to the local governments. One or a combination of the four model building codes has been adopted by most States or municipalities. These codes (and the geographic area) dominated by their association/author are: (1) the Uniform Building Code written by the International Conference of Building Officials (adopted primarily in the West); (2) the Basic Building Code compiled by the Building Officials and Code Administrators, International, Inc. (found in the Northeast and North Central areas); (3) the Southern Standard Building Code enacted by the Southern Building Code Conference (adopted in

the South); and (4) the National Building Code, developed by the National Board of Fire Underwriters. Local variations exist despite the model codes. Some municipalities have adopted selected provisions rather than the entire code. Interpretations of the same code differ from city to city.

Unlike zoning ordinanaces, most building codes apply retroactively. Three types of information are provided in most codes: definition of terms; licensing requirements; and standards. Taken together, the definitions and licensing requirements have the effect of prescribing who is authorized to conduct particular sorts of construction activity. For example, the International Association of Plumbing and Mechanical Officials Code states that only licensed plumbers may do work defined as plumbing. Many codes require that structural design plans be prepared by a state certified engineer.

Two types of code standards exist: technical specifications and performance standards. Codes prescribing technical specifications set out how, and with what materials, a building is to be constructed. Performance standards represent a more progressive and technically more flexible approach. Codes based on these standards state product requirements that do not prescribe designs and materials. For example, "the structural frame of all buildings, signs, tanks and other exposed structures shall be designed to resist the horizontal pressures due to wind in any direction. . ." Typical construction components specified in codes are structural and foundation loads and stresses, construction material, fireproofing, building height (this represents a common duplication of the zoning ordinance), and electrical installation. The developer is likely to be required to comply with the standards for structural and foundation loads and stresses, as standards set out the minimum force measure in pounds-per-square-inch that the design must bear under certain circumstances, e.g., wind or snow. The electrical code regulates the use of all electrical wiring when voltage levels are above 36 volts.

Dissatisfaction with the building inspector's denial of a permit may result in an appeal before the local board of building appeals. The common bases of appeal provided by the codes are: an incorrect interpretation of the code by the building official; the availability of an equally good or better form of construction not specified in the code; and the existence of practical difficulties in carrying out the requirements of the code. The local board members are usually appointed experts in the field of construction. The local board may uphold, modify, or reverse the building official's decision. Further appeals to the state board of building appeals or to the courts are also available.

*As a corollary to this, the rate structures on such transportation are also regulated

GASOLINE ALLOCATIONS

A letter containing the following information is required from the applicant interested in obtaining a gasoline allocation for gasohol:

The office that is final arbiter on increasing gasoline allocation for gasohol usage is:

Office of Hearings and Appeals
U. S. DEPARTMENT OF ENERGY
2000 M Street, N.W.
Washington, D.C. 20461
(202) 254-3008

(a) Investment

1. What is the amount of investment already made?

2. How is/was this investment financed?

3. What would be the value of the investment if approval for reallocation is not granted?

4. Provide photographs of production facilities.

5. For what purpose(s) will the investment be used—gasohol blending, ethanol production, facility construction, feedstock purchase, etc.?

(b) Alcohol

1. Where are the sources of alcohol?

2. What will be the cost of obtaining the alcohol for each source listed?

3. Supply copies of supply contracts, if any exist.

4. What is the efficiency of the alcohol production techniques(s) to be used?

5. What feedstock will be utilized in producing the alcohol?

(c) Unleaded Gasoline

1. What is the availability of unleaded gasoline in the intended marketing area?

2. List refiners in the marketing area.

3. Describe the contacts already made in attempting to obtain unleaded gasoline.

4. List your preferred supplier(s) for unleaded gasoline.

5. Of the gasoline you obtain, what percentage is unleaded?

6. What is your selling price for unleaded gasoline?

7. What quantities of unleaded gasoline do you anticipate needing in the next 12 months? Specify month-by-month.

(d) Marketing

1. Describe the intended marketing area.

2. Describe your plans for the marketing of gasohol.

3. What is the anticipated sales of gasohol for 12-month period?

4. Describe any feasibility studies showing demand for gasohol in marketing area.

5. List retailers in the marketing area that would sell gasohol. Describe any contracts that exist with them.

6. What is the expected selling price for the gasohol?

A copy of this letter of application (with confidential information deleted) should be sent to the base supplier of gasoline.

DEPARTMENT OF THE TREASURY — BUREAU OF ALCOHOL, TOBACCO AND FIREARMS

APPLICATION AND PERMIT FOR ALCOHOL FUEL PRODUCER UNDER 26 U.S.C. 5181

INSTRUCTION SHEET FOR ATF FORM 5110 74

PLEASE READ CAREFULLY AN INCOMPLETE OR INCORRECT APPLICATION WILL DELAY YOUR ALCOHOL FUEL PRODUCER'S PERMIT

1 PURPOSE The application is completed by a person (applicant) who would like to establish a plant to produce, process, and store, and use or distribute distilled spirits to be used exclusively for fuel use under 26 U S C 5181 Distilled spirits means only ethanol or ethyl alcohol The production of methanol does not require a permit from the Bureau of Alcohol, Tobacco and Firearms The production of distilled spirits from petroleum, natural gas, or coal is not allowed by the Alcohol Fuel Producers Permit

2 GENERAL PREPARATION Prepare this form and any attachments in duplicate Use separate sheets of approximately the same size as this form when necessary or as required Identify these separate sheets with your name, and attach to form

3 WHERE TO FILE Submit application to the appropriate regional regulatory administrator (A listing of their offices and the areas they are responsible for is on the reverse)

4 TYPE OF APPLICATION (See item 1 of application) The type of permit you need, small, medium, or large, depends on how many proof gallons of distilled spirits you intend to produce and receive by transfers from other plants during one calendar year The number of proof gallons may be calculated by taking the proof of the spirits multiplied by the wine gallons (your standard American gallon) and dividing by 100 For example

$$50 \text{ gallons of } 190° \text{ proof spirits} =$$
$$190 \times 50 - 100 = 95 \text{ proof gallons}$$

After determining the quantity of spirits to be produced and received by transfers from other plants, check applicable box in item 1

5 AMENDED PERMIT Item 2 is only completed when changing or amending an existing permit

6 INFORMATION ABOUT APPLICANT CURRENTLY ON FILE WITH REGIONAL REGULATORY ADMINISTRATOR NEED NOT BE RESUBMITTED State in item requesting such information the type and the number of the license or permit for which the information was filed

7 CAPACITY OF STILLS Item 9 d asks for the capacity of your still(s) in proof gallons The capacity should be given as the greatest number of proof gallons of spirits that could be distilled in a 24 hour period Capacity of column still may be shown by giving the diameter of the base and the number of plates or packing material The capacity of a pot or kettle still may be shown by giving the volumetric (wine gallon) capacity of the pot or kettle

8 SAMPLE DIAGRAM OF PREMISES Item 12 requires a diagram of your plant premises It can be drawn by hand and does not have to the drawn to scale Below is a sample of such a diagram

9 ITEM 15—SIGNATURE OF/FOR APPLICANT (a) Individual owners sign for themselves, (b) partnerships have all partners sign, or have one partner who has attached an authorization to act on behalf of all the partners (unless this authorization is provided by State law), (c) corporations have an officer, director, or other person who is specifically authorized by the corporate documents sign, (d) any other person who signs on behalf of the applicant must submit ATF F 1534 (5000 8), Power of Attorney or other evidence of their authority

10 ADDITIONAL INSTRUCTIONS FOR APPLICANT OF SMALL ALCOHOL FUEL PLANT Complete all items on the application form Be sure that you sign and date the form in items 15, and 16 NO ADDITIONAL INFORMATION IS REQUIRED

11 APPLICANT FOR PERMIT OF MEDIUM AND LARGE ALCOHOL FUEL PLANT Complete all items on the form Supply in duplicate the additional information and forms as stated below
a The following information is requested for an individual proprietor, each partner, or each director of a corporation or similar entity and each officer of the corporation or similar entity who will have responsibilities and connection with the operations covered by the permit In addition applicants for large plants shall show the same information for each interested person who is an individual as listed in the statement of interest in the organizational documents required by 27 CFR 19 941
(1) full name including middle name,
(2) title in connection with applicants' business,
(3) social security number,
(4) date of birth,
(5) place of birth, and
(6) address of residence
b A statement as to whether the applicant or any person required to be listed by the instructions above has been previously convicted of any violation of Federal or State law (other than minor traffic violations)
c A statement of the maximum quantity of distilled spirits to be produced and received from other plants during a calendar year
d A Distilled Spirits Bond, ATF F 5110 56, as required by 27 CFR 19 956
e Any other information required by the regional regulatory administrator after examination of this application

SPECIAL INSTRUCTIONS FOR ALL APPLICANTS

12 OPERATIONS BEFORE ISSUANCE OF PERMIT Unless otherwise specifically authorized by law or regulations, an applicant for an Alcohol Fuel Producer Permit may not engage in operations until the permit has been issued by the regional regulatory administrator

13 STATE AND LOCAL LAWS This permit does not allow you to operate in violation of State or local laws Applicants should check with appropriate State and local authorities before engaging in alcohol fuel plant operations

14 ATF FORMS AND REGULATIONS FOR ALCOHOL FUEL PLANTS ATF forms and publications including regulations for alcohol fuel producers may be ordered from the ATF Distribution Center, 3800 S Four Mile Run Drive, Arlington, VA 22206

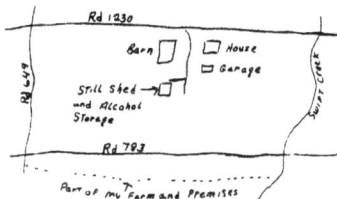

My Farm is 1,200 acres, 1,000 acres are bounded by Rd 649, 1230, 783 and Swift Creek. 200 acres are just the other side of Rd 783.

ATF F 5110 74 (6 80)

Figure B-1.—BATF Form 5110.74

12 DIAGRAM OF PLANT PREMISES *(In the space provided or by attached map or diagram, show the area to be included for the alcohol fuel plant, Identify roads, streams, lakes, railroads, buildings, and other structures or topographical features on the diagram. Show location(s) where alcohol fuel plant operations will occur. The diagram should be in sufficient detail to locate your operations and premises.) (See directions for sample diagram.)*

13 I WILL COMPLY WITH THE CLEAN WATER ACT *(33 U S C 1341(a)) (Will not discharge into navigable waters of the U S)*

☐ YES   ☐ NO

14 IF THIS APPLICATION IS APPROVED AND THE PERMIT IS ISSUED, I CONSENT TO THE DISCLOSURE OF THE NAME AND ADDRESS SHOWN ON THE APPLICATION IN AN ATF PUBLICATION, "ALCOHOL FUEL PRODUCERS", WHICH MAY BE DISTRIBUTED ON REQUEST TO THE GENERAL PUBLIC *(including media, business, civic, government agencies, and others)* UNDER 26 U S C 6103 YOU HAVE A LEGAL RIGHT NOT TO GIVE THIS RELEASE

☐ YES   ☐ NO *(A no response will have no effect on the consideration given this application)*

APPLICANTS FOR MEDIUM AND LARGE ALCOHOL FUEL PLANTS MUST ATTACH THE ADDITIONAL INFORMATION REQUIRED IN INSTRUCTIONS

Under the penalties of perjury, I declare that I have examined this application, including the documents submitted in support thereof or incorporated therein by reference, and, to the best of my knowledge and belief, it is true, correct, and complete

15 SIGNATURE OF/FOR APPLICANT	16 TITLE *(Owner, Partner, Corporate Officer)*

STOP
MAKE NO FURTHER ENTRIES ON THIS FORM

ALCOHOL FUEL PRODUCERS PERMIT UNDER 26 U.S.C. 5181	1 EFFECTIVE DATE
	2 PERMIT NUMBER AFP —

Pursuant to the above application and subject to applicable law and regulations and to the conditions set forth below you are hereby authorized and permitted at the premises described in your application to produce, process and store, and use or distribute distilled spirits (Not including distilled spirits produced from petroleum, natural gas or coal) exclusively for fuel use. The quantity to be produced and received from other plants during the calendar year is limited to the quantity stated in this application

This permit is continuing, and will remain in force until suspended, revoked, voluntarily surrendered, or automatically terminated. This permit does not allow you to operate in violation of State or local laws

THIS PERMIT IS NOT TRANSFERABLE In the event of any lease, sale, or other transfer of the operations authorized, or of any other change in the ownership or control of such operations, this permit shall automatically terminate (See 27 CFR 19 145 AND 19 949)

3 SIGNATURE OF REGIONAL REGULATORY ADMINISTRATOR, BUREAU OF ALCOHOL, TOBACCO AND FIREARMS

CONDITIONS

1 That the permittee in good faith complies with the provisions of Chapter 51 of Title 26 of the United States Code and regulations issued thereunder

2 That the permittee has made no false statement as to any material fact in his application for this permit

3 That the permittee discloses all the material information required by law and regulation

4 That the permittee shall not violate or conspire to violate any law of the United States relating to intoxicating liquor and shall not be convicted of

any offense under the United States Code punishable as a felony or of any conspiracy to commit such an offense

5 That all persons employed by the permittee in good faith observe and conform to all of the terms and conditions of this permit

6 That the permittee engages in the operations authorized by this permit within a 2 year period

7 This permit is conditioned on compliance by you with the Clean Water Act (33 U S C 1341(a))

ATF F 5110 74 (6 80)

Figure B-1.—BATF Form 5110.74 (continued)

<table>
<tr><td colspan="2">DEPARTMENT OF THE TREASURY
BUREAU OF ALCOHOL, TOBACCO AND FIREARMS

DISTILLED SPIRITS BOND

(See instructions on back)</td><td colspan="2">TYPE OF BOND AS PRESCRIBED UNDER 26 U S C 5173
(Check applicable box)

1 OPERATIONS   □ 2 WITHDRAWAL BOND
□ (a) ONE PLANT BOND   □ 3 UNIT BOND
□ (b) ADJACENT WINE CELLAR BOND □ 4 ALCOHOL FUEL
                                            PRODUCER BOND
□ (c) AREA BOND</td></tr>
</table>

PRINCIPAL *(See instructions 2, 3, and 4)*	ADDRESS OF BUSINESS OFFICE *(Number, street, city, State, ZIP Code)*

SURETY (OR SURETIES)		AMOUNT OF BOND	EFFECTIVE DATE

KIND OF BOND *(Check applicable box)*

□ ORIGINAL   □ STRENGTHENING   □ SUPERSEDING

TYPE OF ACTIVITY (DISTILLER, WAREHOUSEMAN, PROCESSOR, ADJACENT WINE CELLAR OR ALCOHOL FUEL PRODUCER) AND THE PREMISES COVERED BY THIS BOND ARE SPECIFIED ON THE BACK OF THIS BOND AND, IF NECESSARY, ON AN ADDITIONAL SHEET APPROPRIATELY IDENTIFIED AND ATTACHED TO THIS BOND

PURPOSE The above principal has filed an application for registration of the distilled spirits plant(s) specified

CONDITIONS The above principal and surety (sureties) are bound independently and jointly for the payment to the United States in the above amount of lawful money of the United States In this bond, the terms principal or surety include the heirs, executors, administrators, successors and assigns of the principal or surety

THE PRINCIPAL SHALL

(1) Comply with all requirements of law and regulations, now or hereafter in force, relating to the activities covered by this bond,

(2) Pay all penalties incurred and fines imposed for violations of law or regulations, now or hereafter in force, relating to the activities covered by this bond,

(3) Pay all taxes (including any penalties and interest in respect of failure to file a timely return, or to pay such tax when due) on distilled spirits withdrawn from bonded premises imposed under 26 U S C Chapter 51,

(4) Pay all taxes (including any penalties and interest) imposed under 26 U S C Chapter 51, including taxes on all unexplained shortages of bottled distilled spirits,

(5) Comply with all requirements of law and regulations, now or hereafter in force, pertaining to all distilled spirits (including denatured spirits, fuel alcohol and articles) removed from or returned to the bonded premises free of tax,

(6) With respect to distilled spirits withdrawn from the bonded premises without payment of tax as authorized by law, (a) comply with all requirements of law and regulations, now or hereafter in force relating thereto, and (b) as to the said distilled spirits or any part thereof withdrawn, for example, for exportation or for use on vessels or aircraft or for transfer to a foreign-trade zone or for transfer to a Customs bonded warehouse, or for research, development, or testing, and not so exported, used or transferred, or otherwise lawfully disposed of or accounted for, pay the tax imposed thereon by law, now or hereafter in force, together with penalties and interest, and

(7) As the proprietor of a bonded wine cellar, pay all taxes imposed by law, now or hereafter in force, (including any penalties and interest) for which he may become liable with respect to operation of the said bonded wine cellar, and all distilled spirits and wine now or hereafter in transit thereto or received thereat, and on all distilled spirits and wine removed therefrom, including wine withdrawn without payment of tax, on notice by the principal, for exportation, or use on vessels or aircraft, or transferred to a foreign-trade zone, and not so exported, used, or transferred, or otherwise lawfully disposed of or accounted for, Provided, that this obligation shall not apply to taxes on wine in excess of $100 which have been determined for deferred payment upon removal of the wine from the premises of the said bonded wine cellar or transfer to a taxpaid wine room thereon

    (If this bond covers only withdrawals, strike out clauses 4, 5, 6, and 7 above)
    (If this bond covers only operations, strike out clause 3 above)
    (If this bond covers only alcohol fuel production operations, strike out clauses 3 and 7 above)

DEFAULT If the principal fails to fulfill any of the terms or conditions of this bond, the United States may seek compensation and pursue its remedies independently from either the principal or surety, or jointly from both The surety hereby waives any right or privilege it may have of requiring, upon notice, or otherwise, that the United States shall first commence action, intervene in any action of any nature whatsoever already commenced, or otherwise exhaust its remedies against the principal

CHANGE OF PREMISES All stipulations, covenants, and agreements of this bond shall extend to and apply to any change in the business address of the premises, the extension or curtailment of such premises, including the buildings thereon, or any equipment or any other change which requires the principal to file a new or amended registration, application, or notice, except where the change constitutes a change in the proprietorship of the business, or in the location of the premises Further, this bond shall continue in effect whenever operation of the plant is resumed from time to time following suspension of operations by an alternate proprietor

EFFECTIVE DATE This bond shall not in any case be effective before the above date, but if accepted by the United States, it shall be effective according to its terms on and after that date without notice to the obligors Provided, that if no effective date is inserted in the space provided, the date of execution shown below shall be the effective date of the bond

Witness our hands and seals this _____ day of _____ , 19_____ Signed, Sealed and Delivered in the

presence of _____

_____ SEAL

_____ SEAL

_____ SEAL

_____ SEAL

ATF F 5110 56 (7-80)   PREVIOUS EDITIONS ARE OBSOLETE

Figure B-2.—BATF Form 5110.56

NAME AND ADDRESS	REGISTRY NUMBER	OPERATIONS COVERAGE (State activities at each premises and in the last block the amount of coverage for such activities)	WITHDRAWAL COVERAGE (State amount allocated to each premises (distilled spirits plants only) and total amount)
			$
			$
			$
			$
			$
			$
			$
AMOUNT OF COVERAGE		$	TOTAL $

APPROVED

SIGNATURE OF REGIONAL REGULATORY ADMINISTRATOR	REGION	DATE

INSTRUCTIONS

1 This bond shall be filed in duplicate with the Regional Regulatory Administrator, Bureau of Alcohol, Tobacco and Firearms, of the region where the premises covered by the bond are located

2. The name, including the full given name, of each party to the bond shall be given in the heading, and each party shall sign the bond, or the bond may be executed in his name by an empowered attorney-in-fact

3 In the case of a partnership, the firm name, followed by the names of all its members, shall be given in the heading In executing the bond, the firm name shall be typed or written followed by the word "by" and the signatures of all partners, or the signature of any partner authorized to sign the bond for the firm, or the signature of an empowered attorney-in-fact

4 If the principal is a corporation, the heading shall give the corporate name, the name of the State under the laws of which it is organized, and the location of the principal office The bond shall be executed in the corporate name, immediately followed by the signature and title of the person authorized to act for the corporation

5 If the bond is signed by an attorney-in-fact for the principal, or by one of the members for a partnership or association, or by an officer or other person for a corporation, there shall be filed with the bond an authenticated copy of the power of attorney, or a resolution of the board of direcors, or an excerpt of the bylaws, or other document, authorizing the person signing the bond to execute it for the principal, unless such authorization has been filed with the Regional Regulatory Administrator, Bureau of Alcohol, Tobacco and Firearms, in which event a statement to that effect shall be attached to the bond

6 The signature for the surety shall be attested under corporate seal The signature for the principal, if a corporation, shall also be attested if the corporation has a corporate seal, if the corporation has no seal, that fact should be stated Each signature shall be made in the presence of two persons (except where corporate seals are affixed), who shall sign their names as witnesses

7 A bond may be given with corporate surety authorized to act as surety by the Secretary of the Treasury, or by the deposit of collateral security consisting of bonds or notes of the United States The Act of July 30, 1947 (Section 15, Title 6 U S C) provides that "the phrase 'bonds or notes of the United States' shall be deemed *** to mean any public debt obligations of the United States and any bonds, notes or other obligations which are unconditionally guaranteed as to both interest and principal by the United States "

8 If any alteration or erasure is made in the bond before its execution, there shall be incorporated in the bond a statement to that effect by the principal and surety or sureties, or if any alteration or erasure is made in the bond after its execution, the consent of all parties thereto shall be written in the bond

9 The penal sum named in the bond shall be in accordance with 27 CFR Part 19

10 If the bond is approved, a copy shall be returned to the principal

11 All correspondence about the filing of this bond or any subsequent action affecting this bond should be addressed to the Regional Regulatory Administrator, Bureau of Alcohol, Tobacco and Firearms with whom the bond is filed

GPO 870 571

ATF F 5110 56 (7-80)

Figure B-2.—BATF Form 5110.56 (continued)

COMMERCIAL SCALE ETHANOL PRODUCTION AND FINANCING

DEPARTMENT OF THE TREASURY
BUREAU OF ALCOHOL, TOBACCO AND FIREARMS

ALCOHOL FUEL PLANT REPORT

1. PERMIT NUMBER

GENERAL INSTRUCTIONS

1. Each proprietor of an alcohol fuel plant shall file this report of plant operations. The period covered by the report depends on the size of the alcohol fuel plant. A report is due even if no operations were conducted during the period

2. Prepare in duplicate. Send the original to the Regional Regulatory Administrator, Bureau of Alcohol, Tobacco and Firearms. Keep the copy with your alcohol fuel plant records

3. The report is due by the 30th day following the end of the reporting period

4. Report spirits in proof gallons and alcohol fuel in wine gallons. Round to the nearest whole number

5. Small and medium plants complete Parts I and II. Large plants complete Parts I and III.

WHEN TO FILE REPORTS

SMALL PLANT

1. Proprietors of small plants will prepare and file a report once a year to cover all operations for the calendar year

2. Report is due by January 30th following the end of the calendar year

MEDIUM PLANT

1. Proprietors of medium plants will prepare and file a report twice a year. Each report will cover all operations for period reported (January 1 thru June 30 or July 1 thru December 31) and will be due 30 days after the end of the reporting period

LARGE PLANT

1. Proprietors of large plants will prepare and file a report 4 times a year. Reports will cover calendar quarters (example Jan - March) and reflect all transactions in that period.

2. Reports are due 30 days after the end of the reporting period.

PART - I

1 SMALL PLANTS *(Annual Report)*	2 MEDIUM PLANTS *(Semi-Annual Report)*	3. LARGE PLANTS *(Quarterly Report)*
19 ____ *(Year)*	____ *(Month)* Through ____ *(Month)* 19 ____ *(Year)*	____ *(Month)* Through ____ *(Month)* 19 ____ *(Year)*

4. NAME OF PROPRIETOR *(If partnership, include name of each partner)*	5 LOCATION *(If no street show rural route)*

PART - II
FOR SMALL AND MEDIUM PLANTS ONLY

SPIRITS TRANSACTIONS	PROOF GALLONS	PROOF GALLONS
1. PRODUCED		
2. RECEIVED		
3. USED ON PREMISES FOR FUEL *(Do not include fuel alcohol)*		
4 USED IN MAKING FUEL ALCOHOL		
5 DESTROYED		
6. TRANSFERRED		
FUEL ALCOHOL TRANSACTIONS	**WINE GALLONS**	**WINE GALLONS**
7. PRODUCED *(Rendered Unfit)*		
8 USED ON PREMISES FOR FUEL		
9. TRANSFERRED		

Under penalties of perjury I declare that I have examined this report, and to the best of my knowledge and belief, it is a true and complete report of operations

10 SIGNATURE	11 DATE

ATF F 5110 75 (7-80)

Figure B-3.—BATF Form 5110.75

PART - III
FOR LARGE PLANTS ONLY

Report the loss (line 11) or the gain (line 4) of spirits found by taking the required physical inventory for the reporting period. Report any losses by theft separately (line 10). Report and identify on line 5 the quantity of any imported spirits received from customs custody

SPIRITS TRANSACTIONS	PROOF GALLONS	PROOF GALLONS
1 ON HAND BEGINNING OF REPORTING PERIOD		
2 PRODUCED BY DISTILLING		
3. RECEIVED FROM OTHER PLANTS		
4 INVENTORY GAIN		
5		
6 TOTAL — *(Lines 1 through 5)*		
7 USED IN MAKING FUEL ALCOHOL		
8 USED ON PLANT PREMISES FOR FUEL *(Do not include fuel alcohol)*		
9 USED AS DISTILLING MATERIAL OR FOR REDISTILLING		
10. LOST BY THEFT		
11 INVENTORY LOSS		
12 TRANSFERRED TO OTHER PLANTS		
13 DESTROYED		
14		
15 TOTAL — *(Lines 7 through 14)*		
16 ON HAND END OF PERIOD *(Subtract line 15 from line 6)*		

FUEL ALCOHOL TRANSACTIONS	WINE GALLONS
17 MANUFACTURED	
18 DISTRIBUTED OR SOLD FOR FUEL PURPOSES	
19. USED ON PLANT PREMISES FOR FUEL PURPOSES	
20 DESTROYED	
21 ALL OTHER DISPOSITIONS	

22 REMARKS

Under penalties of perjury I declare that I have examined this report, and to the best of my knowledge and belief, it is a true and complete report of operations

23 SIGNATURE	24 DATE

ATF F 5110 75 (7-80)

Figure B-3.—BATF Form 5110.75 (continued)

COMMERCIAL SCALE ETHANOL PRODUCTION AND FINANCING

APPENDIX C

Technical Reference Data

Conversion Table Properties of Gasoline, Methanol, and Ethanol

Red Winter Wheat and All Wheat Production Costs, 1978

Barley Production Costs, 1978

Grain Sorghum Production Costs, 1978

Grain Sorghum Silage and Forage Use, 1976 to 1978

Grain Sorghum Area, Yield, and Production, 1977-1978

Corn Area, Yield, and Production, 1977-1978

Sugar Beet Area, Yield, and Production

Sugar Beet Production and Value, 1977

Production of Selected Fruit Crops

Components of Molasses

Nutritional Content of Distillers Feed from Corn in Percentages

Available Energy from Corn in Distillers Feeds

Conversion Table

To convert from	To	Multiply By
barrels (oil)	gallons	42
°C	F	$1.8 \times (°C + 32)$
centimeters	inches	0.394
cubic feet	bushels	0.804
cubic feet	cubic meters	0.028
cubic feet	cubic yards	0.037
cubic meters	bushels	28.377
cubic meters	cubic feet	35.314
cubic meters	cubic yards	1.308
cubic meters	gallons	264.17
cubic meters	liters	1000
cubic meters/kilogram	cubic feet/pound	16.02
cubic meters/second	million gallons/day	22.83
grams/liter	parts per million	1000
grams/liter	pounds/cubic foot	0.062
grams/liter	pounds/1000 gallons	8.35
hectares	acres	2.471
kilograms	pounds	2.205
kilograms/cubic meter	pounds/cubic foot	0.062
kilojoules	Btu	0.948
kilojoules/cubic meter	Btu/cubic foot	33.5
kilojoules/kilogram	Btu/pound	0.447
kilojoules/liter	Btu/gallon	3.589
kilopascals	atmospheres	9.869×10^1
kilopascals	pounds/square inch	0.145
kilowatthours	Btu	3.413
kilowatthours	joules	3.6×10^6
liters	gallons	0.264
meters	feet	3.281
metric tons	pounds	2,205
metric tons	tons	1.102
micrometers	inches	0.000039
millimeters	inches	0.039
pounds/square inch	atmospheres	0.068
proof	% alcohol by volume	0.5

Properties of Gasoline, Methanol, and Ethanol

CHEMICAL PROPERTIES	GASOLINE	METHANOL	ETHANOL
Formula	C_4-C_{12}	CH_3OH	C_2H_5OH
Molecular Weight	Varies	32.04	46.1
% Carbon (by Weight)	85–88	38.70	52.1
% Hydrogen (by Weight)	12–15	9.70	13.1
% Oxygen (by Weight)	Indefinite	51.60	34.7
C/H Ratio	5.6–7.4	3.00	4.0
Stoichiometric Air-to-Fuel Ratio	14.2–15.1	6.45	9.0

PHYSICAL PROPERTIES	GASOLINE	METHANOL	ETHANOL
Specific Gravity	0.70–0.78	0.79	0.794
Liquid Density			
lb/ft^3	43.6 Approx.	49.30	49.30
lb/gal	5.8–6.5	6.59	6.59
Vapor Pressure			
psi at 100°F (Reid)	7–15		2.50
psi at 77°F	0.3 Approx.		0.85
Boiling Point (°F)	80–440	149	173
Freezing Point (°F)	-70 Approx.	-208	-173
Solubility in Water (ppm)	240	Infinite	Infinite
Solubility of Water in Compound (ppm)	88	Infinite	Infinite
Viscosity at 68°F (Centipoise)	0.288		1.17

THERMAL PROPERTIES	GASOLINE	METHANOL	ETHANOL
Lower Heating Value			
Btu/lb	18,900 (Avg.)	9,066	11,500
Btu/gal	115,400 (Avg.)		73,560
Higher Heating Value			
Btu/lb at 68°F	20,260	10,258	12,800
Btu/gal	124,800		84,400
Heat of Vaporization			
Btu/lb	150	506	396
Btu/gal	900	3,340	3,378
Octane Ratings			
Research	91–105	106	106–108
Pump (RON + MON)/2	86–90	92	98–100
Flammability Limits (% by Volume in Air)	1.4–7.6		3.3–19.0
Specific Heat (Btu/lb – °F)	0.48		0.60
Autoignition Temperature (°F)	430–500		685
Flash Point	-50		55
Coefficient of Thermal Expansion at 60°F and 1 atm	0.0006		0.00112

Red Winter Wheat and All Wheat Production Costs, 1978

COST ITEM	CENTRAL PLAINS	SOUTHERN PLAINS	NORTHERN PLAINS	SOUTHWEST	UNITED STATES	ALL WHEAT TOTAL
COSTS PER ACRE						
Variable Items						
Seed	$ 2.26	$ 2.79	$ 2.75	$ 9.19	$ 2.64	$ 3.89
Fertilizer	4.27	5.80	6.18	18.38	5.28	7.81
Lime	NA	NA	NA	NA	NA	0.12
Chemicals[1]	0.61	1.21	1.05	4.74	0.94	1.16
Custom Operations[2]	2.36	3.03	2.66	7.49	2.72	2.92
All Labor	9.14	9.01	7.19	21.60	9.15	8.63
Fuel and Lubrication	5.37	5.80	4.25	6.24	5.40	5.19
Repairs	5.81	5.65	5.91	7.63	5.81	5.89
Purchased Irrigation Water	NA	NA	0.73	33 16	0.80	0.55
Interest	1.28	1.38	1.36	4 66	1.40	1.48
Total Cost, Variable Items	$31.10	$34.67	$32.08	$113.09	$34.14	$37.64
Machinery Ownership						
Replacement	13.52	13 64	14 51	17.69	13.76	14.15
Interest	5.90	5.59	6.71	7.63	5.93	6.17
Taxes and Insurance	1.81	1.68	2.05	2.31	1.80	1.87
Total Cost, Machinery Ownership	$21.23	$20.91	$23.27	$ 27.63	$21.49	$22.19
General Farm Overhead	6.63	5.17	6.16	13.44	6.36	6.55
Management[3]	5.90	6.08	6.15	15.42	6.20	6.64
TOTAL COST PER ACRE, EXCLUDING LAND	$64.86	$66.83	$67.66	$169.58	$68.19	$73.02
Land Allocation, Composite With:						
Current Value[4]	38.72	29.02	44.36	86.28	37.27	43.68
Average Acquisition Value[5]	23.96	15.96	21.71	62.39	21.96	25.07
COSTS PER BUSHEL						
Variable	$ 1.13	$ 1.66	$ 1.00	$ 1.84	$ 1.28	$ 1.27
Machinery Ownership	0.77	1.00	0.73	0 45	0.80	0.75
Farm Overhead	0.24	0.25	0 19	0.22	0.24	0.22
Management	0.21	0.29	0.19	0.25	0.23	0.22
Total, Excluding Land	$ 2.35	$ 3.20	$ 2.11	$ 2.76	$ 2.55	$ 2.46
Land Allocation, Composite With:						
Current Value	1.40	1.39	1.39	1.41	1.40	1.47
Average Acquisition Value	0.87	0.76	0.68	1.02	0.82	0.84
Value of Pasture	0.06	0.35	NA	NA	0.12	0.06
TOTAL PER BUSHEL COSTS OF PRODUCTION TO A RENTER						
Cost to Share Renter[6]	3.48	4.65	3 09	3.66	3.73	3.62
Cost to Cash Renter[7]	3.44	3.73	3 13	3.72	3.57	3.61
Weighted Renter Cost[8]	3.48	4.44	3.10	3.69	3.71	3.62
YIELD PER ACRE (BUSHELS)	27.6	20.9	32.0	61.3	26.7	29.7
PERCENT OF U.S. PRODUCTION	54.7	24.6	13.3	4.9	97.5	96.9

[1] Includes herbicides, insecticides, and rodenticides not otherwise included under custom operations
[2] Includes custom application of crop chemicals, the cost of chemicals in some cases, and custom harvesting and hauling
[3] Based on 10 percent of above costs
[4] Based on prevailing tenure arrangements in 1974, reflecting actual combinations of cash rent, net share rent, and owner-operator land allocations, land values, land tax rates, and cash rents updated to current year
[5] Same as footnote 4, except average value of cropland during the last 35 years is used for owner-operated land instead of current land value
[6] Share-renter portion of cost divided by share-renter portion of crop
[7] Cash-renter costs including cash rent divided by total yield
[8] Weighted average of share renter and cash renter based on prevailing tenure arrangements in 1974
NA — Not applicable

Source   USDA Agricultural Statistics 1979

Barley Production Costs, 1978

COST ITEM	NORTHEAST	NORTHERN PLAINS	SOUTHERN PLAINS	SOUTHWEST	NORTHWEST	UNITED STATES
COSTS PER ACRE						
Variable Items						
Seed	$ 6.23	$ 4.05	$ 4 57	$ 7 32	$ 5 43	$ 4 75
Fertilizer	14.16	5.60	7 07	8.52	10.73	7.01
Lime	2.74	—	—	—	—	0.04
Chemicals[1]	1.78	1.26	2.75	1.50	2.35	1 54
Custom Operations[2]	3.24	2.60	2.52	4.38	3.26	2.95
All Labor	9.76	7.90	13.24	21.38	12 42	10.62
Fuel and Lubrication	4.95	4.83	12.68	5.87	4.12	5.19
Repairs	5.23	5.89	7.60	7.23	5.80	6 11
Purchased Irrigation Water	—	0.97	—	25.35	4.57	4 63
Miscellaneous	—	0.11	—	—	—	0.07
Interest	2.29	0.71	1.57	3.53	1.14	1.20
Total Cost, Variable Items	$50.38	$33.92	$52.00	$ 85.08	$49.82	$44 11
Machinery Ownership						
Replacement	14.38	14.20	21.61	16.63	15.38	15.03
Interest	6.02	6.37	9.04	7.05	7.07	6.68
Taxes and Insurance	1.81	1.93	2 79	2.12	2.13	2.02
Total Cost, Machinery Ownership	$22.21	$22.50	$33.44	$ 25.80	$24.58	$23 73
General Farm Overhead	8.35	5.55	5.60	11.15	7.05	6.56
Management[3]	8.09	6.20	9.10	12.20	8.15	7.44
TOTAL COST PER ACRE, EXCLUDING LAND	$89.03	$68.17	$100.14	$134.13	$89.60	$81.84
Land Allocation, Composite With:						
Current Value[4]	109.79	34.31	40.23	71.92	60.55	44 82
Average Acquisition Value[5]	40.56	18.46	25.12	46.61	30.52	24.66
COSTS PER BUSHEL						
Variable	$ 1.16	$ 0.78	$ 1.09	$ 2.02	$ 0.88	$ 0.97
Machinery Ownership	0.51	0.51	0.70	0.62	0.43	0.52
Farm Overhead	0.19	0.13	0.12	0.26	0.12	0.14
Management	0.19	0.14	0.19	0.29	0.14	0.16
Total, Excluding Land	$ 2.05	$ 1.56	$ 2.10	$ 3.19	$ 1 57	$ 1 79
Land Allocation, Composite With:						
Current Value	2.52	0.79	0.85	1.71	1.06	0.98
Average Acquisition Value	0.93	0.42	0.53	1.11	0.53	0.54
TOTAL PER BUSHEL COSTS OF PRODUCTION TO A RENTER						
Cost to Share Renter[6]	3.59	2.31	2.95	4.89	2.20	2.72
Cost to Cash Renter[7]	2.74	2.17	3.23	3.84	2.11	2.46
Weighted Renter Cost[8]	2.83	2.26	2.98	4.63	2.19	2.64
YIELD PER ACRE (BUSHELS)	43.5	43.7	47.6	42.1	57.1	45.8
PERCENT OF U.S. PRODUCTION	1.3	56.5	4 0	10.8	18 5	91 1

[1] Includes herbicides, insecticides, and rodenticides not otherwise included under custom operations
[2] Includes custom application of crop chemicals, the cost of chemicals in some cases, and custom harvesting and hauling
[3] Based on 10 percent of above costs
[4] Based on prevailing tenure arrangements in 1974, reflecting actual combinations of cash rent, net share rent, and owner-operator land allocations, land values, land tax rates, and cash rents updated to current year
[5] Same as footnote 4, except average value of cropland during the last 35 years is used for owner-operated land instead of current land value
[6] Share-renter portion of cost divided by share-renter portion of crop
[7] Cash-renter costs including cash rent divided by total yield
[8] Weighted average of share renter and cash renter based on prevailing tenure arrangements in 1974

Source   USDA Agricultural Statistics 1979

Grain Sorghum Production Costs, 1978

COST ITEM	CENTRAL PLAINS	SOUTHERN PLAINS	SOUTHWEST	UNITED STATES
COSTS PER ACRE				
Variable Items				
Seed	$ 3.56	$ 3.46	$ 7.57	$ 3.60
Fertilizer	12.14	11.89	29.70	12.37
Lime	0.02	(1)	(1)	0.01
Chemicals[2]	3.83	1.85	2.00	2.95
Custom Operations[3]	4.73	4.82	4.96	4.77
All Labor	10.42	14.07	39.04	12.53
Fuel and Lubrication	7.37	12.51	26.80	9.94
Repairs	6.70	10.75	12.55	8.54
Drying	2.17			1.20
Purchased Irrigation Water			23.28	0.45
Interest	1.29	1.61	5.32	1.50
Total Cost, Variable Items	$52.23	$60.96	$151.22	$57.86
Machinery Ownership				
Replacement	15.82	25.06	25.96	19.95
Interest	6.96	10.13	11.41	8.40
Taxes and Insurance	2.12	3.00	3.39	2.52
Total Cost, Machinery Ownership	$24.90	$38.19	$40.76	$30 87
General Farm Overhead	6.63	6.52	11.56	6.67
Management[4]	8.38	10.57	20.35	9.54
TOTAL COSTS PER ACRE, EXCLUDING LAND	$92.14	$116.24	$223.89	$104.94
Land Allocation, Composite With:				
Current Value[5]	42.00	28.56	82.01	37.04
Average Acquisition Value[6]	27.58	17.74	56.42	23.94
COSTS PER BUSHEL				
Variable	$ 0.91	$ 1.34	$ 2.16	$ 1.10
Machinery Ownership	0.43	0.84	0.58	0.59
Farm Overhead	0.12	0.14	0.16	0.13
Management	0.15	0.23	0.29	0.18
Total Cost, Excluding Land	$ 1.61	$ 2.55	$ 3.19	$ 2.00
Land Allocation, Composite With:				
Current Value	0.73	0.63	1.17	0.71
Average Acquisition Value	0.48	0.39	0.80	0.46
TOTAL PER BUSHEL COSTS OF PRODUCTION TO A RENTER				
Cost to Share Renter[7]	2.36	3.48	4.54	2.83
Cost to Cash Renter[8]	2.05	3.03	3.71	2.63
Weighted Renter Cost[9]	2.33	3.42	4.20	2.81
YIELD PER ACRE (BUSHELS)	57.2	45.5	70.2	52.5
PERCENT OF U.S. PRODUCTION	58.9	36.0	2.5	97.4

[1] Not applicable
[2] Includes herbicides, insecticides, and rodenticides not otherwise included under custom operations
[3] Includes custom application of crop chemicals, the cost of chemicals in some cases, and custom harvesting and hauling
[4] Based on 10 percent of above costs
[5] Based on prevailing tenure arrangements in 1974, reflecting actual combinations of cash rent, net share rent, and owner-operator land allocations, land values, land tax rates, and cash rates updated to current year
[6] Same as footnote 5, except average value of cropland during the last 35 years is used for owner-operated land instead of current land value
[7] Share-renter portion of cost divided by share-renter portion of crop
[8] Cash-renter costs including cash rent divided by total yield
[9] Weighted average of share renter and cash renter based on prevailing tenure arrangements in 1974

Source USDA Agricultural Statistics 1979

COMMERCIAL SCALE ETHANOL PRODUCTION AND FINANCING

Grain Sorghum Silage and Forage Use, 1976 to 1978

STATE	SILAGE									FORAGE AREA HARVESTED		
	AREA HARVESTED			AVERAGE YIELD PER ACRE			PRODUCTION					
	1976	1977	1978*	1976	1977	1978*	1976	1977	1978*	1976	1977	1978*
	1,000 ACRES	1,000 ACRES	1,000 ACRES	TONS	TONS	TONS	1,000 ACRES	1,000 ACRES	1,000 ACRES	1,000 ACRES	1,000 ACRES	1,000 ACRES
AL	18	15	18	11.5	10.5	10.0	207	158	180	11	18	10
AZ	5	6	5	16.5	17.0	17 0	83	102	85	2	2	1
AK	13	17	12	9.5	10.5	10 5	124	179	126	10	11	13
CA	11	7	12	17.0	17.0	17.0	187	123	204	10	10	8
CO	21	21	19	11.0	7.0	11.0	231	147	209	165	146	131
GA	30	18	29	12.0	7.0	12.5	360	126	363	8	11	8
IL	10	11	7	11.5	12.5	11.5	115	138	81	7	5	5
IN	8	6	7	12.5	12.0	13.0	100	72	91	–	–	–
IA	13	11	9	11.0	13.0	14.0	143	143	126	2	1	2
KS	290	300	230	8.6	12.5	10.0	2,494	3,750	2,300	340	430	355
KY	13	11	8	12.5	12.5	12.0	163	138	96	5	5	4
LA	11	9	9	10.0	11.0	10.0	110	99	90	5	5	2
MS	20	27	35	11.5	12.5	13 0	230	338	455	9	4	8
MO	30	44	33	9.5	11.0	11.5	285	484	380	35	56	39
NE	80	90	70	7.5	11.5	13.0	600	1,035	910	60	70	50
NM	5	8	4	11.0	12.0	13.0	55	96	52	41	37	47
NC	25	22	24	11.5	8.0	13 5	288	176	324	7	12	13
OK	41	34	45	8.0	11 0	9 0	328	374	405	139	121	115
SC	12	11	11	9.0	6.5	9.5	108	72	105	2	3	2
SD	45	56	55	3.0	8.0	7.5	135	448	413	115	74	55
TN	12	8	11	11.0	10.0	10 5	132	80	116	8	10	8
TX	50	100	50	12.0	8.0	10.5	600	800	515	900	600	700
VA	9	10	11	10.0	9 0	11.0	90	90	121	2	2	2
U S	772	842	714	9 3	10 9	10 9	7,168	9,168	7,747	1,883	1,633	1,578

* Preliminary.

Source: USDA Agricultural Statistics 1979.

Grain Sorghum Area, Yield, and Production, 1977-1978

STATE	AREA PLANTED FOR ALL PURPOSES		SORGHUM FOR GRAIN				PRODUCTION	
			AREA HARVESTED		AVERAGE YIELD PER HARVESTED ACRE			
	1977	1978	1977	1978	1977	1978	1977	1978*
	THOUSAND ACRES	THOUSAND ACRES	THOUSAND ACRES	THOUSAND ACRES	BUSHELS	BUSHELS	THOUSAND ACRES	THOUSAND ACRES
AL	75	65	27	34	27.0	37.0	729	1,258
AZ	100	80	90	73	80.0	78.0	7,200	5,694
AR	285	230	252	200	52 0	60.0	13,104	12,000
CA	150	210	132	185	73.0	71.0	9,636	13,135
CO	460	490	263	280	31.0	81.0	8,153	8,860
GA	75	85	24	43	28 0	29.0	672	1,247
IL	80	80	64	68	64.0	68.0	4,096	4,624
IN	23	25	15	15	78.0	65.0	1,170	975
IA	45	36	32	24	74.0	75.0	2,368	1,800
KS	4,850	4,700	4,050	4,020	60.0	52.0	243,000	209,040
KY	50	37	32	23	57.0	62.0	1,824	1,426
LA	35	30	20	17	33.0	34.0	660	576
MS	60	65	24	21	32.0	38.0	768	798
MO	1,050	930	930	850	73.0	80.0	67,890	68,000
NE	2,300	2,000	2,070	1,830	71.0	75.0	146,970	137,250
NM	297	336	245	267	48.0	46.0	11,760	12,282
NC	110	125	72	86	37.0	52.0	2,664	4,472
OK	765	700	565	485	38.0	36.0	21,470	17,460
SC	23	26	12	15	16.0	32.0	192	480
SD	29	29	343	340	49.0	50.0	16,807	17,000
TN	40	45	20	24	51.0	51.0	1,020	1,224
TX	5,600	5,700	4,800	4,650	48.0	49.0	230,400	227,850
VA	24	25	10	11	43.0	47.0	430	517
U S	16,526	16,049	14,092	13,561	56 3	55 1	792,983	747,788

* Preliminary.

Source: USDA Agricultural Statistics 1979.

COMMERCIAL SCALE ETHANOL PRODUCTION AND FINANCING

Corn Area, Yield, and Production, 1977-1978

STATE	AREA PLANTED FOR ALL PURPOSES		CORN FOR GRAIN					
			AREA HARVESTED		AVERAGE YIELD PER HARVESTED ACRE		PRODUCTION	
	1977	1978	1977	1978	1977	1978	1977	1978[1]
	THOUSAND ACRES	THOUSAND ACRES	THOUSAND ACRES	THOUSAND ACRES	BUSHELS	BUSHELS	THOUSAND BUSHELS	THOUSAND BUSHELS
AL	840	640	375	544	29.0	50.0	10,875	27,200
AZ	65	70	50	50	100 0	115.0	5,000	5,750
AR	55	40	43	30	53.0	58.0	2,279	1,740
CA	430	420	247	281	116.0	126.0	28,652	35,406
CO	960	1,000	695	720	116.0	110.0	80,620	79,200
CT	54	53	–	–	–	–	–	–
DE	203	187	185	175	56.0	96.0	10,360	16,800
FL	623	430	299	370	35.0	52 0	10,465	19,240
GA	2,240	1,700	1,000	1,500	24.0	50.0	24,000	75,000
ID	120	123	28	39	86.0	87.0	2,408	3,393
IL	11,350	11,000	11,080	10,730	105 0	111 0	1,163,400	1,191,030
IN	6,400	6,100	6,210	5,900	102.0	108.0	633,420	637,200
IA	13,800	13,300	12,700	12,500	86.0	117 0	1,092,200	1,462,500
KS	2,030	1,820	1,680	1,500	96.0	102.0	161,280	153,000
KY	1,650	1,570	1,470	1,410	90 0	85.0	132,300	119,850
LA	86	65	65	47	52.0	59.0	3,380	2,773
ME	51	50	–	–	–	–	–	–
MD	730	690	600	590	72.0	97.0	43,200	57,230
MA	42	43	–	–	–	–	–	–
MI	2,800	2,670	2,320	2,250	85.0	81.0	197,200	182,250
MN	6,900	7,000	6,000	6,190	100 0	104.0	600,000	643,760
MS	250	215	160	135	36.0	56.0	5,760	7,560
MO	2,900	2,400	2,650	2,200	76.0	87.0	201,400	191,400
MT	90	88	11	5	68.0	72.0	748	360
NE	7,150	7,100	6,550	6,550	99.0	113.0	648,450	740,150
NV	3	–	–	–	–	–	–	–
NH[2]	26	27	–	–	–	–	–	–
NJ	149	135	95	95	70 0	91.0	6,650	8,645
NM	135	90	114	72	90.0	105.0	10,260	7,560
NY	1,375	1,300	640	600	80.0	90 0	51,200	47,400
NC	2,000	1,760	1,740	1,600	51.0	76.0	88,740	121,600
ND	620	600	237	253	73.0	79.0	17,301	19,987
OH	3,900	3,850	3,620	3,610	105.0	105.0	380,100	379,050
OK	140	120	95	73	82.0	65.0	7,790	4,745
OP	45	45	12	13	95.0	95 0	1,140	1,235
PA	1,615	1,615	1,160	1,190	92 0	95 0	106,720	113,050
RI	4	4	–	–	–	–	–	–
SC	825	640	690	550	36.0	55 0	24,840	30,250
SD	3,000	3,250	2,150	2,560	59 0	67 0	126,850	171,520
TN	900	820	730	660	65 0	66.0	47,450	43,560
TX	1,800	1,600	1,650	1,440	98.0	100.0	161,700	144,000
UT	80	92	13	16	89.0	90.0	1,157	1,440
VT	110	112	–	–	–	–	–	–
VA	855	825	560	615	55 0	82.0	30,800	50,430
WA	128	130	64	65	119.0	121 0	7,616	7,865
WV	100	93	54	58	74 0	77.0	3,996	4,466
WI	3,850	3,750	2,800	2,750	104 0	98.0	291,200	269,500
WY	89	87	30	34	85.0	81.0	2,550	2,754
U S	83,568	79,719	70,872	69,970	90 7	101 2	6,425,457	7,081,849

[1] Preliminary.

[2] Estimates discontinued after 1977 crop.

Source   USDA Agricultural Statistics 1979.

Production of Selected Fruit Crops

YEAR	APPLES (COMMERCIAL CROP) THOUSAND TONS	PEACHES THOUSAND TONS	PEARS THOUSAND TONS	GRAPES THOUSAND TONS
1964	3,160	1,726	728	3,478
1965	3,070	1,737	500	4,351
1966	2,881	1,695	748	3,734
1967	2,718	1,344	464	3,069
1968	2,735	1,818	624	3,549
1969	3,410	1,842	727	3,899
1970	3,199	1,498	549	3,103
1971	3,187	1,441	749	3,994
1972	2,939	1,186	612	2,579
1973	3,133	1,295	730	4,198
1974	3,290	1,459	742	4,199
1975	3,765	1,419	749	4,366
1976	3,237	1,510	841	4,398
1977	3,336	1,492	787	4,298
1978	3,817	1,351	727	4,567

Source: USDA Agricultural Statistics 1979.

Sugar Beet Production and Value, 1977

STATE	PRODUCTION THOUSAND TONS	SEASON AVERAGE PRICE PER TON RECEIVED BY FARMERS DOLLARS	VALUE OF PRODUCTION THOUSAND DOLLARS
AZ	285	$24 40	$ 6,954
CA	5,664	26 40	149,530
CO	1,404	26 30	36,925
ID	2,094	25 50	53,397
KS	401	21 90	8,792
MI	1,796	21.10	36,100
MN	4,732	20 60	97,479
MT	896	29.10	26,074
NB	1,365	27 00	36,558
NM	23	25 00	575
ND	2,769	21 40	59,257
OH	457	20.20	9,231
OR	206	23 00	4,738
TX	309	23 40	7,231
UT	173	26 50	4,619
WA	1,495	26 50	39,618
WY	949	28 80	27,331
U.S.	25,048	$24.20	$604,409

* Relates to year of harvest, includes some acreage planted previous fall

Source. USDA Agricultural Statistics 1979.

COMMERCIAL SCALE ETHANOL PRODUCTION AND FINANCING

Sugar Beet Area, Yield, and Production

STATE	AREA PLANTED		AREA HARVESTED		AVERAGE YIELD PER HARVESTED ACRE		PRODUCTION	
	1977	1978 [1]	1977	1978 [1]	1977	1978 [1]	1977	1978 [1]
	THOUSAND ACRES	THOUSAND ACRES	THOUSAND ACRES	THOUSAND ACRES	TONS	TONS	THOUSAND ACRES	THOUSAND ACRES
AZ [2]	12.9	15.7	12 8	15.0	22.3	20.5	285	308
CA [2]	227.0	207.0	217.0	195.0	26.1	24.5	5,664	4,778
CO	77.0	89.0	72.0	84 0	19.5	18.3	1,404	1,538
ID	115.4	136.3	107.4	134.1	19.5	20.3	2,094	2,722
KS	26.0	18.0	24 0	26 0	16.7	17.0	401	442
MI	92.3	93.0	85 5	91 0	21.0	19.3	1,796	1,756
MN	264.0	265.0	260.0	263.0	18.2	18.9	4,732	4,971
MT	46.4	45.4	45 0	44.0	19 9	19.8	896	885
NB	75.0	79.0	67.7	67 7	20.0	18.0	1,354	1,368
NM	1.3	2.1	1 2	1 2	19.2	20.6	23	37
ND	157.8	156.2	155.2	155.2	17.8	19.7	2,769	3,056
OH	24.9	24.5	23.3	23.3	20.3	16.9	457	394
OR	8.9	9.2	8.9	8.9	25 1	24.0	206	214
TX	19.9	28.0	23.5	23.5	17.3	17.6	309	414
UT	10.4	14.9	14.7	14.7	17.7	17.0	173	250
WA	63.9	69.2	68.5	68 5	24.3	26.5	1,495	1,815
WY	49.5	49.5	48.8	48.8	19.6	18.9	949	922
U S	1,272 6	1,302 0	1,235 5	1,263 9	20 6	20 3	25,007	25,870

[1] Preliminary.

[2] Relates to year of harvest

Source. USDA Agricultural Statistics 1979.

Components of Molasses

COMPONENTS	COMPOSITION %	
	RANGE	AVERAGE
Water	17 — 25	20
Sucrose	30 — 40	35
Dextrose (Glucose)	4 — 9	7
Levulose (Fructose)	5 — 12	9
Other Reducing Substrates	1 — 5	4
Ash	7 — 15	12
Nitrogenous Compounds	2 — 6	4.5
Non-Nitrogenous Acids	2 — 8	5
Waxes, Sterols, Phospholipids	0 — 1	0.4

Source. Paturau, J. M., By-products of the Cane Sugar Industry. Amsterdam, The Netherlands Elsevier Publishing, 1969.

Nutritional Content of Distillers Feed from Corn in Percentages

CONSTITUENT	DISTILLERS DRIED GRAINS	DISTILLERS DRIED SOLUBLES	DISTILLERS DRIED GRAINS AND SOLUBLES
Moisture	7.50	4.50	9.00
Protein	27.00	28.50	27.00
Fat	7.60	9.00	8.00
Fiber	12.80	4.00	8.50
Ash	2.00	7.00	4.50
Amino Acids			
Lysine	0.60	0.95	(Available 0.60)
Methionine	0.50	0.50	(Available 0.60)
Cystine	0.20	0.40	0.40
Histidine	0.60	0.63	0.60
Arginine	1.10	1.15	1.00
Aspartic Acid	1.68	1.90	1.70
Threonine	0.90	0.98	0.95
Serine	1.00	1.25	1.00
Glutamic Acid	4.00	6.00	4 20
Proline	2.60	2.90	2.80
Glycine	1.00	1.20	1.00
Alanine	2.00	1.75	1.90
Valine	1.30	1.39	1.30
Isoleucine	1.00	1.25	1 00
Leucine	3.00	2.60	2.70
Tyrosine	0.80	0.95	0.80
Phenylalanine	1.20	1.30	1.20
Tryptophan	0.20	0.30	0.20
Fatty Acids			
Linoleic			
Fat	47.20	49.10	48.50
Ingredient	3.60	4.40	3.90
Linolenic			
Fat	5.20	5.10	5.00
Ingredient	0.38	0.46	0.40

Source Distillers Feed Research Council

Available Energy from Corn in Distillers Feeds

CORN ENERGY	DISTILLERS DRIED GRAINS	DISTILLERS DRIED SOLUBLES	DISTILLERS DRIED GRAINS AND SOLUBLES
Cattle			
DE, kcal/kg	3,408	3,608	3,570
ME, kcal/kg	2,794	2,959	2,927
TDN, %	83	80	82
NEmilk	2,150	2,210	2,210
NEm	2,050	2,030	2,035
NEp	1,347	1,335	1,335
Poultry			
ME, kcal/kg	1,631	2,750	2,620
Swine			
DE, kcal/kg	2,030	3,305	3,085
ME, kcal/kg	1,835	2,985	2,790
TDN, %	46	75	70

Source Distillers Feed Research Council

COMMERCIAL SCALE ETHANOL PRODUCTION AND FINANCING

APPENDIX D

Summary of Ethanol Legislation

- **National Legislation**
- **State Legislation**

This appendix provides a summary of the National and State Legislation on alcohol fuel. Significant portions of this material was initially prepared and published by the National Alcohol Fuel Commission.

NATIONAL LEGISLATION

The Federal programs available for increasing our alcohol fuels production capacity are described in this appendix. This appendix identifies 10 existing programs within the Department of Agriculture, the Department of Commerce, the Small Business Administration, the Department of Housing and Urban Development, the Department of Energy, and the Department of the Treasury. It also identifies four items of authorizing legislation for the Departments of Agriculture, Commerce, and Energy.

The 96th Congress has provided new programs for the stimulation of alcohol fuel production. In addition, existing legislation has been interpreted by the agencies affected which will enable and increase the number of commercial scale plants in the near-term.

Two new major laws have been passed: The Alternative Fuels Production Appropriations Act, P.L. 96-126, Title II; and The Biomass Energy and Alcohol Fuels Act of 1980, P.L. 96-294, Title II.

The Alternative Fuels Act is oriented toward the Department of Energy and sets aside $1.5 billion for the development of alternative fuels consisting of $100 million for cooperative agreements with non-Federal entites, and $500 million for a reserve to cover any defaults from loan guarantees. A portion of these funds will be spent on alcohol fuel projects. The Biomass Energy and Alcohol Fuels Act applies to both the Department of Agriculture and the Department of Energy and provides authority to enter into loan guarantees, price guarantees, and purchase agreements for alcohol fuels and other biomass energy projects. Additionally, the Department of Agriculture may provide insured loans for small-scale projects. A total of $1.05 billion has been appropriated for use over the next two years for biomass projects, including assistance for alcohol fuels projects.

Under existing legislation, the Department of Agriculture provides financial assistance for alcohol production through farm ownership loans, farm operating loans, and business and industrial loans under the responsibility of the Farmers Home Administration, and other loans under the responsibility of the Commodity Credit Corporation. For economic development, the Department of Commerce has directed its attention to providing grants, loans, and loan guarantees for creating and retaining jobs by financing the construction and operation of alcohol plants. The Small Business Administration is attempting to finance plant construction, expansion, conversion, or startup and the Department of Housing and Urban Development endeavors to achieve its mission of economic development and neighborhood revitalization.

In addition, the Department of the Treasury encourages the construction of domestic alcohol production facilities through its powers to provide tax exemptions and credits, the Economic Regulatory Administration of DOE encourage alcohol production through its power to regulate cash entitlements.

This section of the Appendix identifies the existing public laws and highlights some currently planned legislation which could impact on the near-term alcohol fuels production outlook. The following legislation is not addressed in this section.

- Current and pending legislation dealing with research, development, and small-scale testing (e.g., Economic Opportunity Act of 1964, Title II, Section 22(a), P.L. 95-568, Community Services Administration; and the Urban Waste Program, P.L. 93-577, Department of Energy).

- Administrative details such as establishing a Federal infrastructure (e.g., the Priority Energy Project Act; Energy Mobilization Board, S.1308; and Surface Transportation Assistance Act of 1978; the National Alcohol Fuel Commission, P.L. 96-106).

- Those appropriation acts which, in effect, deal only with funding the programs already dealt with in authorization bills (e.g., the U.S. Department of Agriculture Appropriations Act for FY 1980, P.L. 96-108; and the Energy and Water Development Appropriations Act of 1980, P.L. 96-69). Table D-1 summarizes the current Federal Programs.

PUBLIC LAWS

DEPARTMENT OF AGRICULTURE, FARMERS HOME ADMINISTRATION

LEGISLATION

RURAL DEVELOPMENT ACT OF 1972; P.L. 92-419, TITLE I, FARM OPERATING LOANS, AND FARM OWNERSHIP LOANS.

ECONOMIC/FINANCIAL BENEFIT(S)

Direct loans; loan guarantees; insured loans.

OBJECTIVE(S)

Interpreted to be appropriate to finance production of alcohol for on-farm use, or for partial off-farm use and sale, only if it is necessary to make the farm operation commercially viable. (See S.985/HR 3683).

Table D.1. Current Federal Alcohol Fuel Production
Economic/Financial Benefit Programs

	GRANT	COOPERATIVE AGREEMENT	DIRECT LOAN	LOAN GUARANTEE	INSURED LOAN	PURCHASE COMMITMENT	PRICE GUARANTEE	JOINT VENTURE AGREEMENT	OTHER INCENTIVES
DEPARTMENT OF AGRICULTURE									
Farm operating loans and farm ownership loans			•	•	•				
Business and industrial loans				•	•				
Commodity credit loans			•						
DEPARTMENT OF COMMERCE									
Economic development assistance	•		•	•					
SMALL BUSINESS ADMINISTRATION									
Energy loans			•	•	•				
DEPARTMENT OF HOUSING AND URBAN DEVELOPMENT									
Urban development action grants	•								
DEPARTMENT OF ENERGY									
Alternative fuels production	•	•		•		•	•		
Crude oil entitlements									•
DEPARTMENT OF ENERGY AND DEPARTMENT OF AGRICULTURE									
Energy Security Act			•	•	•	•	•	•	
DEPARTMENT OF THE TREASURY									
Energy Tax Act of 1978									•
Crude Oil Windfall Profit Tax Act of 1980									•

ELIGIBILITY

After financial assistance is received, must be operators of family farms who are U.S. citizens, are of legal age, cannot obtain sufficient credit elsewhere at reasonable rates and terms, and otherwise meet the requirements for an FmHA farm loan. If the item for which financial assistance is sought is considered chattel property or operating expense (e.g., an alcohol fueled tractor, or a portable still) then the applicant files for a farm operating loan; if the item is considered a permanent installation on the real estate (e.g., an alcohol plant, methane digestor) then the applicant files for a farm ownership loan.

FINANCIAL TERMS

Guranteed/insured loans are made and services by legally organized private lending institutions, such as commercial banks, Federal land banks, production credit associations, insurance companies, and savings and loan associations. FmHA provides the lender with a guarantee to reimburse up to 90 percent of any loss the lender takes on a loan.

Farm operating loans may be repaid in from one to seven years. In some cases, borrowers may be given an additional seven years to repay. The interest rate currently is at 10.5 percent. The limit of an operating loan made directly by FmHA is $100,000; a private loan guarantee by FmHA has a limit of $200,000.

Farm ownership loans have a maximum term of forty years. Interest rates are currently 10 percent annually. The limit on a farm ownership loan made directly by FmHA is $200,000; on a guaranteed loan is $300,000.

DEPARTMENT OF AGRICULTURE, FARMERS HOME ADMINISTRATION

LEGISLATION

RURAL DEVELOPMENT ACT OF 1972; P.L. 92-419, TITLE I, BUSINESS AND INDUSTRIAL LOAN PROGRAM.

ECONOMIC/FINANCIAL BENEFIT(S)

Loan guarantees; insured loans.

OBJECTIVE(S)

To assist public, private, or cooperative organizations organized for profit or nonprofit, Indian tribes or individuals in rural areas to obtain quality loans for the purpose of improving, developing, or financing business, industry, and employment and improving the economic and environmental climate in rural communities.

ELIGIBILITY

Applicant must be located in the fifty states, Puerto Rico, or the Virgin Islands, in areas other than cities having a population of more than 50,000 and its immediately adjacent urbanized areas with a population density of more than 100 persons per square mile. Gasohol projects located on farms are not appropriate for funding here. (See FmHA Farm Operating and Farm Ownership loans.) Non-farm distilleries can qualify under this program.

Guidelines have been established that the operating prototypes for pre-engineered plant designs must meet. Prototype plants producing 500,000 gallons a year or less must have operated for 60 days at the rated annual capacity, produced a minimum 160 proof alcohol, convert 75 percent of the fermentable carbohydrates, be inspected on-site, and have an insurable design.

Pre-engineered alcohol fuel plants qualifying for loan guarantees will have proven successful in prototype plants operating for at least 60 days. Custom- or self-built plants must be based on plans that have been certified by technical experts or on FmHA approved generic plans that are available at no charge.

Plants producing from 500,000 to 5 million gallons a year must produce 190 or higher proof alcohol, must convert 85 percent of the fermentable carbohydrates, have a designer performance guarantee mechanism, and have an insurable design.

FINANCIAL TERMS

FmHA may guarantee up to 90 percent (80 percent in the case of gasohol) of a loan with applicant responsible for furnishing a minimum of 10 to 20 percent of equity. Maximum terms are 30 years for land, buildings, and permanent features; 15 years for machinery or equipment or the life of the machinery or equipment, whichever is shorter; 7 years for working capital. The interest rate for guaranteed loans may be fixed or variable and the amount will be determined between the lender and borrower. No loan guarantee application may exceed $50 million.

Insured loan rates to public bodies, non-profit associations, and Indian tribes are at the rate of 5 percent per annum. The maximum allowable maturity of an FmHA insured loan for community facilities shall not exceed 40 years.

DEPARTMENT OF AGRICULTURE, COMMODITY CREDIT CORPORATION

LEGISLATION

FOOD AND AGRICULTURE ACT OF 1977; P.L. 95-113, SECTION 1420

ECONOMIC/FINANCIAL BENEFIT(S)

Loan guarantees.

OBJECTIVE(S)

To encourage farmers to make additional uses of their commodities in order to achieve greater crop handling and marketing flexibility.

ELIGIBILITY

Any person who as owner, landlord, tenant, or sharecropper produces one or more applicable commodities. Project must be designed to produce industrial hydrocarbons and alcohol from "agricultural commodities and forest products." (Biomass). The industrial hydrocarbons and alcohol must be for nonfood and nonfeed purposes, such as fuels or industrial energy-type raw materials. Total energy content of the products must exceed the total energy input from fossil fuels used in the manufacture of the products.

FINANCIAL TERMS

This act provides $60 million in loan guarantees to build four pilot alcohol plants in the U.S. The four projects are not to exceed $15 million for each. The loan guarantee covers 90 percent of the estimated aggregate cost of the project. Efforts are underway to expand this program (see S.892/HR 3580).

DEPARTMENT OF COMMERCE, ECONOMIC DEVELOPMENT ADMINISTRATION

LEGISLATION

PUBLIC WORKS AND ECONOMIC DEVELOPMENT ACT OF 1965; P.L. 89-136, as amended.

ECONOMIC/FINANCIAL BENEFIT(S)

Grants; direct loans; loan guarantees.

OBJECTIVE(S)

To assist in the construction of public facilities needed to initiate and encourage long-term economic growth and to sustain industrial and commercial viability in designated areas by providing financial assistance to public and private organizations that create or retain permanent jobs by expanding or establishing facilities and plants in redevelopment areas. Financial assistance must not be available from other sources on terms and conditions that would permit accomplishment of the project and further economic development in the area.

ELIGIBILITY

Any individual private or public corporation, or Indian tribe, provided that the project to be funded is physically situated in an area designated as eligible under the Act at the time the application is filed. Neither business development loans nor guarantees of any kind will be extended to applicants who: (1) have, within the previous three years, relocated any or all of their facilities to another city or state; (2) contemplate relocating part or all of their existing facilities with resultant loss of employment at such facilities; and (3) produce a product or service for which there is a sustained and prolonged excess of supply over demand.

For public works grants (Title I), a proposed alcohol plant must be designed to produce less than one million gallons of ethanol per year. For private sector direct loans and loan guarantees, eligibility is generally confined to plants which produce between 500,000 and 50 million gallons per year unless applicants are ineligible for assistance under FmHA and SBA funding programs.

FINANCIAL TERMS

The basic public sector grant rate may be up to 50 percent of the project cost. Severely depressed areas that cannot match Federal funds may receive supplementary grants to bring the Federal contribution up to 80 percent of the project cost, with designated Indian Reservations eligible for 100 percent assistance. EDA's share is usually limited to $300,000 per project.

Direct, long-term business development loans up to 65 percent of the cost may be used for the acquisition of fixed assets only (i.e., land, building, machinery, and equipment, including land preparation and bulding rehabilitation). Direct loans for working capital needs are not limited by statute, but are available only for short periods.

The government will guarantee up to 90 percent of the unpaid balance of loans for the acquisition of fixed assets or for working capital; and up to 90 percent of the rental payments required by guaranteed lease arrangements. The maximum amount of the loan principal eligible for the Federal guarantee is limited to $5 million.

Efforts are underway to extend this program (see S.914).

SMALL BUSINESS ADMINISTRATION

LEGISLATION

SMALL BUSINESS ACT, AS AMENDED; P.L. 95-315, SMALL BUISNESS ENERGY LOANS, SECTION 7(L).

ECONOMIC/FINANCIAL BENEFIT(S)

Direct loans; loan guarantees; insured loans.

OBJECTIVE(S)

To assist small business concerns to finance plant construction expansion, conversion, or startup; and the acquisition of equipment, facilities, machinery, supplies or materials to enable such concerns to manufacture, design, market, install or service specific energy measures.

ELIGIBILITY

Applicant must be a small business concern as described in SBA regulations and must furnish evidence of being engaged in an eligible energy measure.

A direct loan cannot be made if an immediate participation loan is available, and an immediate participation loan cannot be made if a guaranteed loan is available. No loan may be made under this program unless the financial assistance is not otherwise available on reasonable terms from non-Federal sources.

An applicant must pledge collateral and give such personal guarantee as may be required. Since greater risk is associated with energy loans, emphasis is placed on the applicant's technology, capability of its employees, quality control, and the financial status of the firm.

Energy loans are not available for installing or undertaking energy conservation measures; for this purpose, SBA requests that small firms apply under SBA's regular business loan program. Additionally, energy loans generally cannot be used for research and development. However, under special circumstances, up to 30 percent of a loan may be approved for such purposes. Only 25 percent of any direct energy loan may be used for working capital.

FINANCIAL TERMS

$350,000 to any one borrower is the maximum SBA share of an immediate participation loan where SBA and private lending institutions each put up part of loan funds immediately; and is the maximum SBA direct loan made by the agency. For guaranteed loans made by a bank and partially guaranteed by SBA, the maximum is $500,000. SBA's share of an immediate participation loan cannot exceed 75 percent, and not greater than 90 percent for loan guarantees. No more than 30 percent of the funds can be used for research and development and working capital in any combination.

In FY 79, interest rates were set at 8-1/4 percent on direct loans and SBA's share of immediate participation loans. On the bank's share of an immediate participation loan, and on guaranteed loans, the lending institution may set reasonable and legal rates with a maximum ceiling rate set by SBA from time to time. Repayment period is a maximum of 15 years as a rule, however, working capital loans generally are limited to 6 years, while portions of loans for construction and acquisition of real estate may have a maximum of 20 years.

DEPARTMENT OF HOUSING AND URBAN DEVELOPMENT, COMMUNITY PLANNING AND DEVELOPMENT

LEGISLATION

TITLE I OF THE HOUSING AND COMMUNITY DEVELOPMENT ACT OF 1974, P.L. 93-383, AS AMENDED BY TITLE I OF THE HOUSING AND COMMUNITY DEVELOPMENT ACT OF 1977, P.L. 93-128, URBAN DEVELOPMENT ACTION GRANTS.

ECONOMIC/FINANCIAL BENEFIT(S)

Grants.

OBJECTIVE(S)

To assist severely distressed cities and urban counties in alleviating physical and economic deterioration through economic development and neighborhood revitalization. Alcohol fuel production support is an example of such efforts.

ELIGIBILITY

Eligible applicants are distressed cities and distressed urban counties which meet the following criteria specified in Section 470.452 of the regulations: (a) minimum standards of physical and economic distress; (b) demonstrated results in providing housing for persons of low and moderate income; and (c) demonstrated results in providing equal opportunity in housing and employment for low and moderate income persons and members of minority groups.

Projects which include financial assistance from the State or other public entities will receive more favorable consideration. Other public resources may be provided by matching other Federal grants, or by firm commitments or other Federal or local resources. No activities will be funded unless there is a firm commitment of private resources to the proposed project. Typical projects would include modification, reopening of inner city breweries, distilleries.

FINANCIAL TERMS

Assistance is for a discrete project which can be completed in approximately 4 years. No additional funding will be available in subsequent years to complete a project approved in a prior year. Funds are made available through a Letter of Credit.

There are no dollar limits for each award, and they have ranged from less than $100,000 to over $18 million; however, an amount of funding at least two and one-half times the amount requested from HUD must be provided by the private sector to be eligible for an award. Twenty-five percent of the allocated funds must be given to small cities (population less than 50,000). Urban Development Action Grants (UDAG) funds cannot be used for working capital.

DEPARTMENT OF ENERGY, ASSISTANT SECRETARY FOR RESOURCE APPLICATIONS

LEGISLATION

DEPARTMENT OF THE INTERIOR AND

RELATED AGENCIES APPROPRIATIONS FOR FISCAL YEAR 1980; P.L. 96-126, TITLE II, DEPARTMENT OF ENERGY, ALTERNATIVE FUELS PRODUCTION. P.L. 96-304, SUPPLEMENTAL APPROPRIATIONS

ECONOMIC/FINANCIAL BENEFIT(S)

Grants; cooperative agreement (plus others not applicable to alcohol fuels).

ELIGIBILITY

Alternative fuel produced must be derived from one of the following resources: coal/lignite, shale, tar sands, unconventional gases, peat, biomass, solid wastes (industrial and municipal), or other mineral or organic materials. Biomass (alcohol) projects must be commercial scale, i.e., at least one million gallons per year for ethanol production or its energy equivalent (85×10^9 Btu/year). Fuels which are derived from crude oil or derivatives therefrom are not eligible nor would be producing energy from the direct burning of any of the above resources.

FINANCIAL TERMS

This Act established an Energy Security Reserve for alternative fuels production in the amount of $19 billion. $2.2 billion of this reserve have been made available to the Secretary of Energy to stimulate domestic, commercial production of alternative fuels. This appropriation is divided into the following areas pertinent to alcohol fuels:

-- $100 million for project development feasibility studies, not to exceed $4 million each;

-- $100 million for cooperative agreements with non-Federal entitites, the Government share not to exceed $25 million each, to support commercial scale development of alternative fuels facilities (note, however, that $22 million of this amount has been set aside for the Great Plains Gasification Project pursuant to the legislative history of P.L. 96-126); and

-- P.L. 96-304 provides that (a) not to exceed $100 million shall be available for project development feasibility studies, such individual awards not to exceed $10 million; and (b) not to exceed $100 million shall be available for cooperative agreements with non-Federal entitites, such individual agreements not to exceed $25 million, to support commercial scale development of alternative fuels facilities.

The following DOE policies concerning the Government's financial participation in any project selected for award of a cooperative agreement as a result of the solicitation are: Government funds shall only be used for the Federal share of the total estimated cost for design and construction; and Government funds may not be used to pay any profit or fee to the participant.

It is anticipated that the proposer will share at least 50 percent of the costs of the activities specified in the statement of work. The proposer will set forth in this total cost proposal the amount of costs of this project which he proposed that the Government share. This cost share shall be expressed in terms of dollars and percentage of the total cost.

The schedule of the cost sharing shall be also set out.

Prior work, patents or proprietary data will not be valued in determining the proposer's cost sharing in the cooperative agreement project.

In the event that the cost of providing the site for the project shall be borne by the offeror alone, it shall not be included in the estimated cost which will be cost shared by the Government.

Cost participation by the proposer may be accomplished by a contribution to either direct or indirect costs provided such costs are otherwise allowable in accordance with the cost principles of the award. Allowable costs which are absorbed by the participant as its share of cost participation may not be charged directly or indirectly or may not have been charged in the past to the Federal Government under other contracts, agreements, or grants, nor may other Federal funds be used as cost participation unless specifically authorized by statute.

The proposer shall submit in detail his recommended terms and conditions for repayment of the Government's share of the cost of the cooperative agreement. These terms and conditions shall be the subject of negotiation.

In addition, the grants for feasibility studies must also be repaid (with interest as determined by DOE) if the project becomes a successful and profitable operation. In exceptional cases, the repayment of the cooperative agreement loan and the feasibility study grant may be waived.

DEPARTMENT OF ENERGY, ECONOMIC REGULATORY ADMINISTRATION

LEGISLATION

CRUDE OIL ENTITLEMENTS PROGRAM (43 FR 21429, May 18, 1978; EFFECTIVE DATE JULY 1, 1978. 44 FR 63515, NOVEMBER 5, 1979).

ECONOMIC/FINANCIAL BENEFIT(S)

Cash transfers

OBJECTIVE(S)

To remove the regulatory disincentive to the production and use of petroleum substitutes caused by crude oil price regulations.

ELIGIBILITY

The benefits are available for ethanol derived from domestic biomass which is mixed with gasoline for use as fuel in the United States.

FINANCIAL TERMS

The entitlements earned by eligible firms are equivalent, on a Btu basis, to the entitlements earned on uncontrolled crude oil, i.e., approximately 8 cents per gallon of alcohol at current entitlement levels.

DEPARTMENT OF ENERGY OFFICE OF ALCOHOL FUELS

DEPARTMENT OF AGRICULTURE, FARMERS HOME ADMINISTRATION

LEGISLATION

THE ENERGY SECURITY ACT, P.L. 96-294, TITLE II, THE BIOMASS ENERGY AND ALCOHOL FUELS ACT OF 1980.

ECONOMIC/FINANCIAL BENEFIT(S)

Direct loans; loan guarantees; insured loans; purchase commitments; price guarantees; joint venture agreements.

OBJECTIVE(S)

To promote energy conservation and the development of a domestic synthetic fuels industry, thereby reducing our reliance on imported petroleum.

ELIGIBILITY

Biomass energy projects using a primary fuel other than petroleum or natural gas in the production of fuel include geothermal and solar energy, waste heat, coal, wood, bagasse, and corn stover. Primary fuel is the predominant fuel used by the project, and does not include the incidental use of petroleum or natural gas (such as for flame stabilization).

Projects using new technologies that expand possible biomass feedstocks, produce new forms of biomass energy, or produce biomass fuel using improved or new technologies will be given the same priority.

This priority does not exclude from financial assistance a project not using an alternative primary fuel or applying a new technology.

The Btu content of motor fuels used in the project must not be greater than the Btu content of the biomass fuel produced. This applies only to the biomass energy project and excludes from consideration motor fuels used in the production and transportation of feedstocks. Any displacement of motor fuel or other petroleum products which occurs after the biomass fuel is produced is also to be considered.

For a project to receive financial assistance, the Department must find that necessary feedstocks are available and it is reasonable to expect they will continue to be available in the future. A project must extract the protein content of the feedstock for use as food or feed by readily available markets, if to do so is technically and economically practicable.

Alcohol fuel projects eligible for financial assistance from the Department of Energy must have an anticipated annual production capacity of at least 15 million gallons of ethanol, or use aquatic plants as feedstocks. Biomass energy projects eligible for financial assistance from the Department of Agriculture must have an anticipated annual production capacity of less than 15 million gallons of ethanol, or have an anticipated annual production capacity of 15 million gallons or more if the project is (1) owned and operated by a cooperative and will use feedstocks other than aquatic plants, or (2) will use wood or wood wastes or residues as a feedstock (projects under this category may also be funded by the Department of Energy) and do not use aquatic plants as feedstocks.

FINANCIAL TERMS

The Departments of Energy and Agriculture now have authority to enter into loan guarantees, price guarantees, and purchase agreements for alcohol fuels and other biomass energy projects. Additionally, the Department of Agriculture may provide insured loans for small scale projects. A total of $1.05 billion has been appropriated for use over the next two years by the Departments for biomass energy activities, including financial assistance for biomass alcohol fuels projects.

- Loan Guarantees

  The Departments of Energy and Agriculture may guarantee, against loss of principal and interest, loans made to provide funds for the construction of biomass energy projects. The following provisions apply to all loan guarantees made by both departments for alcohol fuels and biomass energy projects:

-- A loan may be for up to ninety percent of the project's estimated construction costs guarantees made for up to 90 percent of the loan.

-- In the event that total estimated construction costs exceed the originally estimated costs, the Department may guarantee an additional loan for an amount up to sixty percent of the difference between the currently estimated costs and the total costs originally estimated.

-- The borrower must establish that without the Department's guarantee the lender is unwilling to extend credit, at reasonable rates and terms, for construction of the project.

-- The lender must bear a reasonable degree of risk in the financing of the project. This is to ensure (1) that the lender fully participates in the financing, (2) that the lender will fully evaluate and scrutinize the loan for viability, and (3) that the lender will fully service the loan during the life of the loan.

-- In the event that the Department determines that the borrower is unable to meet payments but is not in default, then the Department may elect to pay to the lender the amount of principal and interest the borrower is obligated to pay. However, the borrower must first agree to reimburse the Department on terms and conditions the Department deems necessary to protect the financial interests of the United States.

• Price Guarantees

The Departments of Energy and Agriculture may guarantee a sales price for all or part of the production of a biomass energy project.

The following provisions apply to all price guarantees made by the Departments for alcohol fuels or biomass energy projects:

-- The price guaranteed may not be determined on the basis of the cost of production plus a profit, or other similar arrangement that guarantees a profit to the owner or operator.

-- Any price guarantee must specify the maximum dollar amount of liability of the Federal Government.

-- The price guaranteed and the maximum liability of the Federal Government under that guarantee may be renegotiated to ensure con-

tinuation of a project deemed necessary to achieve the purposes of this program.

• Purchase Agreements

The Departments of Energy and Agriculture may make purchase agreements for all or part of the production of a biomass energy project. The following provisions apply to all purchase agreements made by either Department for alcohol fuels or biomass energy projects:

-- The sales price specified in a purchase agreement may not exceed the estimated prevailing market price as of the date of delivery, unless the Department determines that the sales price must be higher in order to ensure the production of alcohol fuels or biomass energy to achieve the purposes of this program, including national production goals.

-- The alcohol fuels or biomass energy purchased must meet quality standards.

-- The Department may take delivery of alcohol fuels or biomass energy pursuant to a purchase agreement only if arrangements have been made for its distribution to and use by Federal agencies.

-- The Department retains the right to refuse delivery of the alcohol fuels or biomass energy upon such terms and conditions as are specified in the purchase agreement.

-- Any purchase agreement must specify the maximum dollar amount of liability of the Federal Government.

-- The sales price in a purchase agreement and the maximum liability of the Federal Government under that agreement may be renegotiated to ensure continuation of a project deemed necessary to achieve the purposes of this program.

• Insured Loans

Insured loans are available only from the Department of Agriculture for small scale alcohol fuels and biomass energy projects. "Small scale" is limited to projects with an anticipated annual production capacity of not more than 1,000,000 gallons. The following provisions apply to all insured loans made by the Department of Agriculture for alcohol fuels and biomass energy projects.

-- They may be for up to $1 million per project.

-- They may be for up to 90 percent of the total estimated construction costs of the project.

-- Loans for cost overruns are limited to 10 percent of original costs.

-- The applicant must establish that without an insured loan, sufficient credit at reasonable rates and terms is unavailable.

-- The interest rate on an insured loan shall be determined by the Department of Agriculture, taking into consideration the current interest rate charged to the Federal Government to borrow money plus up to one percent additional rate of interest.

DEPARTMENT OF THE TREASURY, INTERNAL REVENUE SERVICE

LEGISLATION

ENERGY TAX ACT OF 1978; P.L. 95-618, CRUDE OIL WINDFALL PROFIT TAX ACT OF 1980; P.L. 96-223.

ECONOMIC/FINANCIAL BENEFIT(S)

Excise tax exemption; Income tax credit; Investment tax credit.

OBJECTIVE(S)

To encourage energy conservation and promote industrial and agricultural conversions from oil and gas to alternative forms of energy, specifically, by encouraging the construction of domestic alcohol fuels production facilities.

ELIGIBILITY

Taxpayers who produce and sell gasoline for blending into gasohol, and those who produce and sell gasoline, provided that the blend contains at least 10 percent alcohol made from any product other than petroleum, natural gas, or coal. Any person who purchases tax-paid gasoline and uses it to make a tax-exempt alcohol-gasoline blend may claim a refundable income tax credit equal to the taxes paid on such gasoline.

To be eligible for a regular investment tax credit and a new energy investment tax credit, the property in question must be depreciable property with a useful life of three years or more. In order to qualify for the energy investment tax credit, the property must be new and be placed in service after September 30, 1978 and before January 1, 1983.

FINANCIAL TERMS

The Energy Tax Act provides that gasohol is exempt from its 4-cent-a-gallon Federal excise tax on motor fuels sold after December 31, 1978, and before October

1, 1984. The Crude Oil Windfall Profit Tax Act of 1980 extends the exemption through December 31, 1992.

The Windfall Profit Tax Act further states that a person who blends alcohol fuel with gasoline or any other liquid fuel suitable for use in an internal combustion engine may claim an income tax credit. The credit is 40 cents per gallon of alcohol of at least 190 proof, and 30 cents per gallon of alcohol of at least 150 proof but less than 190 proof. The tax credit is also applicable to a person (1) who uses straight alcohol fuel (100 percent) as a fuel in his trade or business, or (2) who sells straight alcohol fuel at retail to one who uses it directly. In situations where the retail seller is eligible for the credit, no credit is allowable for the user.

The Energy Tax Act also includes a 10 percent energy investment tax credit in addition to the regular 10 percent investment credit already available. The energy credit applies to costs incurred for the period from October 1, 1978 through December 31, 1982. There is no termination date for the regular investment tax credit.

PENDING LEGISLATION

DEPARTMENT OF AGRICULTURE

BILL

Rural Development Authorization Act; S.892. Rural Development Policy Act; HR 3580. (Soon to be P.L. 96-355)

ECONOMIC/FINANCIAL BENEFIT(S)

Loan guarantees

MAJOR PROVISIONS

S.892 - Authorizes $500 million in loan guarantees for production of industrial hydrocarbons and alcohols from agricultural commodities and forest products; $30 million maximum loan guarantee per project; at least 25 percent of all loans guaranteed must be for projects that produce no more than 2.5 million gallons of alcohol per year. HR 3580 - Provides $180 million in loan guarantees for production of alcohol fuels from agricultural and forest products.

DEPARTMENT OF AGRICULTURE

BILL

Consolidated Farm and Rural Development Act; S.985. Consolidated Farm and Rural Development Act Amendments; HR 3683.

ECONOMIC/FINANCIAL BENEFIT(S)

Direct loans; Loan guarantees.

MAJOR PROVISIONS

Makes explicit that alcohol fuel production is eligible under the loan programs of the Farmers Home Administration.

DEPARTMENT OF COMMERCE

BILL

The Public Works and Economic Development Act; S.914.

ECONOMIC/FINANCIAL BENEFIT(S)

Grants; Direct loans.

MAJOR PROVISIONS

Under the House version of S.914, $100 million is authorized in each of two years for EDA grants and loans for construction and operation of facilities that produce alcohol or methane from renewable resources. The Senate version authorizes about $39 million annually in EDA grants for alcohol fuel production facilities when such grants will create or preserve jobs in small communities.

DEPARTMENT OF ENERGY

BILL

Department of Energy Authorization Act, FY 80-81; S.688 and HR 3000.

ECONOMIC/FINANCIAL BENEFIT(S)

Purchase commitments.

MAJOR PROVISIONS

The Senate version (S.688) emphasizes biomass research and development ($57 million); HR 3000 required DOE to seek suppliers of alternative fuels which would be used to fuel DOE motor vehicles.

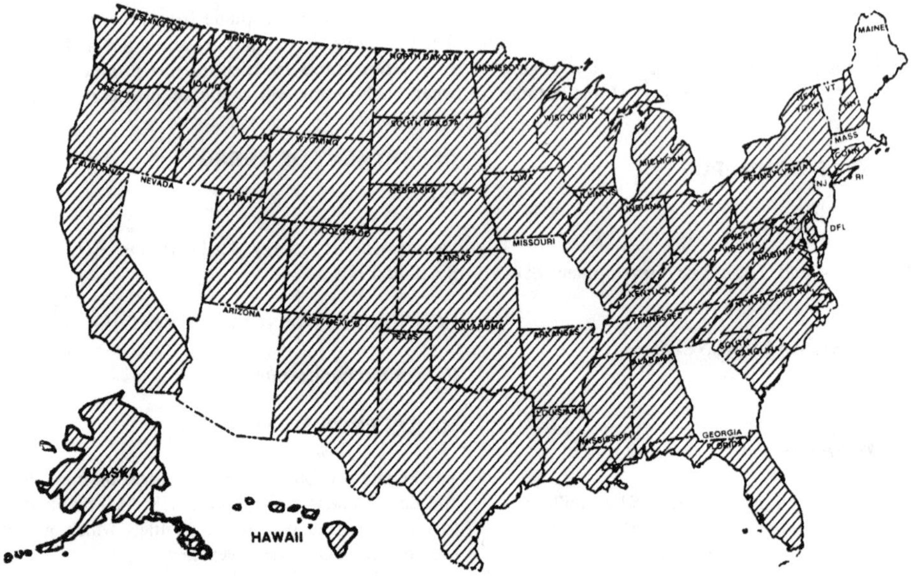

States Which Have Enacted Alcohol Fuels Legislation

▨	States Which Have Enacted Laws
☐	States Which Have Not Enacted Laws

SOURCE U S National Alcohol Fuels Commission, July 1980

Figure D-1. States Which Have Enacted Alcohol Fuels Legislation

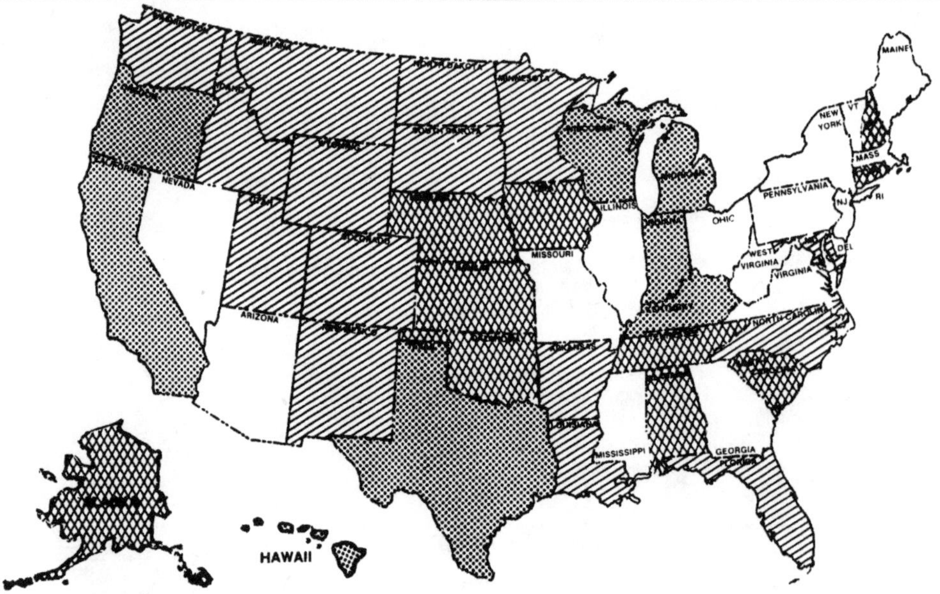

☐ States Which Have Not Enacted Alcohol Fuels Tax Legislation	▨ Excise Tax Exemptions for Gasohol (No other tax incentives)
▨ Both Exemption and Incentives	▨ Property, Sales and/or Income Tax Incentives for Alcohol fuels

SOURCE U S National Alcohol Fuels Commission, July 1980

Figure D-2. States Which Have Enacted Tax Incentives for Alcohol Fuels

COMMERCIAL SCALE ETHANOL PRODUCTION AND FINANCING

STATE LEGISLATION

This section provides a summary of ethanol legislation for each State. The key points included for each state are the current state laws, pending legislation, state regulations, other important alcohol fuel activity, and the name, address, and phone number of a responsible official to contact for further information. Figure D-1 illustrates which states have enacted alcohol fuels legislation. Figure D-2 illustrates which states have enacted tax incentives for alcohol fuels.

ALABAMA

LAWS

SB 354

Exempts gasohol from 3 cents of the 8 cents per gallon state motor fuel tax. The gasohol must contain at least 10 percent ethyl alcohol produced from agricultural, forest or other renewable resources and be at least 99 percent pure. Effective July 1, 1980.

SB 286

Creates the Alabama State Department of Energy which will be responsible for alcohol fuels and other alternative energy sources.

PENDING LEGISLATION

HB 37

Exempts ethyl alcohol produced and sold for motor fuel and the accompanying distillery operation from all municipal, county, and state sales taxes.

HB 198

Exempts alcohol used for fuel purposes produced in Alabama from taxation and from Alabama alcohol regulatory provisions.

HB 952

Exempts gasohol from the 8 cents per gallon state motor fuel tax. The gasohol must be manufactured in Alabama or produced in a state which exempts Alabama produced gasohol.

STATE REGULATION

Permit Procedure—A permit is required by the Alabama Alcoholic Beverage Control Board for the production of fuel alcohol. A $100 fee, payable October 1, is required each year. All alcohol provided for fuel use must be denatured in the final state of distilling.

State Volatility Requirements—Alabama requires that all gasoline (including gasohol) meet ASTM Distillation and Reid Vapor Pressure requirements. ASTM D439 not required.

OTHER IMPORTANT ALCOHOL FUELS ACTIVITY IN ALABAMA

The Johnson Environmental and Energy Center at the University of Alabama at Huntsville is conducting tests on small scale commercial alcohol fuels stills manufactured in Alabama. The center is also conducting research on vacuum distillation and potential end uses of alcohol fuels, and has recently completed a feasibility study on comparative feedstock values of agricultural resources grown in Alabama.

Universities and Colleges offering alcohol fuels workshops, seminars and courses: Auburn University; University of Alabama at Tuscaloosa; University of Alabama at Huntsville; Talledega College.

Responsible Official for Further Information:

Fred Braswell (205) 832-5010
Program Coordinator
Alabama Energy Management Board
3734 Atlanta Highway
Montgomery, Alabama 36130

ALASKA

LAWS

SB 438

Exempts gasohol from the 8 cents per gallon state motor fuel tax. The gasohol must contain at least 10 percent alcohol. Effective January 1, 1981.

HB 687

Authorizes funding for alternative fuels (including gasohol) research and development. Establishes the Alaska Energy Center. Effective July 1, 1980.

PENDING LEGISLATION

The legislative session adjourned June 6, 1980. The legislature will convene on January 12, 1981.

STATE REGULATION

Permit Procedure—No state permit is required for the production of fuel alcohol.

State Volatility Requirements—Alaska has no state volatility requirements for gasoline or gasohol.

OTHER IMPORTANT ALCOHOL FUELS ACTIVITY IN ALASKA

Feasibility studies on alcohol fuels are being conducted by the state and the University of Alaska.

The Alaska Council on Science and Technology has appropriated funds for the construction of a small scale model methanol plant.

The U.S. Department of Energy Appropriate Technology Program has provided funds for the construction in Alaska of a new methane production plant using crab waste. The plant is now operational and the first test is underway.

Universities and Colleges offering alcohol fuels workshops, seminars and courses: University of Alaska at Fairbanks; University of Alaska Kenai Community College at Soldotna; Matanuska-Susitna Community College at Palmer.

Responsible Official for Further Information

Paula Wellen (907) 452-4761
Community Information Center
Fairbanks North Star Borough
P.O. Box 1267
520 Fifth Avenue
Fairbanks, Alaska 99707

ARIZONA

LAWS

No laws have been enacted as of July 1, 1980.

PENDING LEGISLATION

The legislative session adjourned May 1, 1980.

STATE REGULATION

Permit Procedures—A state permit is required for the production of alcohol. All distilling apparatus used in the production of fuel alcohol must be registered with the State Department of Liquor Licenses and Control.

State Volatility Requirements—Arizona requires that gaoline (including gasohol) meet ASTM D439 specifications which include Distillation and Reid Vapor Pressure specifications.

LOCAL GOVERNMENT REQUIREMENTS AND INCENTIVES

The City of Wilcox has set aside property for the development of a municipal industrial park which will lease land at competitive rates for alcohol fuels production facilities.

OTHER IMPORTANT ALCOHOL FUELS ACTIVITY IN ARIZONA

The Arizona Department of Transportation is using gasohol in approximately 750 state vehicles and is also conducting a fleet test on alcohol fuels using gasohol and dieselhol in three vehicles.

A "Biofuels Task Force" was established by the Arizona legislature on January 1, 1980, to research the production and use of alcohol fuels in Arizona.

The Arizona Solar Energy Commission has funded a research project on the production of ethanol from biomass and is operating a small scale solar powered ethanol still.

Universities and Colleges offering alcohol fuels workshops, seminars, or courses: University of Arizona at Tucson; Mohave Community College; Yavapai College.

Responsible Official for Further Information:

Rich Wetzel (602) 255-5705
Arizona Office of Economic Planning & Development
1700 W. Washington
Executive Tower, Room 505
Phoenix, Arizona 85007

LAWS

SB 454

Exempts gasohol from all 9.5 cents per gallon of the state motor fuel tax and special motor fuel tax. The gasohol must contain at least 10 percent alcohol produced in Arkansas from agricultural or forest resources distilled or manufactured in Arkansas, or must be manufactured or distilled in a state which exempts from its motor fuel tax such a mixture produced within the State of Arkansas. Effective March 20, 1979.

SB 28

Authorizes the establishment and maintenance of a state Alcohol Fuels Production Registry and requires all persons manufacturing fuels containing alcohol, or manufacturing alcohol for use in or as a fuel, to register with this new department. Effective April 1, 1980.

HB 742

Authorizes a state income tax deduction for qualified energy conservation and renewable energy source expenditures made between January 1, 1979, and December 31, 1984. (Bioconversion energy devices qualify for this deduction.) Effective April 6, 1979.

SB 545

Appropriates additional funds to the Arkansas Department of Energy (including $100.000 to be used for the creation of an Energy Innovation Grant Program).

PENDING LEGISLATION

The legislative session adjourned January 29, 1980.

STATE REGULATION

Permit Procedure—The state beverage law permit requirement for the production of alcohol has been waived for alcohol used as a motor fuel April 1, 1980, and replaced by the regulations and procedures set forth in the alcohol fuels production registry (see SB 28 above). No fee is required.

State Volatility Requirements—Arkansas requires that gasoline (including gasohol) meet ASTM D439 specifications, which include Distillation and Reid Vapor Pressure specifications.

Environmental Regulations—The by-products of alcohol production are subject to state environmental control regulations.

OTHER IMPORTANT ALCOHOL FUELS ACTIVITY IN ARKANSAS

Stevens Inc. is conducting a feasibility study for five to six county cooperatives on a 40 million gallon per year, $30 million dollar corn/ethanol plant.

The Arkansas State Department of Energy has published a "how to" manual for alcohol fuel producers.

The Arkansas Joint Legislative Interim Committee on Energy appropriated $10,000 for alcohol fuels workshops in 1980.

Delta-Vocational Technical College in Marked Tree, Arkansas, is conducting research on engine retrofitting.

Van Buren, Arkansas Project—Highly automated alcohol fuel production using the chambers ACR advanced distillation process.

University of Arkansas (Gulf Chemical) Project Dr. George Emert is conducting research on the conversion of cellulosic waste to alcohol by enzymatic hydrolysis.

Universities and Colleges offering alcohol fuels workshops, seminars, or courses: Mississippi County Community College; East Arkansas Community College at Forrest City; Phillips County Community College.

Responsible Official for Further Information:

Alford Drinkwater (501) 371-1370
Biomass/Resource Recovery Coordinator
Arkansas Department of Energy
3000 Kavanaugh
Little Rock, Arkansas 72205

LAWS

AB 2004

Exempts from any law or regulations, for the three-year period from June 13, 1980 to June 13, 1983, any blend of gasoline consisting of at least 10 percent ethanol, if the gasoline used in the blend meets the 9 pounds per square inch Reid Vapor Pressure standard. The law requires the Legislative Analyst to submit a report to the Legislature two years after the effective date of the bill on the impact of the use of gasohol on the ambient air quality in California. Effective June 13, 1980.

AB 1401

Authorizes the use of methanol and methanol/gasoline blends in vehicles, and permits vehicles to be modified to run on these fuels as of January 1, 1980. The law also establishes a 10-year experimental methanol program in California, and permits vehicles registered with the California Division of Motor Vehicles to operate exclusively on methanol or methane and a mixture of gasoline or diesel. The law requires manufacturers or distributors of pollution control or engine modification devices to submit such devices to the California Air Resources Board for testing. The state board must report test results to the legislature within 120 days after receiving the device for testing. Effective January 1, 1980.

SB 318

Requires the Department of General Services to develop a gasohol plan for converting 25 percent of the state fleet to be run on alcohol fuel by January 1, 1980.

SB 620

Allocates $10 million for an investigation of the practicality and cost effectiveness of alternative motor vehicle fuels.

SB 1324

Exempts gasohol from the state sales tax on a decreasing scale basis. The exemption for gasohol will be as follows: January 1, 1981 - January 1, 1982 = 5¢, January 1, 1982 - January 1, 1983 = 4¢, January 1, 1983 - January 1, 1984 = 3¢. The gasohol must contain at least 10% methyl or ethyl alcohol distilled from agricultural commodities, renewable resources or coal. The alcohol must be rendered unsuitable for human consumption at the time of its manufacture or immediately thereafter and be dyed a different color from the color of other gasolines. Effective January 1, 1981. Expires January 1, 1984.

PENDING LEGISLATION

SB 1420 (AB 2348)

Allows franchisees to purchase gasoline, diesel, and gasohol from any available source.

AB 2093 (SB 1324)

Exempts the sale of gasohol from sales and use taxes.

AB 2603 (SB 1626)

Provides that any alcohol produced for use as a motor vehicle fuel will be taxed as motor vehicle fuel and not subject to beverage taxes.

AB 2882

Establishes an alcohol fleet in Ventura County and provides for a reimbursement of costs incurred with this project to the Ventura County government.

SB 1205

Provides funding for the development of alternative energy sources, including alcohol fuels.

SB 1922 (SB 3158)

Exempts synthetic fuel manufacturing plants from district Air Quality Board emission standards.

STATE REGULATION

Permit Procedure—A state permit is required for the production of fuel alcohol. A $6.60 fee is required.

Gasohol containing at least 10 percent ethanol is exempted from any Reid Vapor specification laws or regulations from June 13, 1980 to June 13, 1983, if the base gasoline meets the 9 pounds per square inch Reid Vapor Pressure standard. Prior to this ruling, gasohol was required by the California Air Resources Board to meet the same Reid Vapor Pressure specifications as gasoline (9 pounds per square inch) only during the summer months (April-October). This ruling in effect banned the sale of gasohol during these "summer" months.

State Volatility Requirements—California requires that gasoline meet ASTM D439 specifications, which include distillation and Reid Vapor Pressure specifications.

OTHER IMPORTANT ALCOHOL FUELS ACTIVITY IN CALIFORNIA

Los Angeles County is conducting a vehicle fleet test using gasohol and methanol blends in 100 vehicles.

The Bank of America is conducting a vehicle fleet test using pure methanol and methanol blends in approximately 120 vehicles.

The Los Angeles Times sponsored a vehicle fleet test from November 1979 to January 7, 1980, using gasohol in 125 vehicles.

The California General Services Department sponsored a vehicle fleet test during July 1979 using gasohol in 120 vehicles.

Contra Costa County is conducting a vehicle fleet test (August 1, 1979, to August 1, 1980) using gasohol in 30 vehicles.

The Coordination Research Council for the U.S. Department of Energy is conducting a vehicle fleet test in Anaheim, California, using gasohol in 14 vehicles.

The California State Legislature sponsored a vehicle fleet test in 1979 using methanol X (methanol + higher alcohols) in two vehicles.

The University of Santa Clara conducted a vehicle fleet test in December 1978 using ethanol and methanol blends in seven vehicles.

The California Energy Resources and Conservation Development Commission conducted a vehicle fleet test from November 1978 to November 1979 using ethanol and methanol blends in five state vehicles. The Commission will conduct three additional fleet tests using pure methanol and ethanol in 50-150 vehicles altered to achieve maximum efficiency and to meet emission standards.

The California Department of Transportation conducted a vehicle fleet test from November 1979 to July 1980 using gasohol in 500 vehicles.

Oakton Community College in association with University Extension (University of California at San Diego) has received, grants from the National Endowment for the Humanities and the National Science Foundation for a research project evaluating the costs and benefits of synthetic fuels and other related issues.

The California Resources and Conservation Developments Commission is jointly funding with private industry the construction of ethanol plants.

Universities and Colleges offering alcohol fuels workshops, seminars, and courses: University of California at Berkeley; University of California at Davis; University of California at Santa Cruz; LaVerne College; Cabrillo Junior College; Modesto Junior College; College of Siskiyous.

Responsible Official for Further Information:

Dick DeZeeuw (916) 920-6033
California Energy Commission
1111 Howe Avenue, MS 68
Sacramento, California 95825

Peter Ward (916) 920-6031
California Energy Commission
Development Division
1111 Howe Ave.
Sacramento, California 95825

LAWS

HB 1135

Exempts gasohol from 5 cents of the 7 cents per gallon state motor fuel tax. To be eligible gasohol must contain at least 10 percent ethyl or methyl alcohol produced in Colorado from agriculture or forest resources, petroleum coke, waste coke, or coal waste, and be at least 95 percent pure. Originally, this exemption contained a 200,000 maximum population ceiling restriction and 20 million gallon ceiling; however, these provisions were removed by additional legislation in 1979. (See HB 1463 below.) Effective July 1, 1978. The exemption expires July 1, 1985.

HB 1607

Extends the state motor fuel tax reduction to gasohol made from petroleum coke, waste coke, coal waste, coal, etc. The law also extends the property tax reduction to apply to these types of plants. Effective July 1, 1979.

SB 80

Creates a nine-member board, the Colorado Gasohol Promotion Committee, which will promote ethanol fuels. Effective July 1, 1978. The board expires July 1, 1985.

HB 1463

Removes HB 1135's population and gallonage ceilings. It also requires all state and local government agencies and units to use gasohol in their vehicles when such fuel becomes readily available, provided this conversion does not necessitate the mechanical modifications of vehicles.

The law also authorizes a temporary reduction for property tax purposes in an assessed valuation for alcohol plants producing 2.5 million gallons or less of alcohol annually on a decreasing scale basis. The reduction is as follows: 2 percent of actual value for the first year, 9 percent the second year, 16 percent the third year, 23 percent the fourth year, and 30 percent thereafter.

The law establishes a special fund under the Gasohol Promotion Committee to promote gasohol and provides a check off option for off-road users of motor fuel on the refund of the motor fuel tax. Effective July 1, 1979.

LEGISLATION INTRODUCED AND NOT ENACTED

The legislative session adjourned May 2, 1980.

HB 1145

Energy Appropriation Act. Appropriates funds for the creation of a state operated "Energy Farm" demonstration alcohol plant.

STATE REGULATION

Permit Procedure—No state permit required for the production of fuel alcohol.

A retailer planning to sell gasohol in his station must file and receive approval with the State Gasoline and Alcohol Inspection Board of the Colorado Revenue Service.

State Volatility Requirements—Colorado requires that all gasoline (including gasohol) meet only ASTM Distillation requirements. No Reid Vapor Pressure or other ASTM D439 requirements.

OTHER IMPORTANT ALCOHOL FUELS ACTIVITY IN COLORADO

The Colorado Gasohol Promotion Committee has conducted research on alcohol fuels and presented recommendations to the Colorado State Legislature on February 15, 1980.

Public Service Co. in Denver, Colorado, conducted a vehicle fleet test during May 1979, using methanol-gasoline mixture (5-10 percent methanol) in 400 vehicles.

The Police Department of Denver, Colorado conducted a vehicle fleet test from September 1979 to October 1979 using 10 percent methanol in 400 vehicles.

Rockwell Institute of Colorado Springs, Colorado is conducting a vehicle fleet test using gasohol in 106 vehicles.

Universities and Colleges offering alcohol fuels workshops, seminars, and courses: Northeastern Junior College at Sterling (developing one-year and two-year programs in alcohol fuels technology); Lamar Community College (conducting nine intensive three-day workshops in conjunction with an operating on-farm alcohol plant in southeastern Colorado); University of Colorado at Boulder.

Responsible Official for Further Information:

Bob Merten (303) 839-3218
Department of Agriculture
525 Sherman Street, 4th Fl.
Denver, Colorado 80203

CONNECTICUT

LAWS

HB 7188

Exempts gasohol from 1 cent of the 11 cents per gallon state motor fuel tax. The gasohol must contain at least 10 percent ethyl or methyl alcohol. Effective July 1, 1979.

PENDING LEGISLATION

The legislative session adjourned May 7, 1980.

STATE REGULATION

Permit Procedures—No state permit is required for the production of fuel alcohol. However, producers must file a notice of alcohol fuel production with the State Tax Department.

State Volatility Requirements—Connecticut requires that all gasoline (including gasohol) meet only Distillation requirements. No Reid Vapor Pressure or other ASTM D439 requirements.

OTHER IMPORTANT ALCOHOL FUELS ACTIVITY IN CONNECTICUT

Universities and Colleges offering alcohol fuels workshops, seminars, and courses: Wesleyan University.

Responsible Official for Further Information:

Joe Belanger (203) 566-5898
Director, Energy Research & Policy
81 Washington Street
Hartford, Connecticut 06115

DELAWARE

LAWS

No laws have been enacted as of July 1, 1980.

PENDING LEGISLATION

HB 679

Exempts gasohol from 9 cents of the 19 cents per gallon state motor fuel tax.

HB 784

Encourages the State of Delaware to use gasohol in all state vehicles.

STATE REGULATION

Permit Procedure—No state permit is required for the production of fuel alcohol. The Delaware Alcoholic Beverage Control Commission requires fuel alcohol producers submit copies of the Federal Bureau of Alcohol, Tobacco and Firearms permit, permit approval notice, and a copy of the Federal bond.

State Volatility Requirements—Delaware has no state volatility requirements for gasoline or gasohol.

OTHER IMPORTANT ALCOHOL FUELS ACTIVITY IN DELAWARE

Universities and Colleges offering alcohol fuels workshops, seminars, and courses: Delaware Technology and Community College (will also construct three alcohol stills on campus).

Responsible Official for Further Information:

Dan Anstine (302) 736-5647
Delaware Energy Office
P.O. Box 1401
114 W. Walter Street
Dover, Delaware 19901

LAWS

SB 903

(Ch 77) Exempts gasohol from 5 cents of the 8 cents per gallon state motor fuel tax on a decreasing scale basis. The exemption for gasohol will be as follows: July 1, 1980-July 1, 1983 = 5 cents; July 1, 1983-July 1, 1985 = 4 cents; July 1, 1985-July 1, 1987 = 2 cents. The exemption expires July 1, 1987. The gasohol must contain at least 10 percent ethyl alcohol and be 99 percent pure.

Provides for a corporate income tax credit for new or expanded businesses engaged in the distillation of ethyl alcohol for use in motor fuels or in the manufacture of equipment for the processing or distillation of ethyl alcohol for use in motor fuels. The credit is limited to seven years and is available from January 1, 1981 through January 1, 1988.

Requires that oil companies which permit the use of credit cards for the purchase of gasoline must also allow the use of credits for the purchase of gasohol.

Provides an exemption for gasohol from the Florida statute which prohibits the blending or adultering of petroleum products by distributors or retailers, provided that the alcohol blended fuel meets the following temporary specifications (until ASTM approves specifications for gasohol): 90 percent unleaded gasoline; 10 percent ethyl alcohol; the alcohol must be a minimum of 198 proof; a maximum of 50 parts per million acetic acid. The gasoline blend stock must meet Florida volatility specifications until the ASTM gasohol specifications are approved.

PENDING LEGISLATION

None

STATE REGULATION

Permit Procedure—No state permit is required for the production of fuel alcohol.

State Volatility Requirements—Florida requires that gasoline (including gasohol) meet ASTM D439 specifications which include Distillation and Reid Vapor Pressure specifications.

OTHER IMPORTANT ALCOHOL FUELS ACTIVITY IN FLORIDA

Brevard Community College is conducting research on alcohol fuels.

Universities and Colleges offering alcohol fuels workshops, seminars, and courses: University of Florida at Gainsville (Cooperative Extension Service/Institute of Food and Agricultural Services).

Responsible Official for Further Information:

Doug Roberts (904) 488-6146
Governor's Energy Office
Capitol Building
Talahassee, Florida 32301

GEORGIA

LAWS

No laws have been enacted as of July 1, 1980.

PENDING LEGISLATION

The legislative session adjourned May 8, 1980.

STATE REGULATION

Permit Procedures—The State Department of Revenue Motor Fuel Tax Division requires producers of fuel alcohol to submit a copy of the Federal Bureau of Alcohol, Tobacco, and Firearms permit. Producers must also obtain a nonbeverage alcohol manufacturer's license from the State of Georgia, Department of Revenue, Alcohol and Tax Department, and a motor fuel distributor's special fuels license.

State Volatility Requirements—Georgia requires that gasoline (including gasohol) meet ASTM D439 specifications, which include Distillation and Reid Vapor Pressure specifications.

OTHER IMPORTANT ALCOHOL FUELS ACTIVITY IN GEORGIA

The University of Georgia at Athens is conducting research on the use of membrane reactors in enzymatic hydrolysis for alcohol production.

The Lieutenant Governor of Georgia has established a five member committee which will assess the feasibility of alcohol production.

Universities and Colleges offering alcohol fuels workshops, seminars, and courses: University of Georgia (Cooperative Extension Service); Georgia Tech (Continuing Education Department).

Responsible Official for Further Information:

Rob Harvey (404) 656-5176
Office of Energy Resources
270 Washington Street, SW
Room 615
Atlanta, Georgia 30334

HAWAII

LAWS

SB 1906

Exempts gasohol from the 4% state excise tax imposed on the gross proceeds of retail sales. The State Director of Taxation has the authority to terminate the exemption at his discretion. Effective July 1, 1980. The exemption expires June 30, 1985.

PENDING LEGISLATION

The legislative session adjourned April 17, 1980.

STATE REGULATION

Permit Procedure—No permit is required for the production of fuel alcohol.

State Volatility Requirements—Hawaii requires that gasoline (including gasohol) meet ASTM D439 specifications, which include Distillation and Reid Vapor Pressure specifications.

OTHER IMPORTANT ALCOHOL FUELS ACTIVITY IN HAWAII

Maui County conducted a vehicle fleet test during November 1979, using gasohol in county vehicles.

Responsible Official for Further Information:

Takeshi Yoshihara (808) 546-3730
Hawaii Representative
Department of Energy
4322 Prince Kuhio Federal Building
Honolulu, Hawaii 96850

LAWS

SB 1359

Exempts gasohol from 4 cents of the 9 1/2 cents per gallon state motor fuel tax. Gasohol must contain at least 10 percent anhydrous ethyl produced in Idaho from agricultural or forest products grown in Idaho. Effective July 1, 1980.

SB 1247

Provides for a credit against individual or corporate income taxes for any personal property or improvements to real property used or constructed for the purpose of manufacturing ethanol alcohol. The credit is available according to the following schedule. During the first year of use: 0.8 percent of the gross value of ethanol produced in that year; the second year of use: 0.6 percent of the property value, but not to exceed 0.6 percent of the gross ethanol value; the third year of use: 0.4 percent of the property value, but not to exceed 0.4 percent of the gross ethanol value; the fourth year: 0.2 percent of the property value, but not to exceed 0.2 percent of the gross ethanol value; the fifth and succeeding years of use: no credit.

HB 531

Includes gasohol containing 10 percent anhydrous ethanol from agricultural, forest, or waste products in the definition of "motor fuels" for tax purposes. Effective July 1, 1980.

PENDING LEGISLATION

The legislative session adjourned March 31, 1980. The legislature will convene on January 12, 1981.

STATE REGULATION

Permit Procedure—No state permit is required for the production of fuel alcohol.

State Volatility Requirements—Idaho requires that gasoline (including gasohol) meet ASTM D439 specifications, which include Distillation and Reid Vapor Pressure specifications.

OTHER IMPORTANT ALCOHOL FUELS ACTIVITY IN IDAHO

In January 1980, the Governor issued a directive requiring the use of gasohol in all state vehicles where gasohol is available and requested that a fleet test (on 100 percent alcohol) be conducted.

The State Energy Department has issued several alcohol fuels information brochures.

An informal inter-agency committee was created by the state legislature in March 1980 to conduct research on the feasibility of gasohol as a motor vehicle fuel.

Universities and Colleges offering workshops, seminars and courses: University of Idaho at Moscow (Cooperative Extension Service). The College of Southern Idaho in Twin Falls has received a grant from DOE to conduct alcohol fuel workshops.

Responsible Official For Further Information:

Gail Dameworth (208) 334-3800
Research Analyst
Idaho Office of Energy
State House Mail
Boise, Idaho 83720

LAWS

SR 364

Directs the Governor to promote the establishment of an interagency task force, which will focus on disseminating information on alcohol fuels to individuals and businesses.

HD 3403

Directs the Illinois Institute of Natural Resources to serve as a clearinghouse for information on alcohol production technology, develop information packets for the public, and sponsor seminars on alcohol fuels. Effective January 1, 1981.

PENDING LEGISLATION

HB 171

Requires oil companies to mix ethyl alcohol with unleaded gasoline for motor fuel purposes. Requires that motor fuel sold in Illinois must contain 2 percent alcohol by 1980, 5 percent by 1983, and 10 percent by 1984. Makes a "willful violation" of this law a Class A misdeameanor in addition to punishment by forfeiture of the franchise for the petroleum products seller.

HB 464

Exempts gasohol from the motor fuels tax law, Retailers Occupation Tax, Use Tax, Service Occupation Tax, and Service Use Tax Acts.

HB 1599

Requires that all motor fuels contain 10 percent alcohol by July 1, 1980.

HB 1611

Directs the University of Illinois to conduct research on gasohol.

HB 2530

Directs the Illinois Institute of Natural Resources to develop an "alternative fuels incentive program" to investigate tax incentives for gasohol in Illinois.

HB 2731

Amends the state motor fuel tax to provide that such tax does not apply to alcohol manufactured and used by an individual solely for his or her own use as a motor fuel.

HB 2866 (SB 1500) (HB 2920)

Requires retailers of motor fuel to accept credit cards for sales of gasohol if they accept such credit cards for the sale of any form of motor fuel.

HB 2911

Exempts gasohol from 5 cents of the 7 1/2 cents per gallon state motor fuel tax.

SB 1464

Exempts gasohol from 2 cents of the state motor fuel tax of 7 1/2 cents per gallon.

HB 3351

Exempts gasohol from 4 cents of the 7 1/2 cents per gallon state motor fuel tax.

HB 2931

Exempts gasohol from the State Occupation and Use Tax Acts for the period of July 1, 1980, through June 30, 1982.

SB 1518

Exempts gasohol from the State Occupation and Use Tax Acts.

HR 400

Establishes an Alternative Fuels Task Force.

STATE REGULATION

Permit Procedure—That State Department of Revenue requires a permit for the production of fuel alcohol. The Motor Fuels Revenue Act of 1927 allows the Illinois Department of Revenue to tax all fuel producers (including on-farm producers). A $1,000 - $40,000 bond (depending on volume) must be posted and fuel alcohol producers must report all sales on a monthly basis.

State Volatility Requirements—Illinois has no state volatility requirements for gasoline or gasohol.

OTHER IMPORTANT ALCOHOL FUELS ACTIVITY IN ILLINOIS

The Coal and Energy Development Bond Fund was established by the state legislature in 1974. In FY 1977, $5 million of this fund was set aside for alternative energy projects. In 1979, the state General Assembly voted to spend $1 million of this fund. Several alcohol fuels projects have been submitted for funding under this program.

The State Task Force on Alcohol Production is planning to construct at Vienna, Illinois a 500,000 - 1 million gallons per year alcohol fuels plant. The alcohol produced at this plant will be used to fuel state fleet vehicles.

On November 2, 1979, the Governor of Illinois issued an executive order requiring that 10,000 state vehicles operate on gasohol by October 1980.

Pursuant to SR364, May 22, 1980, the Governor established an interagency task force on alcohol fuels.

National Car Rental at Chicago O'Hare airport is conducting a vehicle fleet test using gasohol in shuttle buses.

Illinois Bell Telephone Co. in Peoria, Illinois, conducted a vehicle fleet test from January 1979 to January 1980 using gasohol in 30 vans and trucks.

The Illinois Institute of Technology conducted a vehicle fleet test from July 1979 to September 1979 using gasohol (20 percent ethanol) in vehicles.

Maken Lumber Co. of Wooddale, Illinois, conducted a vehicle fleet test from February 1979 to September 1979 using gasohol in 15 flatbed trucks and private vehicles.

Universities and Colleges offering alcohol fuels workshops, seminars and courses: Parkland Community College; Lincoln Land Community College; Lakeland Community College; Illinois Eastern Community College at Olneu; Illinois Central College at East Peoria; Sangamon State University.

Responsible Officials For Further Information:

John Lehman (217) 782-6675
Alternative Fuels Coordinator
Division of Marketing & Agricultural Services
Department of Agriculture
Emerson Building - State Fairgrounds
Springfield, Illinois 62706

Larry Metzroth (217) 782-8220
Acting Staff Director
Illinois Energy Resources Commission
612 S. 2nd Street, 2nd floor
Springfield, Illinois 62706

Nicholas Hall (217) 785-2800
Illinois Institute of Natural Resources
325 W. Adams
Springfield, Illinois 62706

INDIANA

LAWS

SB 218

Exempts gasohol from the 4 percent state sales tax. The gasohol must contain at least 10 percent ethyl alcohol produced from agricultural resources. Effective July 1, 1979.

HB 1334

Provides that persons allowed a Federal income tax depreciation deduction for a coal conversion system which processes coal into a gaseous or liquid fuel (e.g., methanol) are entitled to a gross income tax deduction for that same year. Effective July 1, 1980. The provison expires January 1, 1990.

SB 315

Establishes vapor pressure ceilings for gasohol. For the months of December, January, February and March: 15.5 psi; for the months of April, October and November: 14.5 psi; for the month of May, June, July, August and September: 12.5 psi. The gasohol must be a blend of 90 percent unleaded gasoline and 10 percent anhydrous alcohol. Effective July 1, 1980.

SB 362

Establishes the "Indiana Energy Development Board" to develop in-state uses of energy resources (including alcohol fuels). Appropriates $1 million for FY 1980-81 for this program. Effective January 1, 1980.

PENDING LEGISLATION

The legislative session adjourned February 26, 1980.

STATE REGULATION

Permit Procedures—No state permit is required for the production of fuel alcohol. The state requires producers of fuel alcohol to submit a copy of the Federal Bureau of Alcohol, Tobacco and Firearms permit.

State Volatility Requirements—Indiana requires that all gasoline (including gasohol) meet Distillation and the Reid Vapor Pressure requirements in the SB 315 legislation. No ASTM D439 requirements.

OTHER IMPORTANT ALCOHOL FUELS ACTIVITY IN INDIANA

1978 - A $750,000 grant was appropriated by the state legislature to Purdue University for the development of alcohol fuels technology (cellulosic conversion).

1979 - $500,000 in additional funds were appropriated by the state legislature to Purdue for the continuation of their cellulosic conversion research.

Vincennes University is conducting research on alcohol still designs.

The U. S. Navy conducted a vehicle fleet test in 1979 using gasohol (12.5 percent ethanol) in 17 vehicles in Crane, Indiana.

The City of Indianapolis conducted a vehicle fleet test from August 1979 to October 1979 using gasohol in 10 vehicles.

Universities and Colleges offering alcohol fuels workshops, seminars and courses: Vincennes University; Purdue University at Lafayette; Taylor University at Upland.

Responsible Official For Further Information:

Ms. Mary Failey (317) 232-8954
Department of Commerce
440 N. Meridian
Indianapolis, Indiana 46201

LAWS

SF 2376

Exempts gasohol from 10 cents of the state road use tax, (but imposes a 3 percent sales tax). At current prices, the net exemption if approximately 6.4 cents. The gasohol must contain at least 10 percent ethyl alcohol distilled from agricultural resources. Effective July 1, 1979, through June 30, 1981. An April 23, 1980 amendment to SF 2376 modifies the exemption for the period May 1, 1981, through June 30, 1983: sales tax is eliminated and gasohol is exempt from 5 cents of the 10 cents road use tax on gasoline. (Net exemption is 5 cents per gallon).

HF 734

Appropriates $50,000 for FY 1979 - FY 1981 for the promotion of gasohol and associated by-products. This money has been granted to the Iowa Development Commission. Effective July 1, 1979. The exemption expires June 30, 1980.

PENDING LEGISLATION

The legislative session adjourned April 15, 1980.

STATE REGULATION

Permit Procedure—The state code requires alcohol producers to obtain a permit for the production of alcohol and pay a $350 fee. This requirement is not enforced when the alcohol produced is used for fuel. The State Attorney General has directed the Iowa Bureau of Liquor Control to continue this non-enforcement policy until the legislature acts to revise the state code. Producers of fuel alcohol must register with the Iowa Bureau of Liquor Control. Fuel alcohol blenders must obtain a blender's license and maintain records of gallonage distributed.

State Volatility Requirements—Iowa requires that all gasoline (including gasohol) meet Distillation and Reid Vapor Pressure requirements. No ASTM D439 requirements.

OTHER IMPORTANT ALCOHOL FUELS ACTIVITY IN IOWA

Iowa State University is conducting research on alcohol fuels and has constructed and is operating a small scale alcohol fuels plant.

Luther College is conducting research on alcohol fuels and has constructed a small scale solar-powered ethanol plant.

Northwest Bell Telephone Co. in Des Moines is conducting a vehicle fleet test using gasohol in 24 vehicles.

Iowa Department of General Services conducted a vehicle fleet test during 1979 and 1980 using gasohol in state vehicles.

The Des Moines Police Department is conducting a vehicle fleet test using gasohol in 12 vehicles.

The Iowa Corn Promotion Board is promoting the development of gasohol through a corn check-off fund system.

Universities and Colleges offering alcohol workshops, seminars and courses: Iowa Central Community College.

Responsible Officials For Further Information:

Mr. Tom Pearson (515) 281-3151
Iowa Department of Commerce
250 Jewett Building
Des Moines, Iowa 50309

Mr. Doug Getter (515) 281-3251
Director, Administrative Services
Iowa Development Commission
250 Jewett Building
Des Moines, Iowa 50309

KANSAS

LAWS

HB 2324

Exempts gasohol from the state motor fuel tax of 8 cents on a decreasing scale basis. The exemption for gasohol will be as follows: July 1, 1979 - July 1, 1980 = 5 cents; July 1, 1980 - July 1, 1981 = 4 cents; July 1, 1981 - July 1, 1982 = 3 cents; July 1, 1982 - July 1, 1983 - 2 cents; July 1, 1983 - July 1, 1984 = 1 cent; July 1, 1985 - no exemption. The bill contains a maximum ceiling provision which provides that if and when the tax revenue loss resulting from the exemption equals $5 million, the exemption will automatically expire. The gasohol must contain at least 10 percent ethyl alcohol produced from grain products grown in Kansas through the use of 10 less energy units than that which would be contained in the converted motor vehicle fuel. Effective July 1, 1979.

HB 2345

Appropriates $55,000 to the Kansas Energy Office for research on the potential energy resource value of grains. Effective July 1, 1979.

PENDING LEGISLATION

The legislative session adjourned April 13, 1980.

STATE REGULATION

Permit Procedure—No state permit is required for the production of fuel alcohol. The state requires producers of fuel alcohol to submit a copy of the Federal Bureau of Alcohol, Tobacco & Firearms permit. A state permit is required to blend and distribute gasohol.

State Volatility Requirements—Kansas requires that all gasoline (including gasohol) meet only Distillation requirements. No Reid Vapor Pressure or other ASTM D439 requirements.

OTHER IMPORTANT ALCOHOL FUELS ACTIVITY IN KANSAS

The Kansas Energy Office has published several alcohol fuels brochures.

Universities and Colleges offering alcohol fuels workshops, seminers and courses: Colby Community College; Kansas State University at Manhattan

Responsible Official For Further Information:

Randy Noon (913) 296-2496
State Energy Office
214 W. 6th Street
Topeka, Kansas 66603

KENTUCKY

HB 838

Exempts gasohol producers from the state 5 percent sales and use tax, the state property tax (except for a one percent minimum), and the state corporate license tax. Exempts gasohol producers from local property taxes. The gasohol must contain at least 10 percent ethyl alcohol produced from grain or other agricultural resources and be at least 198 proof. All provisions expire eight years after the granting of an exemption certificate. Fuel alcohol plants must burn Kentucky coal or convert to such use within two years of certificate receipt in order to qualify for the exemptions.

HR 29

Resolution directing state officials to encourage the U.S. Department of Energy to convert distilleries to alcohol fuel production.

PENDING LEGISLATION

The legislative session adjourned April 1, 1980. The legislature will convene in February 1981.

STATE REGULATION

Permit Procedure—No permit is required by the state for the production of fuel alcohol. The state requires producers of fuel alcohol to submit a copy of the Federal Alcohol, Tobacco & Firearms permit.

The HB 838 exemption certificate is required to receive the financial incentives of HB 838. See HB 838 above.

State Volatility Requirements—Kentucky has no state volatility requirement for gasoline or gasohol.

LOCAL GOVERNMENT REQUIREMENTS AND INCENTIVES

No local property taxes apply to alcohol fuel producers who are eligible for the incentives of HB 838.

OTHER IMPORTANT ALCOHOL FUELS ACTIVITY IN KENTUCKY

Universities and Colleges offering alcohol fuels workshops, seminars and courses: Paducah Community College.

Responsible Official For Further Information:

Bruce Sauer (606) 252-5535
Kentucky Dept. of Energy
P.O. Box 11888/Iron Works Pike
Lexington, Kentucky 40578

LOUISIANA

LAWS

HB 571 (Act 793)

Exempts gasohol from the state sales and use taxes and the entire 8 cents per gallon state motor fuel tax. The gasohol must contain at least 10 percent alcohol distilled in Louisiana from agricultural products of which at least 10 percent must have been grown in Louisiana. The alcohol must also be rendered unsuitable for human consumption at the time of its manufacture or immediately thereafter and be dyed a distinctive color. Effective September 7, 1979, for a period not to exceed 10 years, unless extended by the legislature.

HB 540

Amends Act 793 to exempt gasohol from the 8 cents per gallon state motor fuels tax where the agricultural resources used to produce the alcohol are not available in Louisiana. Effective July 23, 1980

PENDING LEGISLATION

The legislative session adjourned July 14, 1980.

STATE REGULATION

Permit Procedure—No state permit is required for the production of fuel alcohol. Alcohol for industrial use, including fuel, is exempt from state alcohol production regulations and laws.

State Volatility Requirements—Louisiana requires that all gasoline (including gasohol) meet Distillation and Reid Vapor Pressure requirements. No ASTM D439 requirements.

Dye Color—The Louisiana Department of Revenue requires that a green dye must be used to differentiate between gasohol which qualifies for the Louisiana excise tax exemption and all other gasolines.

OTHER IMPORTANT ALCOHOL FUELS ACTIVITY IN LOUISIANA

The Louisiana Department of Natural Resources has spent $150,000 of its $2 million FY 1980 appropriation on alcohol fuels research. This department plans to allocate a similar amount for alcohol fuels research for FY 1981.

Louisiana State University (Audubon Sugar Institute) is conducting research on alcohol fuels, including the use of molasses as an alcohol feedstock, and spillage disposition.

Nicholls State University is conducting research on alcohol fuels.

Universities and Colleges offering alcohol fuels workshops, seminars and courses: Nicholls State University.

Responsible Official For Further Information:

Thomas Landrum (504) 342-4594
Director, Department of Natural Resources
Research and Division
P.O. Box 44156
Capital Station
Baton Rouge, Louisiana 70804

MAINE

LAWS

No laws have been enacted as of July 1, 1980.

PENDING LEGISLATION

The legislative session adjourned on April 3, 1980. The legislature will convene June 5, 1981.

STATE REGULATION

Permit Procedure—No state permit is required for the production of fuel alcohol. The State Bureau of Alcoholic Beverages grants authority to operate an alcohol fuels plant upon receipt of a producer's Federal Bureau of Alcohol, Tobacco & Firearms permit.

State Volatility Requirements—Maine requires that all gasoline (including gasohol) meet only Distillation requirements. No Reid Vapor Pressure or other ASTM D439 requirements.

OTHER IMPORTANT ALCOHOL FUELS ACTIVITY IN MAINE

The State "Alcohol Fuels Task Force" established November 1979 by the Governor will issue recommendations to the State Office of Energy Resources and the Governor on the issue of whether or not the State should support an alcohol fuels industry. This Task Force will issue its recommendations and its final report to the OER and the Governor in November 1980.

The Maine Department of Transportation conducted a vehicle fleet test from June 1979 to September 1979 using gasohol.

Universities and Colleges offering alcohol fuels workshops, seminars and courses: University of Maine at Orono.

Responsible Official For Further Information:

Nancy Holmes (207) 289-3811
Office of Energy Resources
55 Capitol Street
Augusta, Maine 04330

MARYLAND

LAWS

SB 807

Exempts gasohol from one cent of the 9 cents per gallon state motor fuel tax. Effective July 1, 1979. SB 9 (HB 423) enacted May 1, 1980, increased the exemption for gasohol 3 additional cents (or a total net exemption of 4 cents per gallon). The gasohol must contain at least 10 percent ethyl or methyl alcohol. The 2 cent exemption expires May 1, 1981.

Chapter 331, Acts of 1980

Exempts ethyl and methyl alcohol which is not mixed with gasoline and is sold as a motor fuel from all 9 cents of the state 9 cents per gallon motor fuel tax and the state sales tax. Effective July 1, 1980.

SB 823

Authorizes the Maryland Industrial Development Financing Authority to insure loans for the development and production of gasohol. Effective July 1, 1979.

SJR 24

Appoints a Maryland study commission on gasohol through the Maryland Secretary of Agriculture. Effective July 1, 1979.

HB 628

Establishes a gasohol testing program through the Maryland Department of Agriculture. The results of this program were submitted to the legislature January 1, 1980. Effective July 1, 1979.

Chapter 437, Acts of 1980

Includes gasohol in those provisions of the annotated tax code and state regulations which affect gasoline. Effective July 1, 1980.

SJR 7

Urges the United States Congress to pass legislation which promotes the development of alcohol fuels. Effective July 1, 1980.

LEGISLATION INTRODUCED AND NOT ENACTED

The legislative session adjourned April 8, 1980. The legislature will convene January 14, 1981.

SB 492

Requires retailers of motor fuel to accept credit cards for the purchase of gasohol if they accept credit cards for the purchase of any other motor fuel.

STATE REGULATION

Permit Procedure—No state permit is required for the production of fuel alcohol. Producers of fuel alcohol are required to send a copy of the Federal Bureau of Alcohol, Tobacco and Firearms permit to the state. Blenders and distributors are required to register with the state motor fuel tax division.

State Volatility Requirements—Maryland requires that gasoline (including gasohol) meet ASTM D439 specifications, which include Distillation and Reid Vapor Pressure specifications.

LOCAL GOVERNMENT REQUIREMENTS AND INCENTIVES

The government of St. Mary's County has received a grant to build a demonstration still. The Tri-County Council (St. Mary's, Charles and Calvert Counties) sponsored this project.

OTHER GOVERNMENT ALCOHOL FUELS ACTIVITY IN MARYLAND

The Maryland Department of General Services conducted a vehicle fleet test from September 1979 to September 1980 using gasohol in 8 state vehicles.

The Maryland Chapter of AAA is conducting a vehicle fleet test using gasohol in 22 vehicles.

Universities and Colleges offering alcohol fuels workshops, seminars and courses: Cecil Community College.

Responsible Officials For Further Information:

Bruce Williams (301) 787-7307
Maryland Department of Transportation
Public Affairs
P.O. Box 8755
Baltimore-Washington International Airport,
Maryland 21240

Marvin Bond (301) 269-3885
Assistant to the Comptroller
P.O. Box 466
Annapolis, Maryland 21404

MASSACHUSETTS_____

LAWS

No laws have been enacted as of July 1, 1980.

PENDING LEGISLATION

HB 1772

Exempts gasohol from 4.5 cents of the 8.5 cents per gallon state motor fuel tax. The alcohol must be produced from feedstocks indigenous to Massachusetts.

STATE REGULATION

Permit Procedure—No state permit is required for the production of fuel alcohol.

State Volatility Requirements—Massachusetts requires that all gasoline (including gasohol) meet only Distillation requirements. No Reid Vapor Pressure or other ASTM D439 requirements.

OTHER IMPORTANT ALCOHOL FUELS ACTIVITY IN MASSACHUSETTS

Springfield Technical and Community College is conducting research on alcohol fuels.

Clark University is conducting research on alcohol fuels.

The University of Lowell is conducting research on alcohol fuels and is operating an alcohol plant using waste paper as a feedstock.

Universities and Colleges offering alcohol fuels workshops, seminars and courses: University of Lowell.

Responsible Official for Further Information:

Mr. Chris Hansen (617) 727-1990
Executive Office of Energy Resources
73 Tremont Street
Boston, Massachusetts 02108

MICHIGAN_____

LAWS

Public Act 198 of 1974 (as amended)

Tax benefits under the act are granted by the legislative body of the city, township or village in which the facility will be located. Allows producers of new alcohol plant facilities a 50 percent property tax exemption for new plant construction (industrial plants only).

PENDING LEGISLATION

HB 5451

Exempts gasohol from all 11 cents of the state 11 cents per gallon motor fuel tax.

HB 5450

Exempts gasohol from the state sales tax.

SB 414

Requires all motor vehicle fuel sold in Michigan to contain 3 percent alcohol by January 1, 1980, and 10 percent alcohol by January 1, 1984.

SB 480

Imposes a 6 cents per gallon tax on alcohol/gasoline blended fuel for the development of a demonstration still and research on the by-products of alcohol production.

STATE REGULATION

Permit Procedure—No state permit is required for the production of fuel alcohol.

State Volatility Requirements—Michigan has no State Volatility Requirements for gasoline or gasohol.

LOCAL GOVERNMENT REQUIREMENTS AND INCENTIVES

No local information is available.

OTHER IMPORTANT ALCOHOL FUELS ACTIVITY IN MICHIGAN

$195,000 was appropriated by the state to the State Department of Agriculture for research, demonstration, grants and clearinghouse activities. $118,000 of this total has been allocated for a contract with Michigan State University for the development of a demonstration still and related byproducts and a computer research project.

The Michigan Gasohol Committee completed a fact-finding analysis titled "Gasohol, Options and Prospects" which was submitted to the Michigan Legislature on April 15, 1980. This report provides information from which appropriate legislative actions can be taken.

Universities and Colleges offering alcohol fuels workshops, seminars and courses: Jordan College at Cedar Springs; University of Michigan at Flint; Michigan State University at East Lansing; Mott Community College.

Responsible Official For Further Information:

Mr. Randy Harmson (517) 373-1054
Assistant Chief of Marketing and
  International Trade Division
Michigan Department of Agriculture
P.O. Box 30017
Lansing, Michigan 48909

LAWS

HF 1121 "Omnibus Tax Bill"

Exempts gasohol from 4 cents of the 11 cents per gallon state motor fuel tax. The alcohol must contain at least 10 percent ethyl alcohol distilled in Minnesota from agricultural resources produced in Minnesota and be at least 190 proof. Effective May 1, 1980. The exemption expires December 31, 1984.

Minnesota statutes 1978, Section 290.06, Subdivision 13

Provides for a 20 percent income tax deduction on the first $10,000 spent by a producer of renewable energy (including methane, methanol and ethanol) not offered for sale. Effective July 1, 1979. The exemption expires December 31, 1982.

Laws of 1978 - CH 786

Appropriates $84,500 for two alcohol fuels feasibility studies in conjunction with U.S. Department of Agriculture.

Laws of 1979, CH 2, extra session

Appropriates $50,000 to the University of Minnesota at St. Paul Department of Agricultural and Applied Economics for alcohol fuels research.

HF 1710

Appropriates $200,000 to the University of Minnesota at Morris for small scale demonstration plants. Effective April 16, 1980.

PENDING LEGISLATION

The legislative session adjourned April 12, 1980. The legislature will convene January 6, 1981.

STATE REGULATION

Permit Procedure—Producers of fuel alcohol must obtain a permit from the Liquor Control Division of the State Department of Public Safety. A $5.00 fee is required.

The Minnesota Department of Revenue, Petroleum Division, requires that producers of ethanol for use as a non-blended fuel on public highways pay all 11 cents of the state motor fuel tax.

The Boiler License Code of the Department of the Labor & Industry requires that all boilers and boiler operations be licensed. Fees range from $8.00 (30 horsepower) to $15.00 (300 horsepower).

State Volatility Requirements—Minnesota requires that all gasoline (including gasohol) meet only Distillation requirements. No Reid Vapor Pressure or other ASTM D439 requirements.

OTHER IMPORTANT ALCOHOL FUELS ACTIVITY IN MINNESOTA

Canby Atvi, Granite Falls Avti, and Mankato Atvi are experimenting with Alcohol fuels.

The agronomy division of the Minnesota Dept. of Agriculture requires a permit to sell distillers grains.

Universities and Colleges offering alcohol fuels workshops, seminars and courses: Lakewood Community College; University of Minnesota; Southwest State University; Mankato State University; Austin Community College.

Responsible Official for Further Information:

Dennis Devereaux (612) 296-9078
Alternative Energy Projects
Minnesota Energy Agency
980 American Center Building
150 E. Kellogg Blvd.
St. Paul, Minnesota 55101

MISSISSIPPI

LAWS

HB 366

Legalizes the production and blending of alcohol for fuel in Mississippi. Effective July 1, 1980. The intent of this law is to clarify that previous statutes which limited the production and transportation of alcohol do not apply to alcohol fuels.

PENDING LEGISLATION

The legislative session adjourned May 12, 1980.

STATE REGULATION

Permit Procedure-A state license is required for the production of fuel alcohol by the Motor Vehicle Comptroller's Office.

State Volatility Requirements-Mississippi requires that gasoline (including gasohol) meet ASTM D439 specifications, which include Distillation and Reid Vapor Pressure specifications.

OTHER IMPORTANT ALCOHOL FUELS ACTIVITY IN MISSISSIPPI

Universities and Colleges offering alcohol fuels workshops, seminars and courses: Mississippi State; University of Southern Mississippi; University of Mississippi.

Responsible Official for Further Information:

Mr. Robert Smira (601) 961-4403
Conservation Director
Office of Energy
Department of Natural Resources
300 Watkins Building
510 George Street
Jackson, Mississippi 39202

MISSOURI

LAWS

No laws have been enacted as of July 1, 1980.

LEGISLATION INTRODUCED AND NOT ENACTED

The legislative session adjourned April 30, 1980.

HB 1272

Exempts gasohol from 4 cents of the state 7 cents per gallon motor fuel tax. The gasohol must contain at least 10 percent alcohol.

STATE REGULATION

Permit Procedure-All producers of fuel alcohol must obtain a permit from the State Department of Revenue. A $5,000 bond must be posted.

State Volatility Requirements-Missouri requires that all gasoline (including gasohol) meet Distillation and Reid Vapor Pressure requirements. No ASTM D439 requirements.

OTHER IMPORTANT ALCOHOL FUELS ACTIVITY IN MISSOURI

The Governor of Missouri has issued a directive requiring that gasohol be used in the state fleet if the price of gasohol does not exceed the price of gasoline by 3 cents.

Universities and Colleges offering alcohol fuels workshops, seminars, and courses: State Fair Community College; Sedalia Crowder College; and Southern State College at Joplin.

Responsible Official for Further Information:

Ms. Deborah Goldhammer (314) 751-4000
Manager of Program Development
Division of Energy
P.O. Box 176
Jefferson City, Missouri 65102

MONTANA

LAWS

HB 402

Exempts gasohol from the 9 cents per gallon state motor fuel tax on e decreasing scale basis. The gasohol must contain at least 10 percent anhydrous alcohol produced in Montana from Montana resources. The exemption for gasohol will be: April 1, 1979 - April 1, 1985 = 7 cents; April 1, 1985 April 1, 1987 = 5 cents; April 1, 1987 - April 1, 1989 = 3 cents. The governor has the authority to suspend the law at his discretion. The exemption is effective April 12, 1979 and expires April 1, 1989.

SB 523

Authorizes a property tax reduction for property used primarily in the production of gasohol. Qualifying properties are taxed at 3 percent (rather than the nonexempt 100 percent level) of their market value. A decreasing scale rate of reductions is included in the law's provisions. The tax reduction is for the taxable years beginning after December 1, 1979, and applies to the period beginning with construction and the first three years of operation. Effective July 1, 1979.

SB 520

Authorizes funding for the Montana Department of Agriculture to contract for research and development on the feasibility of obtaining fuel from wheat and barley. Approximately $200,000 will be appropriated for this project each year.

SB 86

Establishes a state program to provide grants for research, development and demonstration of renewable energy resources. The program is funded by coal severance tax funds (approximately $800,000 per year).

PENDING LEGISLATION

No regular session was held in 1980. The legislature will convene January 5, 1981.

STATE REGULATION

Permit Procedure—No state permit required for the production of fuel alcohol. Distributors of gasohol must obtain a permit and post a $1,000 bond with the Montana Department of Revenue.

State Volatility Requirements—Montana requires that gasoline (including gasohol) meet ASTM D439 specifications, which include Distillation and Reid Vapor Pressure specifications.

OTHER IMPORTANT ALCOHOL FUELS ACTIVITY IN MONTANA

The U.S. Community Services Administration operates a demonstration still in Butte, Montana.

Montana State University at Bozeman is conducting research on alcohol fuels and has established an inter-departmental Biofuels Committee.

Universities and Colleges offering workshops, seminars and courses: University of Montana at Missoula; Flathead Valley Community College.

Responsible Official for Further Information:

Ms. Georgia Brensbal (406) 449-4624
Program Engineer
Renewable Energy Bureau
State Department of Natural Resources
32 S. Ewing
Helena, Montana 59601

NEBRASKA

LAWS

LB 571

Enacted May 23, 1979, increases the net tax exemption for gasohol to 5 cents. Requires that gasohol contain at least 10 percent ethyl alcohol produced from agricultural resources and be at least 99 percent pure. Beginning February 1, 1982, the 5 cent exemption applies only to agricultural ethanol produced in Nebraska by a plant in operation or under construction by July 1, 1982. The law also authorized the state of Nebraska to underwrite bonds issued by cities and counties which sponsor the construction of alcohol fuel plants, but the state supreme court overruled this bond provision in 1979.

LB 776

Exempts gasohol from 3 cents of the 10 1/2 cents state motor fuel tax. The law contains maximum ceiling provisions. Created a "Grain Alcohol Fuel Tax Fund," which is to be used to promote a grain alcohol industry in Nebraska. This fund contained an initial $40,000 appropriation and 1/8 of one cent of all collected motor fuel tax revenues in the future. Created "Agricultural Products Industrial Utilization Committee" to administer this program. Effective May 26, 1971.

LB 424

Establishes a Nebraska Gasohol and Energy Program to be administered by the Agricultural Products Industrial Utilization Committee. Establishes a matching grant ($500,000 per grant) program for alcohol fuels. No appropriations bill was passed to fund this program. Effective April 21, 1978.

LB 74

Requires the use of gasohol in the Nebraska Department of Roads' vehicles. The department is required to comply with this act and purchase gasohol only if the cost of gasohol does not exceed the cost of other usual fuels by 10 cents. Effective April 5, 1979. Conditions of the bill were to be met by July 1, 1980.

LB 876

Increases the amount of funds transferred to the Agricultural Alcohol Fuel Tax Fund by 1/8 cent (total amount transferred = 1/4 cent of each gallon on which the tax is collected). Effective July 19, 1980.

LB 954

Provides for an exemption from the state sales and use taxes for the purchase of alternative energy equipment including alcohol production equipment. Effective April 23, 1980.

PENDING LEGISLATION

The legislative session adjourned April 18, 1980.

STATE REGULATION

Permit Procedure—A boiler permit is required by the Nebraska Department of Labor for the production of fuel alcohol if the production unit exceeds 15 psi. A $10.00 fee is required.

State Volatility Requirements—Nebraska requires that all gasoline (including gasohol) meet only Distillation requirements. No Reid Vapor Pressure or other ASTM D439 requirements.

OTHER IMPORTANT ALCOHOL FUELS ACTIVITY IN NEBRASKA

The University of Nebraska is conducting research on the potential energy source of distillers dried grains. $10,000 was appropriated for 1980 for this project by the Nebraska Gasohol Commission.

The Nebraska Gasohol Commission is providing equipment to the Southeast Community College in Milford for the construction of an on-farm alcohol plant.

The National Gasohol Commission is conducting a marketing study in Lincoln, Nebraska, on consumers' reactions to gasohol.

Sarpy County is conducting a fleet test using gasohol in 20 county cars.

The Nebraska Agricultural Products Industrial Utilization Committee (NAPIUC) conducted the first major gasohol vehicle fleet test in the U.S., from December 1974 to October 1977, using gasohol in 74 vehicles over two million miles.

In November 1979, the Governor of Nebraska issued an administrative order requiring all state government vehicles to use gasohol whenever possible.

Universities and Colleges offering alcohol fuels workshops, seminars, and courses: Southeast Community College; Midplains Community College; Scottsbluff Western College; Northeast Community College; Central Nebraska Technical College.

Responsible Official For Further Information:

Mr. Steve Sorum (402) 471-2941
Agricultural Products Industrial
Utilization Committee
301 Centennial Mall, South
Third Floor
Lincoln, Nebraska 68509

NEVADA

LAWS

No laws have been enacted as of July 1, 1980

PENDING LEGISLATION

No regular session was held in 1980

STATE REGULATION

Permit Procedure—No state Permit is required for the production of fuel alcohol. A permit is required to become a "dealer" or distributor of gasohol or ethanol.

State Volatility Requirements—Nevada requires that all gasoline (including gasohol) meet only Distillation requirements. No Reid Vapor Pressure or other ASTM D439 requirements.

OTHER IMPORTANT ALCOHOL FUELS ACTIVITY IN NEVADA

$140,000 in state funds was appropriated for matching funds in March 1980 by the state legislature for the construction of an alcohol fuels plant.

Responsible Official for Further Information:

Mr. Kelly Jackson (702) 885-5157
Nevada Department of Energy
400 West King Street
Room 106
Carson City, Nevada 89710

NEW HAMPSHIRE_____

LAWS

HB 201

Exempts gasohol from 5 cents of the 11 cents per gallon state motor fuel tax. The gasohol must contain at least 10 percent alcohol and be produced in New Hampshire. Effective 1979.

PENDING LEGISLATION

No regular session was held in 1980.

STATE REGULATION

Permit Procedure—A state permit is required for commercial production of fuel alcohol by the state Siting Evaluation Committee (Law 162-14).

State Volatility Requirements—New Hampshire has no volatility requirements for gasoline or gasohol.

OTHER IMPORTANT ALCOHOL FUELS ACTIVITY IN NEW HAMPSHIRE

Dartmouth College is conducting a research project on ethanol production from biomass using acid hydrolysis.

The Department of Public Works and Highways conducted a vehicle fleet test on 10 vehicles using gasohol and on two trucks using 100 percent ethanol.

Responsible Official For Further Information:

Ms. Tina Oleson (603) 271-2771
Government's Council on Energy
1/2 Beacon Street
Concord, New Hampshire 03310

NEW JERSEY_____

LAWS

No laws have been enacted as of July 1, 1980.

PENDING LEGISLATION

S 1030

Exempts gasohol from the 8 cents per gallon state gasoline tax.

A 325

Exempts motor vehicles fueled by alternative energy sources from the state sales tax.

A 777

Exempts gasohol from 6 cents of the 8 cents per gallon state gasoline tax.

A 1549

Imposes a one cent per gallon tax on gasohol.

A 136

Provides tax incentives for alcohol fuel production and consumption.

STATE REGULATION

Permit Procedure—No state permit is required for the production of fuel alcohol.

State Volatility Requirements—New Jersey has no state volatility requirements for gasoline or gasohol.

OTHER IMPORTANT ALCOHOL FUELS ACTIVITY IN NEW JERSEY

Glassboro State College plans to build a demonstration still in cooperation with local farmers.

Princeton University is conducting research on alcohol fuels.

The State of New Jersey conducted a vehicle fleet test using gasohol in state vehicles from June 1979 to September 1979.

Universities and Colleges offering alcohol fuels workshops, seminars and courses: Glassboro State College; Ramapo College of New Jersey; Southern New Jersey College.

Responsible Official for Further Information:

Mr. Louis Jarecki (201) 648-6293
New Jersey Department of Energy
Office of Alternate Technology
101 Commerce Street
Newark, New Jersey 07102

LAWS

SB 39

Exempts gasohol from entire 8 cents per gallon state motor fuel tax. The gasohol must contain at least 10 percent ethyl alcohol produced in New Mexico and be at least 190 proof. Exempts gasohol from the state gross receipts and compensation taxes. Effective July 1, 1980. The exemptions expire June 30, 1985.

HM 41

Directs the New Mexico Energy and Mineral Department to undertake research on alcohol fuels. Effective July 1, 1980.

PENDING LEGISLATION

The legislative session adjourned February 14, 1980.

STATE REGULATION

Permit Procedure—No state permit is required for the production of fuel alcohol.

State Volatility Requirements—New Mexico requires that all gasoline (including gasohol) meet only Distillation requirements. No Reid Vapor Pressure or other ASTM D439 requirements.

OTHER IMPORTANT ALCOHOL FUELS ACTIVITY IN NEW MEXICO

In December 1979, $200,000 was appropriated by the state legislature for four alcohol fuels projects.

In November 1979, the Governor appointed an Alcohol Fuels Task Force to research the production and use of alcohol fuels in New Mexico and to prepare recommendations to be submitted to the Governor.

Several alcohol fuels brochures have been issued by the New Mexico Energy & Minerals Department.

The University of New Mexico is conducting research on the enzymatic conversion of cellulose.

Navajo Community College is conducting research on alcohol fuels.

New Mexico State University plans to construct a small scale alcohol fuels still.

The U.S. Department of Defense is conducting a vehicle fleet test at the Los Alamos Lab using gasohol in 1,062 vehicles.

Universities and Colleges offering alcohol fuels workshops, seminars and courses: University of New Mexico at Albuquerque.

Responsible Official for Further Information:

Mr. Gerald Bradley (505) 827-3221, Ext. 217
Tax Research & Statistics Office
  of Taxation Revenue Department
P.O. Box 630
Sante Fe, New Mexico 87509

NEW YORK

LAWS

A 11139

Removes the $625 annual alcohol manufacturer's permit fee if alcohol is produced from biomess feedstock and is used for the producers own fuel use. Reduces the fee to $312.50 for alcohol producers who manufacture less than 100,000 gallons per year if the alcohol is produced for commercial distribution. Effective June 26, 1980.

PENDING LEGISLATION

None

STATE REGULATION

Permit Procedure—Producers and blenders of fuel alcohol must obtain an industrial alcohol manufacturer's permit from the state liquor authority to produce fuel alcohol. Commercial producers of fuel alcohol are required to pay a $625 fee (commercial producers over 100,000 gallons per year) or a $312.50 fee (commercial producers under 100,000 gallons per year).

State Volatility Requirements—New York requires that all gasoline (including gasohol) meet only Distillation requirements. No Reid Vapor Pressure or other ASTM D439 requirements.

OTHER IMPORTANT ALCOHOL FUELS ACTIVITY IN NEW YORK

Cornell University is conducting technical and economic feasibility studies on the production of ethanol in five regions of New York State from agricultural wastes, including apples, beets, potatoes, brewery wastes and cheese whey. The New York State Energy Research and Development Authority has partially funded this program.

New York State's Antonio Ferri Laboratory is conducting research on alcohol fuels.

GOHR Distributing Inc. in Buffalo is conducting a vehicle fleet test using gasohol in 60 trucks.

Brookhaven National Labs conducted a vehicle fleet test from March 1979 to March 1980 using methanol and ethanol blends in 300 vehicles.

New York State conducted a vehicle fleet test from July 1979 to March 1980 using gasohol in 62 vehicles.

The New York City Police Department is conducting a vehicle fleet test (test began August 1979) using methanol and ethanol blends in three vehicles.

New York City conducted a vehicle fleet test in 1979 using gasohol in 80 vehicles.

The NY-NJ Port Authority conducted a vehicle fleet test (beginning in May 1979) using gasohol in airport and police vehicles at Kennedy Airport.

Universities and Colleges offering alcohol fuels workshops, seminars and courses: Onondaga Community College; New York University; CUNY Brooklyn College; SUNY College at Binghamton; SUNY College at Plattsburgh; SUNY Agricultural and Technical College at Delhi.

Responsible Official for Further Inforamtion:

Mr. Mark Bagdon (518) 474-7875
State Energy Office
Agency Building 2
Rockefeller Plaza
Albany, New York 12210

LAW

HB 1556

Exempts gasohol from the state 9 cents per gallon gasoline tax on a decreasing scale basis. The exemption for gasohol will be as follows: January 1, 1981 - June 30, 1981 = 8 cents; July 1, 1981 - June 30, 1982 = 7 cents; July 1, 1982 - June 30, 1983 - 6 cents; July 1, 1983 - June 30, 1984 = 5 cents. The gasohol must contain at least 10 percent anhydrous ethyl alcohol. Exempts non-anhydrous ethyl alcohol from all 9 cents of the 9 cents per gellon state motor fuel tax if the ethanol produced for on-farm use. Effective January 1, 1981. The exemptions expire July 1, 1984.

HB 1554

Establishes a credit against corporate income tax for corporations that construct in North Carolina fuel ethanol distilleries which make ethyl alcohol from agricultural or forestry products for use as a motor fuel. The tax credit equals 20 percent of the installation of equipment costs and an additional 10 percent of those costs if the distillery is powered primarily by use of an alternative fuel source. Establishes a credit against personal income tax for construction of a fuel ethanol distillery for any person who constructs in North Carolina a distillery to make ethanol from agricultural or forestry products for use as a motor fuel. The tax credit equals 20 percent of the installation and equipment costs of construction and an additional 10 percent of those costs if the distillery is powered primarily by use of an alternative fuel source. Effective date: retroactive to January 1, 1980. The exemptions expire January 1, 1985.

PENDING LEGISLATION

None.

STATE REGULATION

Permit Procedures—No state permit is required for the production of fuel alcohol.

State Volatility Requirements—North Carolina requires that gasoline (including gasohol) meet ASTM D439 specifications, which include Distillation and Reid Vapor Pressure specifications.

OTHER IMPORTANT ALCOHOL FUELS ACTIVITY IN NORTH CAROLINA

Pitt Community College in Greenville has received a $10,000 grant from the U.S. Department of Energy for the construction of an alcohol fuels demonstration plant.

Universities and Colleges offering alcohol fuels workshops, seminars and courses: Carteret Technical Institute.

Responsible Official For Further Information:

Mr. John Manuel (919) 733-4493
Energy Division
Dept. of Commerce
P.O. Box 25249
Raleigh, North Carolina 27611

NORTH DAKOTA

LAWS

HB 1384

Exempts gasohol from entire 4 cents per gallon motor fuel tax. The gasohol must contain at least 10 percent ethyl alcohol produced from agricultural resources. Establishes an Agricultural Products Utilization Commission which will promote and research the production and distribution of gasohol. Establishes an agriculturally derived alcohol motor vehicle fuel tax fund which will finance the commission and appropriates $50,000 for this fund. Authorizes a sales tax refund (excepting a 1/8 cent charge) to persons who purchase agriculturally derived alcohol fuels for agriculture or industrial purposes. Effective July 1, 1979.

PENDING LEGISLATION

No regular session held in 1980.

STATE REGULATION

Permit Procedure—A joint cooperative permit between the Federal Bureau of Alcohol, Tobacco and Firearms and the State Department of Treasury is required by the state for the production of fuel alcohol. No fee is required. The Attorney General has waived licensing requirements.

State Volatility Requirements—North Dakota requires that gasoline (including gasohol) meet ASTM D439 specifications, which include Distillation and Reid Vapor Pressure specifications.

OTHER IMPORTANT ALCOHOL FUELS ACTIVITY IN NORTH DAKOTA

Several state end junior colleges are trying to establish demonstration and/or research projects; Lake Region Junior College (Devils Lake, N.D.), State School of Science (Wahpeton, N.D.), Cooperative Extension Service at North Dakota State University (Fargo, N.D.).

Responsible Officials For Further Information:

Mr. Mike Mahlum (701) 224-2250
Federal & Coordinators Office
Energy Management Conservation Programs
1533 W. 12th Street
Bismarck, North Dakota 58501

Mr. Dean McIlroy (701) 224-2232
Executive Director
Agriculture Products Utilization Commission
Department of Agriculture
State Capital
Bismarck, North Dakota 58505

LAWS

HB 602

Establishes an Alcohol Fuels Advisory Council within the Ohio Department of Energy to study the development, production, distribution and use of alcohol fuels in Ohio and the feasibility of using agricultural commodities grown in Ohio as a feedstock for such fuels. The Council will terminate July 1, 1981. It is staffed, funded and chaired by the Ohio Department of Energy. $50,000 has been appropriated for this Council. Effective February 28, 1980.

PENDING LEGISLATION

HB 529 (HB 796, SB 266)

Amends Section 5735.01 of the Revised Code to exempt gasohol from the state 7 cents per gallon motor fuel tax. The gasohol must contain at least 10 percent alcohol.

HB 986

Authorizes the Ohio Department of Energy to design and implement a plan for small-scale alcohol production plants on Ohio farms and appropriates $100,000 for this study. Ohio feedstocks are to be used in this project.

SB 348

Prohibits oil companies from disallowing the purchase of gasohol through credit card mechanisms.

STATE REGULATION

State Code 3717.31

It is illegal to sell, exchange or deliver milk from cows fed on wet distillery waste or starch waste. Effective October 1, 1953.

Permit Procedure—No state permit is required for the production of fuel alcohol.

State Volatility Requirements—Ohio has no state volatility requirements for gasoline and gasohol.

OTHER IMPORTANT ALCOHOL FUELS ACTIVITY IN OHIO

Ohio State University is conducting research on alcohol fuel blends.

Universities and Colleges offering alcohol fuels workshops, seminars and courses: Ohio State University (Cooperative Extension Service).

Responsible Official for Further Information:

Mr. Bob Yaekle (614) 466-4573
Ohio Legislative Service Commission
State House, 5th Floor
Columbus, Ohio 42315

OKLAHOMA

LAWS

SB 248

Exempts gasohol from 6-1/2 cents of the state 6-5/8 cents per gallon motor fuel tax. The gasohol must contain at least 10 percent ethyl alcohol distilled from agricultural resources, and be at least 198°. The distributor of the gasohol must be licensed in Oklahoma and obtain a bond in support of this license. Effective October 1, 1979. Expires October 1, 1984.

SB 428

Establishes permit procedures for fuel alcohol producers. Assigns authority for the enforcement of alcohol fuel regulations to the Oklahoma Department of Agriculture. Effective February 6, 1980.

PENDING LEGISLATION

None.

STATE REGULATION

Permit Procedures—Producers of fuel alcohol must obtain a permit from the Oklahoma Department of Agriculture. A $25.00 fee is required for an experimental plant permit and a $250 fee is required for a distilled spirits plant permit. Whereas distilled spirit plant permit holders may blend alcohol with gasoline and sell the product as a motor fuel, experimental plant permit holders must retain the product for their own use.

State Volatility Requirements—Oklahoma requires that all gasoline (including gasohol) meet only Distillation requirements. No Reid Vapor Pressure or other ASTM D439 requirements.

OTHER IMPORTANT ALCOHOL FUELS ACTIVITY IN OKLAHOMA

The U.S. Department of Energy conducted a vehicle fleet test from November 1978 to December 1979 in Bartlesville, Oklahomas, using methanol blends and neat methanol in 76 vehicles.

Universities and Colleges offering alcohol fuels workshops, seminars and courses: Oklahoma State University at Stillwater. (Agricultural Engineering Dept.)

Responsible Official for Further Information:

Mr. Rex Privett (405) 427-3829
State Gasohol Coordinator
State Department of Energy
4400 N. Lincoln Blvd., Suite 35
Oklahoma City, Oklahoma 73105

OREGON

LAWS

HB 2780

Provides for a property tax exemption for commercial alcohol fuel plant property (100 percent exemption through 1984). Provides for an income tax exemption (100 percent exemption through 1985) for the taxable income from alcohol fuels production if the fuel is methanol, ethanol, or other substitute fuel not produced from petroleum, natural gas, or coal and 75 percent of the total-production yield is used for making gasohol. Effective October 3, 1979.

HB 2779

Requires the Oregon state motor vehicle fleet to use gasohol when it is commercially availeble. Effective January 1, 1980.

SB 515

Provides for a 50 percent investment tax credit over five years (20 percent first year, 10 percent second and third years, 5 percent remaining two years) for alternative energy development projects including alcohol fuels. Effective January 1, 1980. Expires January 1, 1985.

SB 611

Ballot measure 3 was voted on May 20, 1980 (and passed) by the general electorate. Allows the State Department of Energy to issue bonds for alternative energy development projects, including alcohol fuels.

PENDING LEGISLATION

No regular session was held in 1980.

STATE REGULATION

Permit Procedure—No state permit is required for the production of fuel alcohol. Producers of fuel alcohol are required to send a copy of the Federal Bureau of Alcohol, Tobacco and Firearms permit to the Oregon Liquor Control Commission.

State Volatility Requirements—Oregon has no state volatility requirements for gasoline and gasohol.

LOCAL GOVERNMENT REQUIREMENTS AND INCENTIVES

Several Port districts in Oregon issue "development bonds." The Port of Morrow, Port of Umatilla and Klamath Co. Economic Development district in Klamath Falls are involved in alcohol fuel projects.

OTHER IMPORTANT ALCOHOL FUELS ACTIVITY IN OREGON

Oregon General Services Administration conducted a vehicle fleet test (beginning October 1979) using gasohol in 15 vehicles.

Bonneville Power Authority in Portland, Oregon, conducted a vehicle fleet test (beginning July 1979) using methanol blends in seven vehicles.

The University of Oregon is conducting a vehicle fleet test using methanol/gasoline blends.

Universities and Colleges offering alcohol fuels workshops, seminars and courses: Linn-Benton Community College; Eastern Oregon State College at Labrande.

Responsible Official For Further Information:

Mr. Richard Durham (503) 378-4998
Oregon Department of Energy
Labor and Industries Building
Salem, Oregon 97370

PENNSYLVANIA_____

LAWS

SB 111 (Act 129 of 79)

Reduces the state license fee for small scale (on-farm use) ethanol stills from $2,500 to $25.00. Effective December 14, 1979.

PENDING LEGISLATION

HB 1428

Authorizes a tax credit of up to 6 cents on the gasoline pump price of gasohol. The gasohol must contain 10 percent alcohol.

HB 1541

Requires the Pennsylvania Department of General Services to use gasohol in all state vehicles.

HB 1897

Requires the Pennsylvania liquor control board to issue licenses for the limited manufacture and use of ethanol or methanol for liquid fuel purposes and provides for alcohol licensing fees for farm vehicle use only.

HR 2444 (HR 2443)

Establishes a Pennsylvania Energy Development Authority and provides for this agency's powers and duties regarding the implementation and development of energy and energy conservation technologies. Authorizes the issuance of bonds and appropriates funds for this program.

HR 2405

Establishes a Pennsylvania Energy Development Office within the Pennsylvania Industrial Development Authority and gives this office the power to administer grants.

STATE REGULATION

Permit Procedure—Small scale producers of fuel alcohol are required to obtain a permit from the Pennsylvania liquor control board. A $25 fee is required. Large-scale producers and those producers who commercially distribute their product must pay a $2,500 fee.

State Volatility Requirements—Pennsylvania has no state volatility requirements for gasoline and gasohol.

OTHER IMPORTANT ALCOHOL FUELS ACTIVITY IN PENNSYLVANIA

In August 1979 the Governor's office issued a directive encouraging the Pennsylvania Department of General Services to conduct a gasohol vehicle fleet test.

The University of Pennsylvania is conducting research on alcohol fuels (cellulosic conversion).

Pennsylvania Power and Light Company is conducting a vehicle fleet test (November 1979 through the summer of 1980) using gasohol in 26 vehicles.

Universities and Colleges offering alcohol fuels workshops, seminars and courses: Lehigh University; Gettysburg College; University of Pennsylvania at Philadelphia.

Responsible Official For Further Information:

Mr. David L. Krantz (717) 787-3735
Research Analyst
Minority Research Unit
House of Representatives
Box 250
Harrisburg, Pennsylvania 17120

RHODE ISLAND_____

LAWS

No laws have been enacted as of July 1, 1980.

PENDING LEGISLATION

SB 2274

Exempts gasohol from the 10 cents per gallon state motor fuel tax.

STATE REGULATION

Permit Procedure—No state permit required for the production of fuel alcohol.

State Volatility Requirements—Rhode Island requires that gasoline (including gasohol) meet ASTM D439 specifications, which include Distillation and Reid Vapor Pressure specifications.

Responsible Official for Further Information:

Ms. Shelly Greenfield (401) 277-3374
Solar Information Specialist
Governor's Energy Office
80 Dean Street
Providence, Rhode Island 02903

SOUTH CAROLINA_____

LAWS

HB 2442

Exempts gasohol from the 10 cents per gallon state motor fuel tax on a decreasing scale basis. The exemption for gasohol is as follows: October 1, 1979 - July 1, 1985 = 4 cents; July 1, 1985 - July 1, 1987 = 3 cents. The exemptions will expire when the state realizes a $5 million loss of revenue. The gasohol must contain at least 10 percent ethyl alcohol produced from agricultural resources and be at least 99 percent pure. Effective October 1, 1979.

HB 2976 Establishes a $5 million state capital improvement bond program. (Subject to revision in the near future. Guidelines for this money are being proposed.) Effective November 1, 1979. See HB3843 below.

HB 3843

Amends HB 2976 to provide for bond issuance procedures for alcohol fuel development.

PENDING LEGISLATION

None.

STATE REGULATION

Permit Procedure—No state permit is required for the production of fuel alcohol.

State Volatility Requirements—South Carolina requires that all gasoline (including gasohol) meet Distillation and Reid Vapor Pressure requirements. No ASTM D439 requirement.

OTHER IMPORTANT ALCOHOL FUELS ACTIVITY IN SOUTH CAROLINA

Clemson University was awarded a $28,700 grant from the U.S. Department of Agriculture to conduct research on the potential feedstock value of sweet potatoes.

Between March 15, 1980, and March 17, 1980, the Clemson University cooperative extension service canvassed the state for potential alcohol fuels production sites for small scale production.

The state of South Carolina is conducting a vehicle fleet test using gasohol in state vehicles.

Universities and Colleges offering alcohol fuels workshops, seminars and courses: Clemson University; Orangeburg Calhoun Technical College.

Responsible Official For Further Information:

Mrs. Cathy Twilley (803) 758-8110
Administrative Assistant
Governor's Division of Energy Resources
1122 Lady Street
Columbia, South Carolina 29201
(803) 758-8110

LAWS

SDCL 10-49-2.1

Exempts gasohol from 4 cents of the 12 cents per gallon state motor fuel tax. The gasohol must contain at least 10 percent ethyl alcohol and be at least 98 percent pure. Exempts ethyl alcohol mixed with motor fuel used to produce gasohol from the 5 percent state use tax. Effective July 1, 1979. The exemptions expire June 30, 1982.

SB 175

Requires that 98 percent alcohol (196 proof) be used in the blending of gasohol. Effective August 1, 1980.

HB 1008A

Authorizes a property tax assessment credit for alcohol plants that produce alcohol for fuel use on a decreasing scale basis. For on-farm alcohol plants, the credit applies for three continuous years at 100 percent, then (beginning with the fourth year) decreases each succeeding year to 75 percent, 50 percent, 25 percent, 0 percent. Large scale commercial plants receive 50 percent (first three years) then each succeeding year, 30 percent, 25 percent, 12 percent, 0 percent. Effective July 1, 1980.

PENDING LEGISLATION

The legislative session adjourned March 12, 1980. The legislature will convene January 20, 1981.

STATE REGULATION

Permit Procedure—Both producers and blenders of fuel alcohol are required by the South Dakota Department of Revenue to obtain a motor fuel dealer license. The initial bond requirement is $250. Then each succeeding year the bond requirement equals 1/6 of the motor fuel tax for which the dealer is responsible. Motor fuel tax liability for the preceding calendar year is used in calculating the bond requirement.

State Volatility Requirements—South Dakota requires that all gasoline (including gasohol) meet Distillation and Reid Vapor Pressure requirements. No ASTM D 439 requirement.

OTHER IMPORTANT ALCOHOL FUELS ACTIVITY IN SOUTH DAKOTA

The South Dakota Office of Energy Policy has available a handbook entitled "South Dakota Alcohol Fuel Production Handbook - Permits, Regulations, and Assistance".

In November 1979, the Governor of South Dakota established a South Dakota Alcohol Fuels Commission to promote the production of alcohol fuels.

South Dakota State University at Brookings has a demonstration still for public observation on its campus and is conducting research on alcohol fuels.

The South Dakota Farmers Union has established a task force to investigate alcohol fuels.

Responsible Official For Further Information:

Mr. Verne Brakke
Office of Energy Policy
Capitol Lake Plaza
Pierre, South Dakota 57501
(605) 773-3603

LAWS

SB 2081 (HB 1966)

Prevents gasoline retailers from disallowing the use of credit cards for the purchase of gasohol. Prohibits distributors from restricting retailers' voluntary sales of gasohol. July 1, 1980.

SB 1927 (HB 1778)

Legalizes the production of alcohol for fuel purposes, requires the denaturing and coloring of alcohol, adds ethanol and methanol to the list of petroleum products that must be inspected and provides for licensing procedures and accompanying fine schedules. This law does not set up alcohol proof requirements, but provides that the state's provisions be consistent with the requirements of the Federal Bureau of Alcohol, Tobacco and Firearms.

Provides that a manufacturer of alcohol for fuel use who produces less than 1000 gallons annually for his own personal use or for sale to a bonded distributor is exempt from all 7 cents of the state 7 cents per gallon motor fuel tax. The gasohol must contain ethyl or methyl alcohol distilled in accordance with standard Federal Bureau of Alcohol, Tobacco and Firearms levels and be at least 188 proof. Effective April 3, 1980.

PENDING LEGISLATION

The legislative session adjourned April 18, 1980.

STATE REGULATION

Permit Procedures—Producers of fuel alcohol who produce less than 1000 gallons/year of alcohol must register with the State Department of Revenue. A permit is required if more than 1000 gallons is produced annually. A $50.00 fee must be paid to the Department of Revenue if 1000-2500 gallons per year are produced and a $100.00 fee is required if more than 2500 gallons per year are produced (SB 1927).

Fuel alcohol must be denatured and colored pursuant to the quality control provisions in SB 1927 must be followed.

State Volatility Requirements—Tennessee requires that gasoline including gasohol) meet ASTM D439 specifications, which include Distillation and Reid Vapor Pressure specifications.

OTHER IMPORTANT ALCOHOL FUELS ACTIVITY IN TENNESSEE

Pursuant to House Joint Resolution 161 (1979) the state established a one-year Alcohol Fuels Legislative study Committee.

The Tennessee Valley Authority conducted a vehicle fleet test from March 1979 to April 1980 using gasohol in 185 vehicles.

Responsible Official For Further Information:

Ms. Margot Myrick
Research Analyst
Legislative Plaza, Suite 2
Nashville, Tennessee 37219
(615) 741-2354

LAWS

HB 1986

Legalizes the production of ethanol for fuel use in gasohol. The intent of this law is to clarify that previous statutes which limited the production and transportation of alcohol do not apply to fuel alcohol. Effective September 1979.

HB 1803

Grants authority for a loan guarantee program. No money to date has been appropriated for this program, which became effective in September 1979.

Texas revised civil statutes annotated article 7510.1. (The general electorate passed an amendment to the Texas constitution November 1978; the enabling legislation passed the legislature the following year.) Provides for exemptions of property and sales taxes for producers of alcohol fuels for the purchase of equipment used in the production of ethanol for on-site use. In this bill, the equipment must meet the definition of a "solar device" (which includes biomass) as defined in the law. Effective September 1979.

PENDING LEGISLATION

No regular session was held in 1980.

STATE REGULATION

Permit Procedure—Producers of fuel alcohol are required to obtain a Local Industrial Producers Permit from the State Alcoholic Beverage Commission. A fee of $100 per year is required.

Dry county regulations—Alcohol used for fuel can be transported through dry county jurisdictions.

State Volatility Requirements—Texas requires that all gasoline (including gasohol) meet only Distillation requirements. No Reid Vapor Pressure or other ASTM D439 requirements.

OTHER IMPORTANT ALCOHOL FUELS ACTIVITY IN TEXAS

Texas Energy and Natural Resources Advisory Council (TENRAC) is considering funding several small state demonstration projects for alcohol fuels through the Texas Energy Development Fund. It is anticipated that the funding will be authorized during the summer of 1980.

Responsible Official For Further Information:

Mr. Bob Avant (512) 475-5588
Coordinator of Biomass and Wind Programs
TENRAC
411 W. 13th Street, Suite 900
Austin, Texas 78701

UTAH

LAWS

SB 11

Exempts gasohol from 5 cents of the state 9 cents per gallon motor fuel tax. The gasohol must contain at least 10 percent alcohol produced in Utah. No more than 25 percent of the energy value of the alcohol may be derived from electricity, distillable liquid or gaseous fuels in the production of the alcohol. Effective July 1, 1980. The exemption expires June 30, 1985.

SB 10

Provides for an income tax reduction (deduction, and depletion allowances) for alcohol fuel producers. Effective April 16, 1980.

PENDING LEGISLATION

The legislative session adjourned February 2, 1980.

STATE LEGISLATION

Permit Procedure—A state permit is required for the production of alcohol by the State Liquor Control Commission. No more than 25 percent of the energy value of the alcohol in an alcohol plant may be derived from electricity, distillable liquid or gaseous fuels. The Liquor Control Commission requires producers of fuel alcohol to submit a copy of the Federal Bureau of Alcohol, Tobacco, and Firearms permit. A $1.00 fee is required.

State Volatility Requirements—Utah requires that all gasoline (including gasohol) meet Distillation and Reid Vapor Pressure requirements. No ASTM D 439 requirement.

Responsible Official For Further Information:

Mr. Jim Byrne (801) 533-5424
Deputy Director
Utah Energy Office
231 E. 400 South
Salt Lake City, Utah 84111

VERMONT

LAWS

No laws have been enacted as of July 1, 1980.

PENDING LEGISLATION

The legislative session adjourned April 26, 1980.

STATE REGULATION

Permit Procedure—No state permit is required for the production of fuel alcohol.

State Volatility Requirement—Vermont has no volatility requirements for gasoline or gasohol.

Responsible Official For Further Information:

Mr. Larry Ogden
State Energy Office
State Office Building
Montpelier, Vermont 05602
(802) 828-2393

VIRGINIA

LAWS

HB 68

Requires the registration of all industrial alcohol (including fuel alcohol) plants. Effective March 31, 1980.

PENDING LEGISLATION

The legislative session adjourned March 8, 1980. The legislatiure will convene January 14, 1981.

STATE REGULATION

Permit Procedure—A state permit is required for the production of fuel alcohol. No fee is requied.

State Volatility Requirements—Virginia requires that all gasoline (including gasohol) meet Distillation and Reid vapor Pressure requirements. No ASTM D439 requirement.

OTHER IMPROTANT ALCOHOL FUELS ACTIVITY IN VIRGINIA

Virginia State University at Blacksburg has established a research and development alcohol fuels task force and constructed an alcohol fuels demonstration still on its campus.

The Virginia Department of Agriculture and Consumer Services has established a 30-member task force on alcohol fuels production, marketing and regulation.

NASA is conducting a vehicle fleet test at Langley Research Center using gasohol in 100 vehicles.

Responsible Officials for Further Information:

For information and regulation:
Mr. Penn Zentmeyer
Supervisor of Fertilizers,
Time & Motor Fuels
VA Department of Agriculture
203 North Governors Street
Richmond, Virginia 23219
(804) 786-3511

All other:

Mr. I. W. Smith
Director of Rural Afairs/
State Director of Alcohol Fuels
VA Department of Agriculture &
Consumer Services
203 Norther Governors Street
Richmond, Virginia 23219
(804) 786-3519

WASHINGTON

LAWS

ESSB 3629

Exempts gasohol from 1.2 cents of the state 12 cents per gallon motor fuel tax (only the alcohol content of gasohol is exempt). Exempts gasohol from the state retail sales and use taxes. The gasohol must contain at least 9-1/2 percent alcohol. Effective March 1980. The exemption expires December 31, 1986.

HB 1568

Requires the "widest possible use of gasohol and cost-effective alternative fuels" in all motor vehicles. Effective March 19, 1980.

ESSB 3551

Exempts from state property taxation and leasehold taxation property used for manufacturing facilities which produce alcohol for fuel use and other purposes, at an annually determined rate based on the percentage of the total gallons produced per year that is sold or used as an alcohol fuel. No claims under this exemption may be filed after December 31, 1986. Exempts gasohol from state business and occupation taxes through December 31, 1986.

SHB 1630

Exempts alcohol manufactured for fuel from the state liquor control laws. The exemption expires December 31, 1986. Effective June 1, 1980.

PENDING LEGISLATION

The legislative session adjourned March 13, 1980.

STATE REGULATION

Permit Procedure—No state permit is required for the production of fuel alcohol. Gasoline retailers that sell gasohol must register with the State Department of Licensing.

OTHER IMPORTANT ALCOHOL FUELS ACTIVITY IN WASHINGTON

$73,000 was allocated by the Pacific Northwest Regional Commission for research on the potential ethanol production capacity in Washington State. The Washington State Department of Commerce and Economic Development published the results of this study in April 1980.

The Washington State Motor Transport Divison is using gasohol in 100 state vehicles.

The City of Everett is conducting a vehicle fleet test using methanol (10 percent methanol/gasoline blends) in 175 vehicles.

Universities and Colleges offering alcohol fuels workshops, seminars and courses: Tacoma Community College; Skegit North Shore Community College; Washington State University; Central Washington State University.

Responsible Official For Further Information:

Mr. Paul Juhasz
State Energy Office
400 E. Union Street
Olympia, Washington 98504
(206) 754-0700

WEST VIRGINIA_____

LAWS

SB 112E

Eliminates the $250 state fee for production of ethanol if the alcohol fuel produced is used for personal fuel use. Effective June 5, 1980.

PENDING LEGISLATION

The legislative session adjourned March 11, 1980.

STATE REGULATION

Permit Procedure—A state permit is required by the Alcohol Beverage Control Commission (SB 112) for the production of fuel alcohol.

State Volatility Requirements—West Virginia has no volatility requirements for gasoline or gasohol.

OTHER IMPORTANT ALCOHOL FUELS ACTIVITY IN WEST VIRGINIA

West Virginia University is conducting research on coal liquefaction and gasification with state funds.

Universities and colleges offering alcohol fuels workshops, seminars and courses: West Virginia University.

Responsible Official For Further Information:

Ms. Rebecca Scott
Information Representative
West Virginia Fuel & Energy Office
1262 1/2 Greenbrier Street
Charleston, West Virginia 25311
(304) 348-8860

LAWS

Budget Bill, CH 34, Laws of 1979

Appropriated $75,000 to the Wisconsin Department of Administration for feasibility studies on alcohol plant facilities to be built in Wisconsin. Appropriates $225,000 for an alternative energy grant program which will include projects or waste. Lowers the annual state permit fee for small scale ethanol fuel producers from $750 to $10.00. Effective July 29, 1979.

AB 456

Requires that gasoline-alcohol fuel blends be regulated as other fuels by the Department of Industry, Labor and Human Relations. Effective March 29, 1980.

AB 777

Expands the use of the reduced state permit fee for production of ethanol for gasohol to all fuel uses. Allows alcohol production facilities to qualify for municipal revenue bonds. Allows alcohol fuel production systems and coopertives to qualify for the same benefits as do corporations for renewable energy systems as of March 1, 1980. Directs the State Division of Energy to compile and distribute information on gasohol and the availability of existing financial assistance for alcohol fuel producers, and report to the legislature on the desirability of state financial assistance programs. Effective May 22, 1980.

Budget Review Bill, CH 221

Allows alcohol fuel production systems to qualify for individual and corporate income tax credits. Requires that all state fleet cars be run on fuel containing at least 10 percent ethanol (unless ethanol is unavailable) by January 1, 1984. Exempts alcohol fuel producers who manufacture and use alcohol fuels for their own use from the requirement to obtain water discharge, air pollution, and solid waste permits if they dispose of the waste on their own property in an environmentally accepted way. Effective March 30, 1980.

Budget Review Bill, CH 144

Exempts alcohol fuel production systems which consume all the alcohol fuel produced from permits, license, and plant approval requirements.

PENDING LEGISLATION

The legislative session adjourned April 13, 1980.

STATE REGULATION

Permit Procedure—A state permit is required for the production of fuel alcohol. A $10.00 fee is required.

If the distillers by-product is to be sold as animal feed, a permit is required.

State Volatility Requirements—Wisconsin requires that all gasoline (including gasohol) meet Distillation and Reid Vapor Pressure requirements. No ASTM D439 requirement.

OTHER IMPORTANT ALCOHOL FUELS ACTIVITY IN WISCONSIN

The Alcohol Fuels Task Force of the State Division of Energy is conducting research on alcohol fuel use.

On April 1, 1980, $75,000 was allocated by the Wisconsin Department of Administration (Divison of State) for feasibility studies of alcohol fuels production facilities.

$225,000 was appropriated by the state legislature for 1980 biomass research and production grants.

Universities and colleges offering alcohol fuels workshops, seminars and courses: University of Wisconsin (Extension Division); Wisconsin Vocational Technical Adult Education Service; Univers University of Wisconsin at Green Bay; University of Wisconsin at Whitewater.

Responsible Officials For Further Information:

Mr. George Plaza
Biomass Analyst
Divison of State Energy
101 S. Webster Street
Madison, Wisconsin 53702
(608) 266-0985

Ms. Ann Bogar-Rieck
Science Analyst
Wisconsin Legislative Council
State Capitol
Room 147 North
Madison, Wisconsin 53702
(608) 266-0985

LAWS

HB 114

Exempts gasohol from 4 cents of the state 8 cents per gallon motor fuel tax. The gasohol must contain at least renewable resources. Effective July 1, 1979. The exemption expires July 1, 1984.

HB 91

Exempts gasohol from the state sales tax. Effective March 5, 1980.

PENDING LEGISLATION

The legislative session adjourned March 5, 1980.

STATE REGULATION

Permit Procedure—No state permit is required for the production of fuel alcohol.

State Volatility Requirements—Wyoming requires that all gasoline (including gasohol) meet Distillation and Reid Vapor Pressure requirements. No ASTM D439 requirement.

OTHER IMPORTANT ALCOHOL FUELS ACTIVITY IN WYOMING

Universities and Colleges offering alcohol fuels workshops, seminars and courses: Powell Junior College; Farrington Junior College, Western Wyoming Community College.

Responsible Official For Further Information:

Mr. Butch Keadle
Fuel Allocation Specialist
Energy Conservation Office
Capitol Hill Office Building
25th & Pioneer
Cheyenne, Wyoming 80002
(307) 777-7284

Appendixes D-1 through D-4 summarize state ethanol legislation. Appendix D-1 provides a summary of state tax exemptions. Restrictions on state tax incentives are shown in Appendix D-2. A summary of state permit requirements is delineated in Appendix D-3. Appendix D-4 identifies the volatility specifications by state.

APPENDIX D-1

Summary of State Tax Exemptions

State	Excise Tax		Sales, use or gross receipts for exemptions for gasohol*	Property tax reduction**	Income tax reduction***
	Exemption for gasohol/gal.	Total tax/ gal. motor fuel			
Alabama	3¢	11¢	no	no	no
Alaska	8¢	8¢	no	no	no
Arizona	0¢	8¢	no	no	no
Arkansas	9.5¢[1 2 6]	9.5¢	no	no	yes[1]
California	0¢	7¢	yes	no	no
Colorado	5¢[1 2]	7¢	no	yes (98%)[1]	no
Connecticut	1¢	11¢	no	no	no
Delaware	0¢	9¢	no	no	no
Florida	5¢[1]	8¢	no	no	yes[1 7]
Georgia	0¢	7.5¢	no	no	no
Hawaii	0¢	11.5-15.0¢	yes (sales) (4%)[1]	no	no
Idaho	4¢[2]	9.5¢	no	no	yes (0.8%)[1]
Illinois	0¢	7.5¢	no	no	no
Indiana	0¢	8.5¢	yes (sales) (4%)	no	yes[1 10]
Iowa	10¢[1 2]	10¢	no	no	no
Kansas	4¢[1 2]	8¢	no	no	no
Kentucky	0%	9% of the avg. wholesale price	yes (sales) (5%)[1]	yes (99%)[1 2]	no
Louisiana	8¢[1 2]	8¢	yes (sales) 3%[1 2 3]	no	no
Maine	0¢	9¢	no	no	no
Maryland	4¢[2] 9¢[2 4]	9¢	yes	no	no
Massachusetts	0¢	9.8¢	yes	no	no
Michigan	0¢[1 2]	11¢	no	no	no
Minnesota	4¢	11¢	no	no	yes (20%)
Missippi	0¢	9¢	no	no	no
Missouri	0¢	7¢	no	no	no
Montana	7¢[1 2]	9¢	no	yes (97%)[1 2]	no
Nebraska	5¢[1 2]	10.4¢ and 2% of the average statewide cost of fuel.[20]	no	no	no
Nevada	0¢	6¢	no	no	no
New Hampshire	5¢[2]	11¢	no	no	no
New Jersey	0¢	8¢	no	no	no
New Mexico	8¢[1 2]	8¢	yes (3%)(gross receipts)[1 2]	no	no
New York	0¢	8¢	no	no	no
North Carolina	4¢[1 6]	9¢[1 13]	no	no	yes (20%)[1 8 12]
North Dakota	4¢	8¢	yes (sales)[3 9 14]	no	no
Ohio	0¢	7¢	no	no	no
Oklahoma	6.50¢[1 2]	6.58¢	no	no	no 1,15
Oregon	0¢	7¢	no	yes (100%)[1]	yes (100%)[1 15] (20%)[1 16]
Pennsylvania	0¢	11¢	no	no	no
Rhode Island	0¢	10¢	no	no	no
South Carolina	5¢[1]	11¢[20]	no	no	no

Summary of State Tax Exemptions · Continued

| State | Excise Tax | | Sales, use or gross receipts for exemptions for gasohol* | Property tax reduction** | Income tax reduction*** |
	Exemption for gasohol/gal.	Total tax/ gal. motor fuel			
South Dakota	4-5c[1]	12-13c	yes (use) (4%)[1]	yes (100%)[1]	no
Tennessee	7c[5]	7c	no	no	no
Texas	0c	5c	yes (4%)[17]	yes, (100%)[17]	no
Utah	5c[1][2]	9c	no	no	yes[10]
Vermont	0c	9c	no	no	no
Virginia	0c	11c[19]	no	no	no
Washington	1.8c	12c	yes (sales & use) (4.5%)[1]	yes[1][18]	no
West Virginia	0c	10.5c	no	no	no
Wisconsin	0c	9c	no	no	yes[12]
Wyoming	4c[1]	8c	yes (sales) (3%)	no	no

* Sales, use and gross receipts tax exemptions—Percent listed is the sales tax rate in that state. In all cases, the entire amount of the tax is exempted

** Property tax reduction—The percent listed equals the percent of the state property tax applicable to qualifying fuel alcohol production facilities which is exempted.

*** Income tax exemptions, credits and deductions—The percent listed equals the percent of the exemption, deduction, or credit with respect to taxable income attributable to qualifying fuel alcohol production.

[1] The enabling legislation contains a decreasing scale, sunset provision, an expiration date and/or a maximum ceiling provision which limits the total amount of excise tax refunds which the state can lose as a result of the gasohol excise tax exemption

[2] Gasohol's exemption is restricted in some manner to alcohol produced from state crops, produced within the state or blended within the state

[3] Applies to agricultural use only

[4] Includes only unmixed ethyl and methyl alcohol.

[5] If production does not exceed 1000 gallons annually

[6] Effective in 1981.

[7] The corporate income tax reduction is based on the amount of property taxes paid by producers of fuel alcohol for production costs incurred in the manufacture of fuel alcohol.

[8] An additional 10% income tax reduction is credited to those producers of fuel alcohol whose distilleries are powered primarily by the use of an alternative fuel source.

[9] A 1/8 cent sales tax charge is levied on gasohol exempted from the 3% state sales tax

[10] No standard amount of reduction is stated in the law. Certain business expenses incurred in the production of fuel alcohol qualify as deductions in the calculation of the producer's next corporate income tax.

[11] A gross income tax deduction for a coal conversion system to process coal into liquid fuel applies only to those who receive the federal depreciation deductions for a coal conversion system.

[12] The law provides for both coprorate and personal income tax reductions.

[13] A 9 cent excise tax exemption is provided for nonanhydrous alcohol which is not used for sale or distribution.

[14] The exemption only applies when the gasohol is used for agricultural or industrial purposes.

[15] 100% income tax exemption—applies only to alcohol fuels production

[16] 50% investment tax credit—applies to all alternative energy development projects.

[17] The exemption only applies to the purchase of equipment used in the production of ethyl alcohol.

[18] This exemption includes leasehold tax exemptions.

[19] A 2 cent increase in the gasoline excise tax is imposed in Northern Virginia.

[20] Effective October 1, 1980.

Source U S. National Alcohol Fuels Commission, July 1980

APPENDIX D-2

Restrictions on State Tax Incentives

States	Incentive	%Alcohol That Must Be Contained in Gasohol	*Required Alcohol Proof Level and Alcohol Purity	Required Type Of Alcohol	Other Restrictions
ALABAMA	3¢ excise tax exemption	at least 10%	at least 99% purity	Ethyl alcohol produced from agricultural, forest, or other renewable resources	----
ALASKA	8¢ excise tax exemption	at least 10%	----	Alcohol	----
ARIZONA	none	----	----	----	----
ARKANSAS	9-1/2¢ excise tax exemption Income tax deduction	at least 10%	----	Alcohol produced from agricultural or forest resources	Alcohol must be distilled in Arkansas and contain at least 10% Arkansas resources or be distilled in a state which has reciprocity deduction with the Arkansas Excise Tax exemption. The income tax deduction expires December 31, 1984.
CALIFORNIA	sales tax exemption	at least 10%	----	ethyl or methyl alcohol distilled from agricultural commodities, renewable resources or coal.	"Other Restrictions" The alcohol must be distilled from California resources, be rendered unsuitable for human consumption and dyed a distinctive color. The exemption expires January 1, 1984.
COLORADO	5¢ excise tax exemption 98% property tax deduction	at least 10%	at least 95% purity	Ethyl or methyl alcohol produced from agricultural or forest resources, petroleum coke, waste coke, or coal waste	Alcohol must be produced in Colorado. The 5¢ excise tax exemption expires July 1, 1985. The 98¢ property tax reduction is temporary and has a decreasing scale rate.
CONNECTICUT	1¢ excise tax exemption	at least 10%	----	Ethyl or methyl alcohol	----
DELAWARE	none	----	----	----	----
FLORIDA	5¢ excise tax exemption Income tax credit	at least 10%		Ethyl alcohol produced from biomass	The income tax credit expires July 1, 1982. The 5¢ excise tax exemption expires July 1, 1987 and has a decreasing scale rate. credit

*In many cases the terms "proof" and "purity" are used interchangeably. However, technically, "proof" refers only to water content, and "purity" refers to the presence in the alcohol of all non-ethanol matter. This chart uses the statutory terms for each state. In some instances the exact meaning is not specified.

COMMERCIAL SCALE ETHANOL PRODUCTION AND FINANCING

States	Incentive	%Alcohol That Must Be Contained in Gasohol	*Required Alcohol Proof Level and Alcohol Purity	Required Type Of Alcohol	Other Restrictions
GEORGIA	none	----	----	----	----
HAWAII	4% sales tax exemption	----	----	Alcohol	The 4% sales tax exemption expires June 30, 1985. The state Director of Taxation has the authority to terminate the exemption at his discretion.
IDAHO	4¢ excise tax exemption 8% income tax credit	at least 10%	----	Anhydrous ethyl alcohol produced from agricultural or forest resources	The alcohol must be produced in Idaho from Idaho products. The 8% income tax credit expires July 1, 1985 and has a decreasing credit scale rate.
ILLINOIS	none	----	----	----	----
INDIANA	4% sales tax exemption Income tax deduction	at least 10%	----	Anhydrous ethyl alcohol produced from agricultural or forest resources	Income tax deduction applies only to coal conversion, liquid fuels and expires January 1, 1990.
IOWA	10¢ excise tax exemption	at least 10%	----	Ethyl alcohol produced from agricultural or forest resources	A 3% sales tax charge accompanies the 10¢ excise tax exemption through May 1, 1981. May 1, 1981 through June 30, 1983 the excise tax exemption has a decreasing scale rate and the exemption expires June 30, 1983.
KANSAS	3¢ excise tax exemption	at least 10%	at least 190 proof	Ethyl alcohol produced from agricultural or forest resources	The alcohol must be produced from grain products grown in Kansas. Production of alcohol must utilize 10 less energy units than would be contained in the converted motor vehicle fuel. The 3¢ excise tax exemption has a decreasing scale rate, a maximum ceiling, and expires July 1, 1985.
KENTUCKY	5¢ sales tax exemption 99% state property tax reduction 99% local property tax reduction	at least 10%	at least 198 proof	Ethyl alcohol produced from grain or other renewable resources	The exemptions expire eight years after the granting of exemption certificates. Alcohol plants must burn coal produced in KY. or convert to such use within two years of certificate receipt to quality for the 99¢ local exemptions.

States	Incentive	%Alcohol That Must Be Contained in Gasohol	*Required Alcohol Proof Level and Alcohol Purity	Required Type Of Alcohol	Other Restrictions
LOUISIANA	8¢ excise tax exemption 3% sale tax exemption	at least 10%	----	Alcohol distilled from agricultural resources	The alcohol must be distilled in Louisiana and contain at least 10% agricultural resources produced in LA unless such products are not available. The alcohol must be "rendered unsuitable for human consumption at the time of its manufacture or immediately thereafter" and be dyed a distinctive color. The exemptions expire after ten years unless it is extended by the legislature.
MAINE	none	----	----	----	----
MARYLAND	4¢ excise tax exemption 9¢ excise tax exemption	at least 10%	----	Ethyl or methyl alcohol	The 4¢ excise tax exemption expires May 1, 1981. The 9¢ excise tax exemption applies only to alcohol not mixed with gasoline and expires June 1, 1982.
MASSACHUSETTS	none	----	----	----	----
MICHIGAN	50% property tax exmption	----	----	----	The 50% property tax exemption applies to new construction of industrial size plants.
MINNESOTA	4¢ excise tax exemption 20% income tax credit	at least 10%	at least 190 proof	Ethyl alcohol produced from agricultural resources	The alcohol must be distilled from Minnesota resources. Fuel must be blended to quality for the 4¢ excise tax exemption.
MISSISSIPPI	none	----	----	----	----
MISSOURI	none	----	----	----	----
MONTANA	7¢ excise tax exemption 97% property tax reduction	at least 10%	----	Anhydrous alcohol	Alcohol must be produced in Montana from MT resources. The 7¢ excise tax exemption has a decreasing scale rate and expires April 1, 1989. The Governor of MT has the authority to suspend the excise tax exemption. The 97% property tax reduction can be claimed during construction and for the first three years of the fuel alcohol plant's operation. This reduction is applicable to taxable years beginning after December 31, 1979.

States	Incentive	%Alcohol That Must Be Contained in Gasohol	*Required Alcohol Proof Level and Alcohol Purity	Required Type Of Alcohol	Other Restrictions
NEBRASKA	5¢ excise tax exemption	at least 10%	at least 99% purity	Ethyl alcohol produced from agricultural or forest resources	The alcohol must be produced in Nebraska. Beginning in 1982, the 5¢ excise tax exemption applies only to alcohol produced in a plant under construction or in operation by July 1, 1980. The 5¢ excise tax exemption has a decreasing scale rate and maximum ceiling levels.
NEVADA	none	----	----	----	----
NEW HAMPSHIRE	5¢ excise tax exemption	at least 10%	-----	Alcohol	Alcohol must be produced in New Hampshire.
NEW JERSEY	none	----	----	----	----
NEW MEXICO	8¢ excise tax exemption 5% sale tax exemption	at least 10%	at least 190 proof	Ethyl alcohol	Alcohol must be produced in New Mexico. The exemptions expire June 30, 1985.
NEW YORK	none	----	----	----	----
NORTH CAROLINA	4¢ excise tax exemption 9¢ excise tax exemption 20% corporate and personal income tax	at least 10%	----	To qualify for the 4¢excise tax exemption the alcohol must be anhydrous ethyl alcohol. To qualify for the 9¢ excise tax exemption the alcohol must be non-produced for on-farm use and may be non-anhydrous alcohol. To qualify for the income tax credits the alcohol must be produced from agricultural or forest resources.	The 4¢ excise tax exemption has a decreasing scale rate and expires June 30, 1984. The income tax credit expires January 1, 1985.
NORTH DAKOTA	4¢ excise tax exemption 3% sales tax exemption	at least 10%	at least 199 proof	Ethyl alcohol produced from agricultural resources	The 3% sales tax exemption only applies when the gasohol is used for exemption agricultural or industrial purposes.
OHIO	none	----	----	----	----
OKLAHOMA	6.5¢ excise tax exemption	at least 10%	at least 99% at least 198°	Ethyl alcohol distilled from agricultural resources	To qualify for the 6.5¢ excise tax exemption the distributor of the gasohol must be licensed in Oklahoma. The 6.5¢excise tax exemption expires October 1, 1984.

States	Incentive	%Alcohol That Must Be Contained in Gasohol	*Required Alcohol Proof Level and Alcohol Purity	Required Type Of Alcohol	Other Restrictions
OREGON	100% income tax exemption 50% investment tax credit 100% property tax reduction	at least 10%	----	To qualify for the 100% income tax exemption or 100% property tax reduction alcohol must be ethyl or methyl alcohol or other substitute fuel not produced from petroleum, natural gas or coal. Several types of alternative fuels qualify for the 50% investment tax credit.	The 50% investment credit has a decreasing scale rate and expires January 1, 1985. The 100% property tax reduction applies only to commercial plants and expires October 3, 1985.
PENNSYLVANIA	none	----	----	----	----
RHODE ISLAND	none	° ----	----	----	
SOUTH CAROLINA	4¢ excise tax exemption	at least 10%	at least 99% purity	Ethyl or methyl alcohol produced from agricultural	The 4¢ excise tax exemption has a decreasing scale rate and expires when the total revenue loss realized reaches a maximum of $5 million.
SOUTH DAKOTA	4-5¢ excise tax exemption 100% property tax assessment credit 4% use tax exemption	at least 10%	at least 98% purity	Ethyl alcohol	The 4-5¢ excise tax exemption expires June 30, 1985. The 4% use tax exemption expires June 30, 1985. The 100% property tax credit has a decreasing scale rate and has differing rates for small- and large-scale plants. The 100% property tax credit expires July 1, 1986.
TENNESSEE	7¢ excise tax exemption	----	at least 188 proof	Ethyl or methyl alcohol	Alcohol used in gasohol must be dyed a distinctive color and be denatured. Only manufacturers of alcohol who produce less 1000 gallons annually and use this gasohol for their own personal use or for sale to a bonded distributor qualify for the 7¢ excise tax exemption.
TEXAS	100% property tax exemption 4% sales tax exemption	----	----	Ethyl alcohol	The sales and property tax exemptions only apply to the purchase of equipment used in the production of ethyl alcohol.

States	Incentive	%Alcohol That Must Be Contained in Gasohol	*Required Alcohol Proof Level and Alcohol Purity	Required Type Of Alcohol	Other Restrictions
UTAH	5¢ excise tax exemption Income tax reduction	at least 10%	----	Alcohol	Alcohol must be produced in Utah. The 5¢ excise tax exemption expires June 30, 1985. No more than 25% of the energy value of the alcohol may be derived from electricity, distillable liquid, or gaseous fuels in the production of alcohol.
VERMONT	none	----	----	----	----
VIRGINIA	none	----	----	----	----
WASHINGTON	1.2¢ excise tax exemption, 4.5% sales tax exemption 4.5% use tax exemption property tax exemption state business and occupation taxes exemption	at least 9 1/2%	----	Alcohol	The 1.2¢ excise tax exemption, 4.5% sales tax exemption and 4.5% use tax exemption expire December 31, 1986. The property tax exemption expires December 31, 1986.
WEST VIRGINIA	none	----	----	----	----
WISCONSIN	corporate income tax credit personal income tax credit	----	----	Alcohol	----
WYOMING	4¢ excise tax 3% sale tax exemption	at least 10%	----	Alcohol produced from agricultural or other renewable resources	The 4¢ excise tax exemption expires July 1, 1984.

Source: U.S. National Alcohol Fuels Commission, July 1980

APPENDIX D-3

STATE	PERMIT REQUIRED	PERMIT FEE & BONDING REQUIREMENT
Alabama	yes	$100/yr.
Alaska	no	0
Arizona	yes	0
Arkansas	yes	0
California	yes	$6.60
Colorado	no	0
Connecticut	no	0
Delaware	yes	no
Florida	no	0
Georgia	yes	Post $1,000-$150,000 bond (depending on size of plant)
Hawaii	no	0
Idaho	no	0
Illinois	yes	Post $1,000-$40,000 bond (depending on size of plant)
Indiana	no	0
Iowa	no	0
Kansas	no	0
Kentucky	no	0
Lousiana	no	0
Maine	no	0
Maryland	no	0
Massachusetts	no	0
Michigan	no	0
Minnesota	yes	$5
Mississippi	yes	$2,500 for large scale production - $25 for small scale production
Missouri	yes	Post $5,000 bond.
Montana	no	0
Nebraska	no	0
Nevada	no	0
New Hampshire	yes	0
New Jersey	no	0
New Mexico	no	0
New York	yes	$625
North Carolina	no	0
North Dakota	no	0
Ohio	no	0
Oklahoma	yes	$25 (experimental plants) $250 (distilled spirits plants)
Oregon	no	0
Pennsylvania	yes	$25 (on-farm production and use) $2,500 (commercial production and use)
Rhode Island	no	0
South Carolina	no	0
South Dakota	yes	Post $250 bond and obtain a Dept. of Revenue motor fuel dealer license.
Tennessee	yes	$50 (production of 1000-2500 gal/yr.)—$1000 (Production of more than 2500 gal/yr.) Notification required only if production less than 1000 gallon/year.

STATE	PERMIT REQUIRED	PERMIT FEE & BONDING REQUIREMENT
Texas	yes	$100
Utah	yes	$1
Vermont	no	0
Virginia	yes	0
Washington	no	0
West Virginia	yes	0
Wisconsin	yes	$10
Wyoming	no	0

Source: U.S. National Alcohol Fuels Commission, July 1980

APPENDIX D-4

State Volatility Specifications

STATE	D439[1]	DISTILLATION[2]	REID VAPOR PRESSURE[3]
Alabama	No	Yes	Yes
Alaska	No	No	No
Arizona	Yes	Yes*	Yes*
Arkansas	Yes	Yes*	Yes*
California	Yes	Yes*	No**
Colorado	No	Yes	No
Connecticut	No	Yes	No
Delaware	No	No	No
Florida	Yes	Yes*	Yes*
Georgia	Yes	Yes*	Yes*
Hawaii	Yes	Yes*	Yes*
Idaho	Yes	Yes*	Yes*
Illinois	No	No	No
Indiana	No	Yes	Yes
Iowa	No	Yes	Yes
Kansas	No	Yes	No
Kentucky	No	No	No
Louisiana	No	Yes	Yes
Maine	No	Yes	No
Maryland	Yes	Yes*	Yes*
Massachusetts	No	Yes	No
Michigan	No	No	No
Minnesota	No	Yes	No
Mississippi	Yes	Yes*	Yes*
Missouri	No	Yes	Yes
Montana	Yes	Yes*	Yes*
Nebraska	No	Yes	No
Nevada	No	Yes	No
New Hampshire	No	No	No
New Jersey	No	Yes	No
New Mexico	No	Yes	No
New York	No	Yes	No
North Carolina	Yes	Yes*	Yes*
North Dakota	Yes	Yes*	Yes*
Ohio	No	No	No
Oklahoma	No	Yes	No
Oregon	No	No	No
Pennsylvania	No	No	No
Rhode Island	Yes	Yes	Yes
South Carolina	No	Yes	Yes
South Dakota	No	Yes	Yes
Tennessee	Yes	No	No
Texas	No	Yes	No
Utah	No	Yes	Yes
Vermont	No	No	No
Virginia	No	Yes	Yes
Washington	No	No	No
West Virginia	No	No	No
Wisconsin	No	Yes	Yes
Wyoming	No	Yes	Yes

*Part of D439

**Gasohol exempted from RVP from June 13, 1980 to June 13, 1983

[1] D439 - Specifications recommended by the American Society for Testing and for automotive gasoline governing volatility, distillation requirements, Reid Vapor Pressure, lead content, corrosion resistance, gum content, sulfur content, oxidation stability and anti-knock index

[2] Distillation - Specifications which control the *vaporization characteristic* of gasoline

[3] Reid Vapor Pressure - Recommended pounds per square inch of pressure of the gasoline. While designed by ASTM primarily to prevent vapor lock, some recent State ceilings on RVP are designed to reduce pollution

Source: U.S. National Alcohol Fuels Commission, July 1980

COMMERCIAL SCALE ETHANOL PRODUCTION AND FINANCING

APPENDIX E

Resource People and Organizations

- **Associations and Organizations**
- **Biologists**
- **Blending Terminals**
- **Chemists**
- **Consultants for Report**
- **Consultants/Design Engineers**
- **DDG Purchasers**
- **Economists**
- **Engineering Firms**
- **Enzyme and Yeast Producers**
- **Ethanol Producers/Distributors**
- **Feedstock Sources**
- **Plant Operation Consultants**
- **BATF Regional Offices**
- **DOE Regional Offices**
- **SBA Regional Offices**
- **FmHA State Offices**
- **HUD Regional Offices**
- **EDA Regional Offices**

NOTICE

The following list of resource people and organizations is provided for your information. Neither DOE nor SERI recommends or vouches for these sources. We would appreciate receiving additions or revisions to this information.

ASSOCIATIONS AND ORGANIZATIONS

American Agriculture
Movement
308 Second Street, SE
Washington, DC 20515
(202) 544-5750

The Bio-Energy Council
1625 Eye Street, NW
Suite 825A
Washington, DC 20006
Contact: Carol Canelio
(202) 833-5656

Brewers Grain Institute
1750 K Street, NW
Washington, DC 20006

Corn Development Commission
Route 2
Holdrege, NE 68949

Distillers Feed Research
Council
1535 Enquirer Building
Cincinnati, OH 45202
Contact: Dr. William Ingrigg
(513) 621-5985

International Biomass Institute
1522 K Street, NW
Suite 600
Washington, DC 20005
Contact: Dr. Darold Albright
(202) 783-1133

Mid-American Solar Energy Complex
8140 26th Ave. South
Minneapolis, MN 55420
(612) 853-0400

National Farmers Organization
Surprise, NE 68667

National Farmers Union
Denver, CO 80251

National Alcohol Fuel Institute
5716 Jonathan Mitchell Rd.
Fairfax Station, VA 22039
(703) 250-5136

National Alcohol Fuels
Commission
492 1st St., S.E.
Washington, DC 20003
Contact: Jim Childress
(202)426-6490

National Alcohol Fuels
Information Center
Solar Energy Research Institute
1617 Cole Boulevard
Golden, CO 80401
(800) 525-5555 - Continental U.S.
(800) 332-8339 - Colorado

National Gasohol Commission,
Inc.
521 South 14th St., Suite 5
Lincoln, NE 68508
Contact: Myron Reamon
(402) 475-8044 or 8055

Northeast Solar Energy Center
470 Atlantic Ave
Boston, MA 02110
(617) 292-9250

Solar Energy Research Institute
1617 Cole Boulevard
Golden, CO 80401
Contact: Cecil Jones

Southern Solar Energy Center
61 Perimeter Park
Atlanta, GA 30341
(404) 458-8765

Western Solar Utilization Network
Pioneer Park Building
715 S.W. Morrison
Portland, OR 97205
(503) 241-1222

The Wheat Growers
Route #1, Box 27
Hemingford, NE 69438
Contact: Vic Haas
(308) 487-3794

BIOLOGISTS

Dr. Pearse Lyons
ALLTECH, INC.
271 Goldrush Rd.
Lexington, KY 40503
(606) 276-3414

Micro-TEC Lab, Inc.
Route 2, Box 19L
Logan, IA 51546
Contact: John W. Rago
(712) 644-2193

Leo Spano
The Army/Navy Lab
Natick, MA 01760
(617) 653-1000, Ext. 2914

BLENDING TERMINALS

GATX Terminals Corporation
P.O. Box 409
Argo, IL 60501
(312) 458-1330

CHEMISTS

Dr. Harry P. Gregor
Columbia University
353 Seeley West Mudd Building
New York, NY 10027
(212) 280-4716

Antonia R. Moreira
Colorado State University
Fort Collins, CO 80523
(303) 491-5252

Dr. Richard Spencer
Southwest State University
Marshall, MN 56258
(507) 537-7217

CONSULTANTS FOR REPORT

TRW, Inc.
Energy Systems Group
TRW, Inc.
8301 Greensboro Dr.
McLean, VA 22102
Contacts: Mr. V. Daniel Hunt
      Mr. Warren Standley
      Dr. Ed Goretsky
      Dr. Jean-Francois Henry
      Ms. Ann Heywood
      Dr. Jean Simons
      Mr. Arch Wood
(703) 734-6554

Development Planning and
Research Associates
200 Research Drive
P.O. Box 727
Manhattan, KS 66502
Contact: Mr. Milton David
(913) 539-3565

The Eakin Corporation
401 Delphine St.
Baton Rouge, LA 70806
Contact: Mr. Sam Eakin
(504) 346-0453

Center for the Biology of Natural
  Systems
Washington University
St. Louis, MO 63130
Contact: Mr. David Freedman
(314) 889-5317

A. T. Kearney, Inc.
699 Prince St.
Alexandria, VA 22314
Contact: Dr. Cathryn Goddard et al.
(703) 836-6210

Battelle Memorial Institute
505 King Ave.
Columbus, OH 43201
Contact: Mr. David Jenkins
(614) 424-6424

Raphael Katzen Associates
1050 Delta Ave.
Cincinnati, OH 45208
Contact: Dr. Raphael Katzen
(513) 871-7500

Quaintance and Swanson
Suite 210, Van Brunt Building
226 N. Phillips Ave.
Sioux Falls, SD 57101
Contact: Mr. Robert Mabee,
  Attorney
(605) 335-1777

E.F. Hutton
3340 Peachtree St. Rd., N.E.
Atlanta, GA 30026
Contact: Mr. Strud Nash
(404) 262-2110

Solar Energy Research Institute
1617 Cole Blvd.
Golden, CO 80401
Contact: Cecil Jones
(303) 231-1205

PEDCo International
1149 Chester Rd.
Cincinnati, OH 45246
Contact: Dr. William Stark
(513) 782-4717

Office of Alcohol Fuels
U.S. Department of Energy
Room 6A-211
1000 Independence Ave.
Washington, DC 20585
Contact: Mr. Ted Tarr
(202) 252-9487

Arthur Young and Company
235 Peachtree St., N.E.
Atlanta, GA 30303
Contact: Mr. Michael Thomas
(404) 577-8773

CONSULTANTS/DESIGN ENGINEERS

ACR Process Corp.
808 South Lincoln, #14
Irbana, IL 61801
Contact: Robert Chambers
(217) 384-8003

Alltech, Inc.
271 Goldrush Rd.
Lexington, KY 40503
Contact: Dr. Pearse Lyons
(606) 276-3414

Bartlesville Energy
  Technology Center
Bartlesville, OK 74003
Contact: Jerry Allsup
(918) 336-4268

Center for Biology of Natural
  Systems
Washington University
St. Louis, MO 63130
Contact: David Freedman
(314) 889-5317

Chemapec, Inc.
230 Crossways Park Drive
Woodbury, NY 11797
Contact: Dr. Ing. Hans Mueller
(516) 364-2100

Development Planning and
  Research Associates
200 Research Drive
P.O. Box 727
Manhattan, KS 66502
Contact: Milton David
(913) 539-3565

Raphael Katzen Assoc.
1050 Delta Avenue
Cincinnati, OH 45208
Contact: George Moon
(513) 871-7500

Resource Planning Associates Inc.
1901 L Street, NW
Washington DC 20036
Contact:
  Mr. Robert T. McWhinney Jr.
(202) 452-9770

TRW Energy Systems Group
8301 Greensboro Drive
McLean, VA 22102
Contact: Mr. V. Daniel Hunt
(703) 734-6554

Vulcan Cincinnati, Inc.
2900 Vernon Place
Cincinnati, OH 45219
Contact: Donald Miller
(513) 281-2800

DDG PURCHASERS

The Pillsbury Company
3333 South Broadway
St. Louis, MO 63118
(314) 772-5150

ECONOMISTS

Resource Planning Associates Inc
1901 L Street, NW
Washington DC 20036
Contact:
  Mr. Robert T. McWhinney Jr.
(202) 452-9770

ENGINEERING FIRMS

Bohler/Vogelbusch
1625 West Belt North
Houston, TX 77043
Contact: Jerry Korff
(713) 465-3373

Chemapec Inc.
230 Crossways Park Drive
Woodbury, NY 11797
Contact: Rene Loser
(516) 364-2100

Hydrocarbon Research Inc.
134 Franklin Corner Rd.
P.O. Box 6047
Lawrenceville, NJ 08648
Contact: Maurice Jones
(609) 896-1300

Raphael Katzen Associates
1050 Delta Ave.
Cincinnati, OH 45208
Contact: George Moon
(513) 871-7500

Keep Chemical Company
Box 441
Cornwall, NY 12518
Contact: Peter J. Ferrara
(914) 534-4755

A. G. McKee Corporation
10 South Riverside Plaza
Chicago, IL 60606
Contact: Edward A. Kirchner
(312) 454-3685

PEDCo International, Inc.
11499 Chester Rd.
Cincinnati, OH 45246
Contact: Timothy Devitt
(513) 782-4717

Power Engineering Co.
1313 S.W. 27th Ave.
Miami, FL 33145
Contact: John Scopetta

Vulcan Cincinnati, Inc.
2900 Vernon Plaza
Cincinnati, OH 45219
Contact: Donald Miller
(513) 281-2800

ENZYME AND YEAST PRODUCERS

Alltech, Inc.
271 Goldrush Road
Lexington, KY 40503
(606) 276-3414

Anheuser-Busch, Inc.
721 Pestalozzi Street
St. Louis, MO 63118
(314) 577-2000

Biocon, Inc.
261 Midland Ave.
Lexington, KY 40507
(606) 254-0517

Chemapec, Inc.
230 Crossways Park Drive
Woodbury, NY 11797
Contact: Rene Loser
(516) 364-2100

Norbert Haverkamp
Compost Making Enzymes
Rural Route 1, Box 114
Horton, KY 66439
(913) 486-3302

Miles Laboratories, Inc.
Enzyme Products Division
P.O. Box 932
Elkhart, IN 56515
(219) 564-8111

Novo Laboratory, Inc.
59 Danbury Road
Wilton, CT 06897
(203) 762-2401

Scientific Products Co.
North Kansas City, MO 64116
(816) 221-2533

ETHANOL PRODUCERS/ DISTRIBUTORS

Amoco Production Company
P.O. Box 5340a
Chicago, IL 60680
(312) 856-2222

APCO Oil Corporation
Houston Natural Gas Building
Houston, TX 77002
(713) 658-0610

Archer Daniels Midland Company
Box 1470
Decatur, IL 62525
(217) 424-5700

Encore Energy Resources,
   Incorporated
11951 Mitchell Road,
   Mitchell Island
Richmond, British Columbia
(604) 327-8394

Exxon Corporation
Exxon Research and Engineering
Public Relations
P. O. Box 639
Linden, NJ 07036

Georgia-Pacific
900 Southwest 5th Avenue
Portland, OR 97204
(503) 222-5561

Georgia Pacific Company
Bellingham Division
P.O. Box 1236
Bellingham, WA 98225
(206) 733-4410

Grain Processing Corporation
Muscatine, IA 52761
(918) 264-4211

Hiram Walker
31275 Northwestern Highway
Farmington Hills
Detroit, MI 48018

Marcam Industries
527 North Easton Road
Glenside, PA 19038
(215) 885-5400

Midwest Solvents Company
1300 Main Street
Atchison, KS 66002
(913) 367-1480

Milbrew, Incorporated
330 South Mill Street
Juneau, WI 53039
(414) 462-3700

Mode, Ronald C.
Box 682
Glen Alpine, NC 28628
(704) 584-1432

Publicker Industries
777 W. Putnam Avenue
Greenwich, CT 06830
(203) 531-4500

Quaternoin Chemical Industries
72026 Livingston Street
Oakland, CA 94604
Contact: Louis Nagel
(415) 535-2311

Sigmor Corporation
P.O. Box 20267
San Antonio, TX 78220
(512) 223-2631

Texaco, Incorporated
135 East 42nd Street
New York, NY 10017
(212) 953-6000

TIPCO
9000 N. Pioneer Rd.
Peoria, IL 61614
(309) 692-6543

Worum Chemical Company
2130 Kasoto Avenue
St. Paul, MN 55108
Contact: Mr. Ritt
(612) 645-9224

FEEDSTOCK SOURCES

General

Distillers Feed Research Council
1435 Enquirer Bldg.
Cincinnati, OH 45202
(513) 621-5985

Sorghum

Grain Sorghum Producers
   Association
1708 - A 15th Street
Lubbock, TX 79401
Contact: Elbert Harp
          Executive Director
(806) 763-4425

Producers Grain Corp.
P.O. 111
Amarillo, TX 79105
(806) 374-0331

Potatoes

National Potato Council
45th & Peoria Denver, CO 80239
(303) 373-5639

Grains

National Council of Farmer Co-
ops
1800 Massachusetts Ave. NW
Washington, DC 20036
(202) 659-1525

PLANT OPERATION
CONSULTANTS

Alltech, Inc.
271 Goldrush Road
Lexington, KY 40503
Contact: Dr. Pearse Lyons
(606) 276-3414

Chemapec, Inc.
230 Crosways Park Drive
Woodbury, NY 11797
Contact: Dr. Ing. Hans Mueller
(516) 364-2100

PEDCo International
1149 Chester Road
Cincinnati, OH 45246
Contact: Dr. William Stark
(513) 782-4717

BUREAU OF ALCOHOL,
TOBACCO, AND FIREARMS
REGIONAL OFFICES

Central Region

(Indiana, Kentucky, Michigan,
Ohio, West Virginia)
Regional Regulatory Administrator
Bureau of Alcohol, Tobacco,
   and Firearms
550 Main Street
Cincinnati, OH 45202
(513) 684-3334

Mid-Atlantic Region

(Delaware, District of Columbia,
Maryland, New Jersey,
Pennsylvania, Virginia)
Regional Regulatory Administrator
Bureau of Alcohol, Tobacco,
   and Firearms
2 Penn Center Plaza, Room 360
Philadelphia, PA 19102
(215) 597-2248

Midwest Region

(Illinois, Iowa, Kansas, Minnesota,
Missouri, Nebraska, North
Dakota, South Dakota, Wisconsin)
Regional Regulatory Administrator
Bureau of Alcohol, Tobacco,
   and Firearms
230 S. Dearborn Street, 15th Floor
Chicago, IL 60604
(312) 353-3883

North-Atlantic Region

(Connecticut, Maine,
Massachusetts, New Hampshire,
New York, Rhode Island,
Vermont, Puerto Rico,
Virgin Islands)
Regional Regulatory Administrator
Bureau of Alcohol, Tobacco,
   and Firearms
6 World Trade Center, 6th Floor
For letter mail:
P.O. Box 15,
Church Street Station
New York, NY 10008
(212) 264-1095

Southeast Region

(Alabama, Florida, Georgia,
Mississippi, North Carolina, South
Carolina, Tennessee)
Regional Regulatory Administrator
Bureau of Alcohol, Tobacco,
   and Firearms
3835 Northeast Expressway
For letter mail:
P.O. Box 2994
Atlanta, GA 30301
(404) 455-2670

Southwest Region

(Arkansas, Colorado, Louisiana,
New Mexico, Oklahoma, Texas,
Wyoming)
Regional Regulatory Administrator
Bureau of Alcohol, Tobacco,
   and Firearms
Main Tower, Room 345
1200 Main Street
Dallas, TX 75202
(214) 767-2285

Western Region

(Alaska, Arizona, California,
Hawaii, Idaho, Montana, Nevada,
Oregon, Utah, Washington)
Regional Regulatory Administrator
Bureau of Alcohol, Tobacco,
   and Firearms
525 Market Street, 34th Floor
San Francisco, CA 94105
(415) 556-0226

DEPARTMENT OF ENERGY REGIONAL OFFICES

Region 1
Regional Representative
150 Causeway Street
Analex Building, Room 700
Boston, MA 02114
(617) 223-3701
(Use same 7-digit number for FTS)

Region 2
Regional Representative
26 Federal Plaza
Room 3206
New York, NY 10007
(212) 264-1021
(Use same 7-digit number for FTS)

Region 3
Regional Representative
1421 Cherry Street
Room 1001
Philadelphia, PA 19102
(215) 597-3890
(Use same 7-digit number for FTS)

Region 4
Regional Representative
1655 Peachtree Street N.E.
8th Floor
Atlanta, GA 30309
(404) 881-2838
FTS 257-2838

Region 5
N. Regional Representative
175 West Jackson Boulevard
Room A-333
Chicago, IL 60604
(312) 353-0540
(Use same 7-digit number for FTS)

Region 6
Regional Representative
P.O. Box 35228
2626 West Mockingbird Lane
Dallas, TX 75235
(214) 749-7345
(Use same 7-digit number for FTS)

Region 7
Regional Representative
Twelve Grand Building
P.O. Box 2208
112 East 12th Street
Kansas City, MO 64142
(816) 374-2061
FTS 758-2061

Region 8
Regional Representative
P.O. Box 26247,
Belmar Branch
Lakewood, CO 80226
(303) 234-2420
(Use same 7-digit number for FTS)

Region 9
Regional Representative
111 Pine St., Third Floor
San Francisco, CA 94111
(415) 556-7216
(Use same 7-digit number for FTS)

Region 10
Regional Representative
1992 Federal Bldg.
915 Second Ave.
Seattle, WA 98174
(206) 442-7285
FTS 399-7280

SMALL BUSINESS ADMINISTRATION REGIONAL OFFICES

Region 1
(Connecticut, Maine,
Massachusetts, New Hampshire,
Rhode Island, Vermont)
John F. Kennedy Federal Bldg.,
Room 2113
Boston, MA 02203
(617) 223-2100

Region 2
(New Jersey, New York, Puerto
Rico, Virgin Islands)
26 Federal Plaza, Room 3930
New York, NY 10007
(212) 460-0100

Region 3
(Delaware, District of Columbia,
Maryland, Pennsylvania, Virginia,
West Virginia)
1 Decker Square, East Lobby,
Suite 400
Bala Cynwyd, PA 19004
(215) 597-3311

Region 4
(Alabama, Florida, Georgia,
Kentucky, Mississippi, North
Carolina, South Carolina,
Tennessee)
1401 Peachtree St. N.E., Room 441
Atlanta, GA 30309
(404) 526-0111

Region 5
(Illinois, Indiana, Michigan,
Minnesota, Ohio, Wisconsin)
Federal Bldg.
219 South Dearborn St., Room 437
Chicago, IL 60604
(312) 353-4400

Region 6
(Arkansas, Louisiana, New
Mexico, Oklahoma, Texas)
1100 Commerce St., Room 300
Dallas, TX 75202
(214) 749-5611

Region 7
(Iowa, Kansas, Missouri,
Nebraska)
911 Walnut St., 24th Floor
Kansas City, MO 64106
(816) 374-7000

Region 8
(Colorado, Montana, North
Dakota, South Dakota, Utah,
Wyoming)
721 19th St., Room 426A
Denver, CO 80202
(303) 837-0111

Region 9
(Arizona, California, Hawaii,
Nevada, Pacific Islands)
Federal Bldg.
450 Golden Gate Ave.
San Francisco, CA 94102
(415) 556-9000

Region 10
(Alaska, Idaho, Oregon,
Washington)
710 2nd Ave., 5th Floor
Dexter Horton Bldg.
Seattle, WA 98104
(206) 442-0111

COMMERCIAL SCALE ETHANOL PRODUCTION AND FINANCING

DISTRICT OFFICES

Region 1

1326 Appleton St.
Holyoke, MA 01040
(413) 536-8770

Federal Bldg.
40 Western Ave., Room 512
Augusta, ME 04330
(207) 622-6171

55 Pleasant St., Room 213
Concord, NH 03301
(603) 224-4041

Federal Bldg.
450 Main St., Room 710
Hartford, CT 06103
(203) 244-2000

Federal Bldg.
87 State St., Room 210
Montpelier, VT 05602
(802) 223-7472

57 Eddy St., Room 710
Providence, RI 02903
(401) 528-1000

Region 2

225 Ponce de Leon Ave.
Hato Rey, PR 00919
(809) 765-0404

970 Broad St., Room 1635
Newark, NJ 07102
(201) 645-3000

Hunter Plaza
Fayette and Salina Sts., Room 308
Syracuse, NY 13202
(315) 473-3350

Chamber of Commerce Bldg.
55 St. Paul St.
Rochester, NY 14604
(716) 546-4900

Region 3

109 North 3rd St.
Room 301, Lowndes Bldg.
Clarksburg, WV 26301
(304) 624-3461

Federal Bldg.
1000 Liberty Ave., Room 1401
Pittsburgh, PA 15222
(412) 644-3311

Federal Bldg.
400 North 8th St., Room 3015
Richmond, VA 23240
(703) 782-2000

1310 L St., N.W., Room 720
Washington, DC 20417
(202) 382-3731

Region 4

908 South 20th St., Room 202
Birmingham, AL 35205
(205) 325-3011

222 South Church St.
Room 500, Addison Bldg.
Charlotte, NC 28202
(704) 372-0711

1801 Assembly St., Room 117
Columbia, SC 29201
(803) 765-5376

Petroleum Bldg., Suite 690
Pascagoula and Amite Sts.
Jackson, MS 39205
(601) 948-7821

Federal Bldg.
400 West Bay St., Room 261
Jacksonville, FL 32202
(904) 791-2011

Federal Bldg.
600 Federal Pl., Room 188
Louisville, KY 40202
(502) 582-5011

Federal Bldg.
51 Southwest 1st Ave., Room 912
Miami, FL 33130
(305) 350-5011

500 Union St., Room 301
Nashville, TN 37219
(615) 749-9300

502 South Gay St.
Room 307, Fidelity Bankers Bldg.
Knoxville, TN 37902
(615) 524-4011

Region 5

502 East Monroe St.
Ridgely Bldg., Room 816
Springfield, IL 62701
(217) 525-4200

1240 East 9th St., Room 5524
Cleveland, OH 44199
(216) 522-3131

34 North High St.
Columbus, OH 43215
(614) 469-6600

Federal Bldg.
550 Main St.
Cincinnati, OH 45202
(513) 684-2200

1249 Washington Blvd.
Room 1200, Book Bldg.
Detroit, MI 48226
(313) 226-6000

36 South Pennsylvania St.
Room 108, Century Bldg.
Indianapolis, IN 46204
(317) 633-7000

122 West Washington Ave.,
Room 713
Madison, WI 53703
(608) 256-4441

12 South 6th St., Plymouth Bldg.
Minneapolis, MN 55402
(612) 725-4242

Region 6

Federal Bldg. and Courthouse
500 Gold Ave., S.W.
Albuquerque, NM 87101
(505) 843-0311

808 Travis St., Room 1219
Niels Esperson Bldg.
Houston, TX 77002
(713) 226-4011

Post Office and Court House
Bldg.
West Capital Ave., Room 377
Little Rock, AK 72201
(501) 378-5871

1205 Texas Ave.
Lubbock, TX 79408
(806) 747-311

219 East Jackson St.
(Lower Rio Grande Valley)
Harlingen, TX 78550
(512) 423-8933

505 East Travis St.
Room 201, Travis Terrace Bldg.
Marshall, TX 75670
(214) 935-5257

Plaza Tower, 17th Floor
1001 Howard Ave.
New Orleans, LA 70113
(504) 527-2611

30 North Hudson St.
Room 501, Mercantile Bldg.
Oklahoma City, OK 73102
(405) 231-4011

301 Broadway
Room 300, Manion Bldg.
San Antonio, TX 78205
(512) 225-5511

Region 7

New Federal Bldg.
210 Walnut St., Room 749
Des Moines, IA 50309
(515) 284-4000

Federal Bldg.
215 North 17th St., Room 7419
Omaha, NE 68102
(402) 221-1221

Federal Bldg.
210 North 12th St., Room 520
St. Louis, MO 63101
(314) 622-8100

120 South Market St., Room 301
Wichita, KS 67202
(316) 267-6311

Region 8

Federal Bldg.
Room 4001, 100 East B St.
Casper, WY 82601
(307) 265-5550

Federal Bldg.
653 2nd Ave., North, Room 218
Fargo, ND 58102
(701) 237-5771

Power Block Bldg.
Corner Main and 6th Ave.,
Room 208
Helena, MT 59601
(406) 442-9040

Federal Bldg.
125 South State St., Room 2237
Salt Lake City, UT 84111
(801) 524-5500

National Bank Bldg.
8th and Maine Ave., Room 402
Sioux Falls, SD 57102
(605) 336-2980

Region 9

149 Bethel St., Room 402
Honolulu, HI 96813
(808) 546-8950

849 South Broadway
Los Angeles, CA 90014
(213) 688-2121

112 North Central Ave.
Phoenix, AZ 85004
(602) 261-3900

110 West C St.
San Diego, CA 92101
(714) 293-5000

Region 10

1016 West 6th Ave., Suite 200
Anchorage Legal Center
Anchorage, AK 99501
(907) 272-5561

503 3d Ave.
Fairbanks, AK 99701
(907) 452-5561

216 North 8th St., Room 408
Boise, ID 83701
(208) 342-2711

921 Southwest Washington St.
Portland, OR 97205
(503) 221-2000

Courthouse Bldg., Room 651
Spokane, WA 99210
(509) 456-0111

FARMERS HOME ADMINISTRATION STATE OFFICES

ALABAMA

Elizabeth Wright
Room 717, Aronov Bldg.
474 South Court St.
Montgomery, AL 36102
(205) 832-7077
FTS 534-7077

ALASKA

John R. Roderick
P.O. Box 1289
Palmer, AK 99645
(907) 745-2176

ARIZONA

Manual O. Dominquez
Room 3433, Federal Bldg.
230 North First Ave.
Phoenix, AZ 85025
(602) 271-6701
(Use same 7-digit number for FTS)

ARKANSAS

Sherman Williams
5529 Federal Office Bldg.
700 W. Capitol
For letter mail:
P.O. Box 2778
Little Rock, AR 72203
(501) 378-6281

CALIFORNIA

Lowell Pannell
459 Cleveland St.
Woodland, CA 95695
(916) 666-2650
FTS 448-3223

COLORADO

Ernest C. Phillips
Room 231, No. 1 Diamond Plaza
2490 West 26th Ave.
Denver, CO 80211
(303) 837-4347
FTS 327-4347

DELAWARE

(Delaware, District of Columbia, Maryland)
John D. Daniello
151 E. Chestnut Hill Rd., Suite 2
Newark, DE 19713
(302) 573-6694
FTS 487-6694

FLORIDA

Michael R. Hightower
Federal Bldg.
401 S.E. 1st Ave., Room 214
For letter mail:
P.O. Box 1088
Gainesville, FL 32602
(904) 376-3218
FTS 946-7221

GEORGIA

Robert L. Bialock
355 E. Hancock Ave.
Athens, GA 30601
(404) 546-2162
FTS 250-2162

HAWAII

Megumi Kin
345 Kekuanaoa St.
Hilo, HI 96720
(808) 961-4781

IDAHO

Joe T. McCarter
Room 429, Federal Bldg.
304 N. Eighth St.
Boise, ID 83702
(208) 384-1730
FTS 554-1318

ILLINOIS

John W. Lindfield
2106 W. Springfield Ave.
Champaign, IL 61820
(217) 356-1891
FTS 958-9149

INDIANA

James E. Posey
Suite 1700,
5610 Crawfordsville Rd.
Indianapolis, IN 46224
(317) 269-6415
FTS 331-6415

IOWA

Max L. McCord
Room 873, Federal Bldg.
210 Walnut St.
Des Moines, IA 50309
(515) 284-4663
FTS 862-4663

KANSAS

John T. Denyer
444 S.E. Quincy St.
Topeka, KS 66683
(913) 295-2870
FTS 752-2870

KENTUCKY

William E. Burnette
333 Waller Ave.
Lexington, KY 40504
(606) 233-2733
FTS 355-2733

LOUISIANA

Nimrod T. Andrews
3727 Government St.
Alexandria, LA 71301
(318) 448-3421
FTS 497-6611

MAINE

Seth H. Bradstreet
USDA Office Bldg.
Orono, ME 04473
(207) 866-4929
FTS 833-7445

MASSACHUSETTS

(Connecticut, Massachusetts, Rhode Island)
William E. Curry
358 N. Pleasant Street
Amherst, MA 01002
(413) 549-2820

MICHIGAN

Robert L. Mitchell
Room 209
1405 South Harrison Rd.
East Lansing, MI 48823
(517) 372-1910, Ext. 272
FTS 374-4272

MINNESOTA

John Apitz
252 Federal Office Bldg. and
    Courthouse
316 N. Robert St.
St. Paul, MN 55101
(612) 725-5842
(Use same 7-digit number for FTS)

MISSISSIPPI

Mark G. Hazard
Room 830, Milner Bldg.
Jackson, MS 39201
(601) 969-4316
FTS 490-4316

MISSOURI

Allan H. Brock
555 Vandiver Dr.
Columbia, MO 65201
(314) 442-2271, Ext. 3241
FTS 276-3241

MONTANA

Wallace B. Edland
Federal Bldg.
P.O. Box 850
Bozeman, MT 59715
(406) 587-5271, Ext. 4211
FTS 585-4211

NEBRASKA

Leonard T. Hanks
Room 308, Federal Bldg.
100 Centennial Mall North
Lincoln, NE 68508
(402) 471-5551
FTS 541-5551

NEW JERSEY

Lawrence E. Suydam
1 Vahlsing Center
Robbinsville, NJ 08691
(609) 259-3076
FTS 342-0232

NEW MEXICO

David W. King
Room 3414, Federal Bldg.
517 Gold Ave., S.W.
Albuquerque, NM 87102
(505) 766-2462
FTS 474-2462

NEW YORK

(New York, Virgin Islands)
Karen N. Hanson
Room 871, U.S. Courthouse
   and Federal Bldg.
100 So. Clinton St.
Syracuse, NY 13202
(315) 432-5290
FTS 950-5290

NORTH CAROLINA

James T. Johnson
Room 514, Federal Bldg.
310 New Bern Ave.
Raleigh, NC 27601
(191) 755-4740
FTS 672-4640

NORTH DAKOTA

Frederick S. Gengler
Federal Bldg., Room 208
For letter mail:
P.O. Box 1737
Bismarck, ND 58501
(701) 255-4011, Ext. 4237
FTS 783-4781

OHIO

Gene R. Abercrombie
Federal Bldg., Room 507
200 N. High St.
Columbus, OH 43215
(614) 469-5606
FTS 943-5606

OKLAHOMA

Gene F. Earnest
Agricultural Center Office Bldg.
Stillwater, OH 74074
(405) 624-4250
FTS 728-4250

OREGON

Kenneth Keith Keudell
Room 1590, Federal Bldg.
1220 S.W. 3rd Ave.
Portland, OR 97204
(503) 221-2731
FTS 423-2731

PENNSYLVANIA

J. Fred King
Federal Bldg.
Room 728, 228 Walnut St.
For letter mail:
P.O. Box 905
Harrisburg, PA 17108
(717) 782-4476
FTS 590-4476

PUERTO RICO

Juan Jose Jimenez
Federal Bldg.
Carlos Chardon Street
Hato Rey, PR 00918
For letter mail:
G.P.O. Box 6106G
San Juan, PR 00936
(809) 753-4308
(Use same 7-digit number for FTS)

SOUTH CAROLINA

Karl G. Smith
240 Stoneridge Rd.
For letter mail:
P.O. Box 21607
Columbia, SC 29221
(803) 765-5876
FTS 677-5876

TENNESSEE

Earl Wayne Avery
538 U.S. Court House Bldg.
801 Broadway
Nashville, TN 37203
(615) 251-7341
FTS 852-7341

TEXAS

Willaim H. Pieratt
W. R. Poage Bldg.
101 S. Main
Temple, TX 76501
(817) 774-1301
FTS 736-1301

UTAH

(Nevada, Utah)
Reed J. Page
Room 5311, Federal Bldg.
125 South State St.
Salt Lake City, UT 84138
(801) 524-5027
FTS 588-5057

VERMONT

(New Hampshire, Vermont)
Brian D. Burns
141 Main St.
P.O. Box 588
Montpelier, VT 05802
(802) 223-2371
FTS 832-4454

VIRGINIA

Edward A. Ragland
Federal Bldg., Room 8213
400 N. 8th St.
For letter mail:
P.O. Box 20206
Richmond, VA 23240
(804) 782-2451
FTS 925-2451

WASHINGTON

Keith P. Sattler
Room 319, Federal Office Bldg.
301 Yakima St.
Wenatchee, WA 98801
(509) 662-4353
FTS 390-0353

WEST VIRGINIA

James Vacemire
Room 320, Federal Bldg.
For letter mail:
P.O. Box 678
Morgantown, WV 26505
(304) 559-7791
FTS 923-7791

WISCONSIN

Lawrence E. Dahl
P.O. Box 639
Suite 209, First Financial Plaza
Stevens Point, WI 54481
(715) 341-5900
FTS 360-3889

WYOMING

Rudolph W. Knoll
Federal Bldg.
100 East B St.
For letter mail:
P.O. Box 820
Casper, WY 82601
(307) 265-5550, Ext. 3272
FTS 328-5271

To locate County Farmers Home Administration Officers, consult your telephone directory under U.S. Department of Agriculture, or the State Office of the Farmers Home Administration listed above.

DEPARTMENT OF HOUSING AND URBAN DEVELOPMENT REGIONAL OFFICES

Region 1

(Connecticut, Maine Massachusetts, New Hampshire, Rhode Island, Vermont)
Boston Regional Office
Carlton H. Hovey
800 John F. Kennedy Federal Bldg.
Boston, MA 02203
(617) 223-4066 *(Use same 7-digit number for FTS)*

Boston Area Office (Massachusetts, Rhode Island)
Bulfinch Bldg.
15 New Chardon St.
Boston, MA 02114
(617) 223-4111

Hartford Area Office (Connecticut)
Lawrence L. Thompson
999 Asylum Ave., 1 Financial Plaza
Hartford, CT 06103
(203) 244-3638 *(Use same 7-digit number for FTS)*

Manchester Area Office (Maine, New Hampshire, Vermont)
Creeley S. Buchanan
New Federal Bldg.
Chestnut St.
Manchester, NH 03101
(603) 666-7681
FTS 834-7681

Bangor Insuring Office (Maine)
Federal Bldg. and Post Office
202 Harlow St.
Bangor, ME 04401
(207) 942-8271

Burlington Insuring Office (Vermont)
Federal Bldg.
Elmwood Ave.
P.O. Box 989
Burlington, VT 05401
(802) 862-6501

Providence Insuring Office (Rhode Island)
330 Post Office Annex
Providence, RI 02903
(401) 528-4351

Region 2

(New York, New Jersey, Panama Canal Zone, Puerto Rico, Virgin Islands)
New York Regional Office
S. William Green
26 Federal Plaza, Room 3541
New York, NY 10007
(212) 264-8068 *(Use same 7-digit number for FTS)*
(212) 264-8086 *(Use same digit number for FTS)*

Buffalo Area Office (western New York State)
Frank C. Cerabone
Grant Bldg., 560 Main St.
Buffalo, NY 14202
(716) 842-3510
FTS 432-3510

Camden Area Office (southern New Jersey)
Robert E. Hazelwood
The Parkade Bldg., 519 Federal St.
Camden, NJ 08103
(609) 757-5081
FTS 488-5081
(609) 757-5085
FTS 488-5085

Newark Area Office (northern New Jersey)
Thomas Verdon Gateway 1 Bldg., Raymond Plaza
Newark, NJ 07102
(201) 645-3010 or 3899
FTS 341-3899

New York Area Office (eastern New York State)
J. Nugent Lopes
666 Fifth Ave.
New York, NY 10019
(212) 399-5290
FTS 662-5290
(716) 399-5283
FTS 662-5283

San Juan, Commonwealth Area Office
(Panama Canal Zone, Puerto Rico, Virgin Islands)
New Pan Am. Bldg., 255 Ponce de Leon Ave.
Hato Rey, PR
Mailing Address:
P.O. Box 3869, GPO
San Juan, PR 00936
(809) 763-6363

Albany Insuring Office (northern New York State)
30 Russell Rd., Westgate North
Albany, NY 12206
(518) 472-3567

Region 3

(Delaware, District of Columbia, Maryland, Pennsylvania, Virginia, West Virginia)
Philadelphia Regional Office
General Information Center
Curtis Bldg, 625 Walnut St.
Philadelphia, PA 19106
(215) 597-2560 or 2528
(Use same 7-digit number for FTS)

Baltimore Area Office (Maryland, except Montgomery and Price Georges Counties)
Everett Rothschild
Mercantile Bank and Trust Bldg.
Two Hopkins Plaza
Baltimore, MD 21201
(301) 962-2121 *(Use same 7-digit number for FTS)*

District of Columbia Area Office (District of Columbia, Montgomery and Prince Georges Counties in Maryland, and northern Virginia State)
James E. Clay
Universal North Bldg.
1875 Connecticut Ave., N.W.
Washington, DC 20009
(202) 673-5837

Philadelphia Area Office (eastern Pennsylvania, Delaware)
Abner Rappoport
Curtis Bldg., Room 892
625 Walnut St.
Philadelphia, PA 19106
(215) 597-2665 or 2633
(Use same 7-digit number for FTS)

Pittsburgh Area Office (western Pennsylvania, West Virginia)
Debra Krol
Two Allegheny Center, Room 1100
Pittsburgh, PA 15212
(412) 644-2802
FTS 722-2802
(412) 644-2818
FTS 722-2818

Richmond Area Office (southern Virginia State)
701 East Franklin St.
Richmond, VA 23219
(804) 782-2721

Charleston Insuring Office (West Virginia)
New Federal Bldg.
500 Quarrier St.
Mailing address:
P.O. Box 2948
Charleston, WV 25330
(304) 343-6181

Wilmington Insuring Office (Delaware)
Farmers Bank Bldg.
919 Market St.
Wilmington, DE 19801
(302) 571-6330

Region 4

(Alabama, Florida, Georgia, Kentucky, Mississippi, North Carolina, South Carolina, Tennessee)
Atlanta Regional Office (Georgia)
Room 211, Pershing Point Plaza
1371 Peachtree St., N.E.
Atlanta, GA 30309
(404) 526-5585

Atlanta Area Office (Georgia)
1100 Peachtree Center Bldg.
230 Peachtree St., N.W.
Atlanta, GA 30303
(404) 526-4576

Birmingham Area Office (Alabama)
Daniel Bldg., 15 South 20th St.
Birmingham, AL 35233
(205) 325-3264

Columbia Area Office (South Carolina)
1801 Main St., Jefferson Square
Columbia, SC 29202
(803) 765-5591

Greensboro Area Office (North Carolina)
2309 West Cone Blvd., Northwest Plaza
Greensboro, NC 27408
(919) 275-9111

Jackson Area Office (Mississippi)
101-C Third Floor Jackson Mall
300 Woodrow Wilson Ave., West
Jackson, MS 39213
(601) 366-2634

Jacksonville Area Office (Florida)
Peninsular Plaza, 661 Riverside Ave.
Jacksonville, FL 32204
(904) 791-1616

Knoxville Area Office (Tennessee)
1 Northshore Bldg., 1111 Northshore Dr.
Knoxville, TN 37919
(615) 584-8527

Louisville Area Office (Kentucky)
Virgil G. Kinarel
Children's Hospital Foundation Bldg.
601 South Floyd St., P.O. Box 1044
Louisville, KY 40201
(502) 582-5251

Memphis Insuring Office (western Tennessee State)
100 N. Main St.
Memphis, TN 38103
(901) 534-3143

Nashville Insuring Office (central Tennessee State)
801 Broadway
Nashville, TN 37203
(615) 749-5521

Tampa Insuring Office (central Florida State)
224028 Henderson Blvd.
P.O. Box 18165
Tampa, FL 33679
(813) 228-1501

Region 5

(Illinois, Indiana, Michigan, Minnesota, Ohio,
Wisconsin)
Chicago Regional Office
300 South Wacker Dr.
Chicago, IL 60606
Referral No. (312) 353-5680
FTS 353-5680

Chicago Area Office (Illinois)
Edward Bush
17 N. Dearborn St., Room 1201
Chicago, IL 60602
(312) 353-7660 or 6979
(Use same 7-digit number for FTS)

Columbus Area Office (Ohio)
60 East Main St.
Columbus, OH 43215
(614) 469-7345

Detroit Area Office (Michigan)
Kenneth Barnard
McNamara Federal Bldg.
Michigan Ave., 17th Floor
Detroit, MI 48226
(313) 226-7900 *(Use same 7-digit number for FTS)*

Indianapolis Area Office (Indiana)
Choice Edwards
Willowbrook 5 Bldg.
Room 301, 4720 Kingsway Dr.
Indianapolis, IN 46205
(317) 269-6303
FTS 331-6303

Milwaukee Area Office (Wisconsin)
744 North 4th St.
Milwaukee, WI 53203
(414) 224-3221

Minneapolis-St. Paul Area Office (Minnesota)
Margaret Wolszon
Griggs-Midway Bldg.,
1821 University Ave.
St. Paul, MN 55104
(612) 725-4701 or 4801
(Use same 7-digit number for FTS)

Cincinnati Insuring Office (southwestern Ohio State)
9009 Federal Office Bldg., 550 Main St.
Cincinnati, OH 45202
(513) 684-2884

Cleveland Insuring Office (northern Ohio State)
777 Rockwell
Cleveland, OH 44114
(216) 552-4065

Grand Rapids Insuring Office (western and
northern Michigan State)
Northbrook Bldg., No. 11
2922 Fuller Ave., N.E.
Grand Rapids, MI 49505
(616) 456-2225

Springfield Insuring Office (central and
southern Illinois State)
Lincoln Tower Plaza
542 South Second St., Room 600
Springfield, IL 62704
(217) 525-4414

Region 6

(Arkansas, Louisiana, New Mexico, Oklahoma,
Texas)
Fort Worth Regional Office
Jackie Bransford
1100 Commerce St.
Dallas, TX 75242
(214)749-7401 or 7406
(Use same 7-digit number for FTS)

Dallas Area Office (New Mexico; eastern, northern, and
western Texas)
2001 Bryan Tower, 4th Floor
Dallas, TX 75201
(214) 749-1601
FTS 740-1601
Public Information: (214) 749-1625
(Use same 7-digit number for FTS)

Little Rock Area Office (Arkansas and
Bowie County, Texas)
Sterling R. Cockrill, Jr.
Union National Bank Bldg., One Union National Plaza
Little Rock, AR 72201
(501) 378-5401
FTS 740-5401

New Orleans Area Office (Louisiana)
Bruno Lohrmann
Plaza Tower, 1001 Howard Ave.
New Orleans, LA 70113
(504) 589-2063
FTS 682-2063

Oklahoma City Area Office (Oklahoma)
Maxwell D. Harris
301 North Hudson St.
Oklahoma City, OK 73102
(405) 231-4891 or 4168
FTS 736-4168

San Antonio Area Office (Southwest Texas)
Finnias E. Jolly (Area Director)
James Byam (Deputy Director)
Kallison Bldg., 410 South Main Ave.
P.O. Box 9163
San Antonio, TX 78285
(512) 229-6800
FTS 730-6800

Albuquerque Insuring Office (New Mexico)
625 Truman St., N.E.
Albuquerque, NM 87110
(505) 766-3251

Fort Worth Insuring Office (northcentral Texas State)
13A01 Federal Bldg.
819 Taylor St.
Fort Worth, TX 76102
(817) 334-3233

Houston Insuring Office (East Central Texas State)
Two Greenway Plaza East
Houston, TX 77046
(713) 226-4335

Lubbock Insuring Office (Northwest Texas State)
514 Courthouse and Federal Office Bldg.
1205 Texas Ave., P.O. Box 1647
Lubbock, TX 79408
(806) 762-7265

Shreveport Insuring Office (northern Louisiana and 5
counties in eastern Texas)
New Federal Bldg.
500 Fannin, 62nd Floor
Shreveport, LA 71120
(318) 226-5011

Tulsa Insuring Office (eastern Oklahoma State)
1708 Utica Square, P.O. Box 52554
Tulsa, OK 74152
(918) 581-7435

Region 7

(Iowa, Kansas, Missouri, Nebraska)
Kansas City Regional Office
300 Federal Office Bldg.
911 Walnut St.
Kansas City, MO 64106
(816) 374-2661
FTS 758-2661
Public Information Desk: (816) 374-4391
FTS 758-4391

Kansas City Area Office (Kansas, western Missouri)
Carleta Foltz
Two Gateway Center, 4th and State St.
Kansas City, KS 66101
(816) 374-4355
FTS 758-4355
(816) 374-4220
FTS 758-4220

Omaha Area Office (Iowa, Nebraska)
Sue Burkett
Univac Bldg., 7100 West Center Rd., 3rd Floor
Omaha, NE 68106
(402) 221-9301
FTS 864-9301
(402) 221-9345
FTS 864-9345

St. Louis Area Office (eastern Missouri)
Craig Rydgig
210 North 12th St.
St. Louis, MO 63101
(314) 622-4761
FTS 279-4761

Des Moines Insuring Office (Iowa)
259 Federal Bldg.
210 Walnut St.
Des Moines, IA 50309
(515) 284-4512

Topeka Insuring Office (Kansas, except Johnson and
Wyandotte Counties)700 Kansas Ave.
Topeka, KS 66603
(913) 234-8241

Region 8

(Colorado, Montana, North Dakota, South Dakota,
Utah, Wyoming)
Denver Regional Office
2500 Executive Towers, 1405 Curtis St.
Denver, CO 80202
(303) 837-4881
FTS 327-2891

Casper Insuring Office (Wyoming)
Federal Office Bldg.
100 East B St.
Casper, WY 82601
(307) 265-5550

Denver Insuring Office (Colorado)
4th Floor Title Bldg., 909 17th St.
Denver, CO 80202
(303) 837-2441

Fargo Insuring Office (North Dakota)
Federal Bldg.
653 2nd Ave., N.
P.O. Box 2483
Fargo, ND 58102
(701) 237-5136

Helena Insuring Office (Montana)
616 Helena Ave.
Helena, MT 59601
(406) 442-3237

Salt Lake City Insuring Office (Utah)
125 South State St.
P.O. Box 11009
Salt Lake City, UT 84111
(801) 524-5237

Sioux Falls Insuring Office (South Dakota)
119 Federal Bldg., U.S. Courthouse
400 S. Phillips Ave.
Sioux Falls, SD 57102
(605) 336-2223

Region 9

(Arizona, California, Hawaii, Nevada, Guam,
American Samoa, Pacific Trust Territories)
San Francisco Regional Office
450 Golden Gate Ave., P.O. Box 36003
San Francisco, CA 94102
(415) 556-4752

Los Angeles Area Office (Arizona, southern
California)
Gilbert Meza
2500 Wilshire Blvd.
Los Angeles, CA 90057
(213) 688-5973 or 3836
FTS 798-3836

San Francisco Area Office (northern California,
Hawaii, Nevada, Guam, American Samoa, Pacific
Trust Territories)
Suite 1600, 1 Embarcadero Center
San Francisco, CA 94111
(415) 556-2238 *(Use same 7-digit number for FTS)*
Public Information: (415) 556-5900
(Use same 7-digit number for FTS)

Honolulu Insuring Office (Hawaii, Guam, American
Samoa)
1000 Bishop St., P.O. Box 3377
Honolulu, HI 96813
(808) 546-2136

Phoenix Insuring Office (Arizona)
244 W. Osborn Rd., P.O. Box 13468
Phoenix, AZ 85002
(602) 261-3900

Reno Insuring Office (Nevada)
1050 Bible Way, P.O. Box 4700
Reno, NV 89505
(702) 784-5356

Sacramento Insuring Office (northeastern California)
801 I St.
P.O. Box 1978
Sacramento, CA 95809
(916) 449-3471

San Diego Insuring Office (Imperial and
San Diego Counties, California)
110 West C St., P.O. Box 2648
San Diego, CA 92112
(714) 293-5310

Santa Ana Insuring Office (Orange, Riverside, and
San Bernardino Counties, California, for home
mortgages)
1440 East First St.
Santa Ana, CA 92701
(714) 836-2451

Region 10

(Alaska, Idaho, Oregon, Washington)
Seattle Regional Office
Merrill Ash
Arcade Plaza Bldg., 1321 Second Ave., Stop 329
Seattle, WA 98101
(206) 442-5415
FTS 399-5415
(206) 442-0934
FTS 399-0934

COMMERCIAL SCALE ETHANOL PRODUCTION AND FINANCING

Portland Area Office (southern Idaho, Oregon, Washington Counties of Clar, Klickitat, and Skamania)
Harold Stephens
520 Southwest 6th Ave.
Portland, OR 97204
(503) 221-2561
FTS 423-2561
(503) 221-2552
FTS 423-2552

Seattle Area Office (Alaska, northern Idaho, Washington except Clark, Klickitat, and Skamania Counties)
Richard Berinck
Arcade Plaza Bldg.
1321 Second Ave., Stop 409
Seattle, WA 98101
(206) 442-7456
FTS 399-7456

Anchorage Insuring Office (Alaska)
334 W. 5th Ave.
Anchorage, AK 99501
(907) 272-5561

Boise Insuring Office (west-central Idaho, Baker and Malheur Counties in Oregon)
331 Idaho St., P.O. Box 32
Boise, ID 83707
(208) 342-2711

Spokane Insuring Office (northern Idaho, eastern Washington)
920 Riverside Ave., West
Spokane, WA 99201
(509) 456-2510

ECONOMIC DEVELOPMENT ADMINISTRATION REGIONAL OFFICES

Colorado
(Rocky Mountain: Colorado, Iowa, Kansas, Missouri, Montana, Nebraska, North Dakota, South Dakota, Utah, Wyoming)
Craig Smith, Regional Director
Suite 505, Title Bldg., 909 17th St.
Denver, CO 80202
(303) 837-4717
FTS 327-4717

Georgia
(Southeastern: Alabama, Florida, Georgia, Kentucky, Mississippi, North Carolina, South Carolina, Tennessee)
Ann Crighton, Regional Director
Suite 700
1365 Peachtree St., N.E.
Atlanta, GA 30309
(404) 881-7401
FTS 257-7401

Illinois
(Midwestern: Illinois, Indiana, Michigan, Minnesota, Ohio, Wisconsin)
Edward Jeep, Regional Director
175 W. Jackson Blvd., Suite A 1630
Chicago, IL 60604
(312) 353-7706 (Use same 7-digit number for FTS)

Pennsylvania
(Atlantic: Connecticut, Delaware, District of Columbia, Maine, Maryland, Massachusetts, New Hampshire, New Jersey, New York, Pennsylvania, Puerto Rico, Rhode Island, Vermont, Virgin Islands, Virginia, West Virginia)
John E. Corrigan, Regional Director
10424 Federal Bldg.
600 Arch St.
Philadelphia, PA 19106
(215) 597-4603 (Use same 7-digit number for FTS)

Texas
(Southwestern: Arkansas, Louisiana, New Mexico, Oklahoma, Texas)
Joseph B. Swanner, Regional Director
Suite 600, American Bank Tower
221 West Sixth St.
Austin, TX 78701
(512) 397-5461
FTS 734-5461

Washington
(Western: Alaska, American Samoa, Arizona, California, Guam, Hawaii, Idaho, Nevada, Oregon, Washington)
Phyllis Lamphere, Regional Director
1700 Westlake Ave., North, Suite 500
Seattle, WA 98109
(206) 442-0596
FTS 399-0596

APPENDIX F

Bibliography

- General

- Conversion

- Coproducts

- Design

- Distillation

- Economics

- Energy Balance

- Environmental Considerations

- Feedstocks

- Fermentation

- International

- Regulatory

- Transportation Use

GENERAL

Introductory

Pimental, D. et al.; 1975 (November 21) "Energy and Land Constraint in Food Production; *Science*. Vol. 190 (no. 4126): pp. 754-761.

Solar Energy Research Institute. 1980. *Fuel from Farms: A Guide to Small-Scale Ethanol Production*. Golden, Co. SERI. Stock No. 061-000-00372-0. Available from: Superintendent of Documents of U.S. Government Printing Office, Washington, DC 20402.

U.S. Department of Agriculture. 1980. *Small-Scale Fuel Alcohol Production*. Washington, DC: USDA. Stock no. 001-000-04124-0. Available from: Superintendent of Documents, U.S. Government Printing Office, Washington, DC 20402.

Reports

Baratz B., Oullette, R., Park, W., Stokes, B. 1975 (November). *Survey of Alcohol Fuel Technology. Volume I.* McLean, VA: Mitre Corporation. Report No. PB-256007. 117p. Available from: National Technical Information Service, 5285 Port Royal Road, Springfield, VA 22161

Freeman, J. H. et al. 1976 (July). *Alcohols — A Technical Assessment of Their Application as Fuels.* Washington, DC: American Petroleum Institute. Publication no. 4261. 32 p. Available from: American Petroleum Institute, 2101 L Street, N.W., Washington, DC 20037

Office of Technology Assessment. 1979. Gasohol — A Technical Memorandum. Washington, DC: OTA, U.S. Congress. Stock No. 052-003-00706-1. Available from: Superintendent of Documents, U.S. Government Printing Office, Washington, DC 20402

Park, W., Price, E., Salo, D. 1978 (August) *Biomass-Based Alcohol Fuels: The Near Term Potential for Use with Gasoline* McLean, VA: Mitre Corporation. Report no. HCP/T4101-3. 84 p. Available from: National Technical Information Service, 5285 Port Royal Road, Springfield, VA 22161.

Proceedings of the Third International Symposium on Alcohol Fuels Technology. 1980. Asilomar, California; May 29-31, 1979. Washington, DC: U.S. Department of Energy. Report no. CONF-790520. Available from: National Technical Information Service, 5285 Port Royal Road, Springfield, VA 22161.

U.S. Department of Energy. 1979 (June) *The Report of the Alcohol Fuels Policy Review:* Washington, DC: US DOE. Report no. DOE/PE-0012.119p. Available from National Technical Information Service, 5285 Port Royal Road, Springfield, VA 22161.

Books

Cheremisnoff, N.P. 1979 *Gasohol for Energy Production.* Ann Arbor, MI: Ann Arbor Science. 140 p.

Paul, J.K. 1979 *Ethyl Alcohol Production and Use as A Motor Fuel.* Chemical Technology Review. No. 144. Park Ridge, NJ: Noyes Data Corporation. 354 p.

Magazines

Alcohol Update. Semi-monthly. August 1980+. $25.00/yr. Available from: Alcohol Update, P.O. Box 35211, Minneapolis, MN 55435.

Biomass Digest. Monthly $87.00/yr. Available from: Technical Insights, P.O. Box 1304, Fort Lee, NJ 07024.

Alcohol Week. Weekly-December 1980+. $245.00/yr. Available from;*Alcohol Week*, P.O. Box 7167, Benjamin Franklin Station, Washington, DC. 20044

Biotimes. Bimonthly, January 1979+; 10.00/yr. Available from: International Biomass Institute, 1522 K St.; N.W., Suite 600, Washington, DC 20005.

Gasohol USA Monthly; June 1979+, $12.00/yr. Available from: Box 9547, Kansas City, MO 64133.

Congressional Hearings.

U.S. House of Representatives. Ninety-fifth Congress, Second Session. *Alcohol Fuels: Hearings Before the Subcommittee on Advanced Energy Technologies and Energy Conservation Research, Development and Demonstration of the Committee on Science and Technology.* Washington, DC: U.S. House of Representatives; 11-13 July 1978. Stock no. 35-520. Available from: House Committee on Science and Technology. Room 3154, House Annex #2, Washington, DC 20515.

U.S. House of Representatives. Ninety-sixth Congress, First Session. *National Fuel Alcohol and Farm Commodity Production Act of 1979: Hearings Before the Subcommittees on Conservation and Credit Department Investigations Oversight and Research, and Livestock and Grains, of the Committee on Agriculture.* Washington, DC: U.S. House of Representatives; 15-16 May 1979. Stock No. 052-070-05071-3. Available from: Superintendent of Documents, U.S. Government Printing Office, Washington, DC 20402.

U.S. House of Representatives. Ninety-sixth Congress, First Session. *Oversight - Alcohol Fuel Options and Federal Policies. Hearings Before the Subcommittee on Energy Development and Applications of the Committee on Science and Technology.* Washington, DC: U.S. House of Representatives. 4 May, 12 June 1979. Stock No. 49-650. Available from: Superintendent of Documents, U.S. Government Printing Office, Washington, DC 20402.

U.S. Senate. Ninety-fifth Congress, Second Session. *Alcohol Fuels: Hearing Before the Committee on Appropriations.* Washington, D.C. U.S. Senate. 13 January 1978. Stock no. 052-070-04679-1. Available from: Superintendent of Documents, U.S. Government Printing Office, Washington, DC 20402.

U.S. Senate. Ninety-fifth Congrss, *Second* Session. *The Gasohol Motor Fuel Act of 1978: Hearings Before the Subcommittee on Energy Research and Development of the Committee on Energy and Natural Resources.* Washington, DC: U.S. Senate. 7-8 August 1978. Publication No. 95-165. Available from: Superintendent of Documents, U.S. Government Printing Office, Washington, D C 20402.

BIBLIOGRAPHY

National Technical Information Service. 1979 (July). *Alcohol Fuels: (Citations from the NTIS Data Base), Volume 1 1964-1977.* Washington, DC: NTIS Report no. NTIS/PS-79/0712. 170p. Available from: National Technical Information Service, 5285 Port Royal Road, Springfield, VA 22161.

National Technical Information Service. 1979 (July): *Alcohol Fuels (Citations from the NTIS Data Base), Volume 2, 1978—June 1979.* Washington, DC: NTIS. Report No. NTIS/PS-79/0713. 144 p. Available from: National Technical Information Service, 5285 Port Royal Road, Springfield, VA 22161.

National Technical Information Service. 1979 (July). *Alcohol Fuels (Citations from the NTIS Data Base), Report for 1970—June 1979.* Washington, DC: NTIS. Report no. NTIS/PS-79/0714. 247p. Available from: National Technical Information Service, 5285 Port Royal Road, Springfield, VA 22161.

NATIONAL ALCOHOL FUELS COMMISSION REPORTS

U.S. National Alcohol Fuels Commission "Farm and Cooperative Alcohol Plant Study: Technical and Economic Assessment as a Commercial Venture." 1980, 200 pp. Prepared by Raphael Katzen Associates International, Inc. Available from NTIS.

U.S. National Alcohol Fuels Commission. "Fuel Alcohol: Report and Analysis of Plant Conversion Potential to Fuel Alcohol Production." 1980, 123 pp. Prepared by Davy McKee Corporation. Available from NTIS.

U.S. National Alcohol Fuels Commission. "Ethanol: Farm and Fuel Issues." 1980, 145 pp. Prepared by Schnittker Associates. Available from NTIS, Report No. PB 80-215692, $11.00.

U.S. National Alcohol Fuels Commission. "Energy Balances in the Production and End-Use of Alcohols Derived from Biomass." 1980, 75 pp. Prepared by TRW.

Conversion

Brushke, H. "Direct Processing of Sugarcane into Ethanol" in: *Proceedings of the International Symposium of Alcohol Fuel Technology: Methanol and Ethanol.* Wolfsburg, Federal Republic of Germany. November 21-23, 1977 Report No. CONF-771175, paper 5-5. Entire report available from: National Technical Information Service, 5285 Port Royal Road, Springfield, VA 22161

Nathan, R. A. 1978. *Fuels from Sugar Crops: Systems Study for Sugarcane, Sweet Sorghum, and Sugar Beets.* Oak Ridge, TN: Technical Information Center, U.S. Department of Energy. Report no. TID-22781. 137p. Available from: National Technical Information Service, 5285 Port Royal Road, Springfield, VA 22161.

Coproducts

Colorado State University, 1979. *Analysis of Alcohol Fermentation By-Products for Livestock* and *Poultry Feeding in Colorado.* Fort Collins, CO: Colorado State University. Available from: Department of Animal Sciences, Colorado State University, Fort Collins, CO 80523.

Paturau, J. M., 1969. *By-products of the Cane Sugar Industry.,* Amsterdam, The Netherlands: Elsevier Publishing Co., 274p.

Reilly, P. J. 1978 *Conversion of Agricultural By-Products to Sugars, Progress Report* Ames, IA: Iowa State University, Available from: Department of Chemical Engineering, Engineering Research Institute, Iowa State University, Ames, Iowa. 50010.

Wisner, R. N. and Gidel, J. O. 1977 (June) *Economics Aspects of Using Grain Alcohol as a Motor Fuel, with Emphasis on By-product Feed Markets.* Ames, IA: Iowa Agriculture Experiment Station. Report no. 9. Available from: Agriculture Engineering, Extension, Davidson Hall, Ames, IA, 50010.

Design

Brackett, A. T. et al. 1978 *Indiana Grain Fermentation Alcohol Plant.* Indianapolis, IN: Department of Commerce. 80p. Available from: Indiana Department of Commerce, State House, Room 336, Indianapolis, IN 46204.

Chambers, R.S. 1979 *The Small Fuel-Alcohol Distillery: General Description and Economic Feasibility Workbook.* Urbana, IL: ACR Process Corporation. 21p. Available from: ACR Process Corporation, 808 S. Lincoln Ave., Urbana, IL 61801.

Grain Motor Fuel Alcohol Technical and Economic Assessment Study. 1978 (December). Cincinnati, OH: Raphael Katzen Associates. Report no. HCP/J6639-01. 341p. Available from: National Technical Information Service, 5285 Port Royal Road, Springfield, VA 22161.

Distillation

King, C. J. 1971. *Separation Process.* New York, NY: McGraw Hill Book Publishing Co. 809 p.

Tassios, D. P. 1972. "Rapid Screening of Extractive Distillation Solvents." *Extractive and Azeotropic Distillation.* Advances in Chemistry Series, No. 115. Washington, DC: American Chemical Society. pp. 46-63.

McCabe, W. and Smith, J. C. 1976 *Unit Operatons in Chemical Engineering* 3rd Edition. New York, NY: McGraw Hill Book Publishing Co. 1028 p.

ECONOMICS

David, M. L., et al. 1978 (July) *Gasohol: Economic Feasibility Study—Final Report.* Manhattan, KS: Development Planning and Research Associates, Inc. Report no. SAN-1681-T1. 280p. Available from: National Technical Information Service, 5285 Port Royal Road, Springfield, VA 22161.

Gasohol from Grain—The Economic Issues. 1978 (January 19). Washington, DC: Economics, Statistics and Cooperative Service. Report no. PB-280120/7ST. 23p. Available from: National Technical Information Service, 5285 Port Royal Road, Springfield, VA 22161.

ENERGY BALANCE

Alich, J. A., et al. 1978 (January). *An Evaluation of the Use of Agricultural Residues as an Energy Feedstock: A Ten Site Survey.* Palo Alto, CA: Stanford Research Institute - International. Report no. TID-27904/2. 402p. Available from: National Technical Information Service, 5285 Port Royal Road, Springfield, VA 22161.

Commoner, Barry. 1979 (July 23) *Testimony before United States Senate Committee on Agriculture, Nutrition and Forestry, Subcommittee on Agricultural Research and General Legislation on "The Potential for Energy Production by U.S. Agriculture."* St. Louis, MO: Center for the Biology of Natural Systems, Washington University. Report No. CBNS-AEP-5. Available from: Center for the Biology of Natural Systems, Washington University, St. Louis, Mo.

Ladisch, M. R., Dyck, K. 1979 (August 3). "Dehydration of Ethanol: New Approach Gives Positive Energy Balance." *Science.* Vol. 205 (no. 31): pp. 898-900.

Lewis, C. W. 1977 (September). "Fuels from Biomass—Energy Outlays vs. Energy Returns: A Critical Appraisal." *Energy* Vol. 2 (no. 3): pp. 241-248.

ENVIRONMENTAL CONSIDERATIONS

Aerospace Corporation. 1980 (June). *Environmental Control Perspective for Ethanol Production from Biomass.* (Draft). Germantown, MD: Report no. ATR-80 (7848-01)-1.

Brown, D., McKay, R., and Weir, W. 1976. "Some Problems Associated with the Treatment of Effluents from Malt Whiskey Distilleries. *Progress in Water Technology*. Vol. 8 (no. 2/3): pp. 291-300.

Council on Environmental Quality. 1979 (December). *Environmental Quality: The Tenth Annual Report of the Council of Environmental Quality*. Washington, DC.

Hagey, G., et al. "Methanol and Ethanol Fuels: Environmental, Health and Safety Issues," *in: Proccedings of the International Symposium on Alcohol Fuel Technology: Methanol and Ethanol*. Wolfsburg, Federal Republic of Germany. November 21-23, 1977. Report no. CONF-771175, paper 8-2. Entire report available from: National Technical Information Service, 5285 Port Royal Road, Springfield, VA 22161.

Jackson, E. A. 1977 (April). "Distillery Effluent Treatment in the Brazilian National Alcohol Programme" Chemical Engineer. no. 319: pp. 239-242.

Kant, F.H. et al.1974. *Feasibility Studies of Alternative Fuels for Automotive Transportation*. Washington, DC: U.S. Environmental Protection Agency. Report no. EPA-460/374-009. Available from: National Technical Information Service. 5285 Port Royal Road, Springfield, VA 22161.

Grain Motor Fuel Alcohol Technical and Economic Assessment Study. 1978 (December). Cincinnati, OH: Raphael Katzen Associates. Report no. HCP/J6639-01. 341p. Available from: National Technical Information Service, 5285 Port Royal Road, Springfield, VA 22161.

Lee, Linda K. *1978. A Perspective on Cropland Availability*. Washington, DC: Economics, Statistics, and Cooperative Services, U.S. Department of Agriculture. Report no. 406. 23 p. Available from: Superintendent of Documents, U.S. Government Printing Office, Washington, DC 20402.

Lowrey, S. P. and Deroto, R. S. 1976. "Exhaust Emissions from a Single Cylinder Engine Fueled with Gasoline, Methanol, and Ethanol." *Combustion Science and Technology*. Vol. 12 (no. 4, 5, 6): pp. 177-182.

A. J. Moriarty. Toxicological Aspects of Alcohol Fuel Utilization *in: Proceedings of the International Symposium on Alcohol Fuel Technology: Methanol and Ethanol*. Wolfsburg, Federal

Republic of Germany. November 21-23, 1977. Report no. CONF-771175, paper 8-1. Entire report available from: National Technical Information Service, 5285 Port Royal Road, Springfield, VA 22161.

Scarberry, R.M. and Papai, M.P, 1980 (January). *Source Test and Evaluation Report: Alcohol Synthesis Facilty for Gasohol Production*. McLean, VA: Radian Corporation. Available from Radian Corporation.

Sittig, M. 1979. Hazardous and Toxic Effects of Industrial Chemicals. Park Ridge, NJ: Noyse Data Corporation.

Office of Technology Assessment. 1979. *Gasohol—A Technical Memorandum*. Washington, DC: OTA, U.S. Congress. Stock no. 052-003-00706-1. Available from: Superintendent of Documents, U.S. Government Printing Office, Washington, DC 20402.

Unger, S. G. 1977 (October). *Environmental Implications of Trends in Agriculture and Silviculture, Volume 1: Trend Identification and Evaluation*. Manhattan, KS: Development Planning and Research Associates. Report no. PB-274-233. 232 p. Available from: National Technical Information Service, 5285 Port Royal Road, Springfield, VA 22161.

U.S. Environmental Protection Agency. 1978 (December). *Report to Congress, Industrial Cost Recovery*. Washington, DC: Office of Water Program Operations. Report no. PR80-204746. 61p. Available from: National Technical Information Service, 5285 Port Royal Road, Springfield, VA 22161.

FEEDSTOCKS

Atchison, J. E. 1977. *Preliminary Investigation of New Process for Separation of Components of Sugarcane Sweet Sorghum, and Other Plant Stalks*. Columbus, OH: Battelle. Report no. TID-28734. 315 p. Available from: National Technical Information Service, 5285 Port Royal Road, Springfield, VA 22161.

Atlas of Nutritional Data of United States and Canadian Feeds. 1971. Washington, DC: National Academy of Science. 772 p.

Chubey, B. B. and Dorrell, D. G. 1974. "Jerusalem Artichoke A Potential Fructose Crop for the Prairies" *Journal of the Canadian Institute of Food Science Technology*. Vol. 7 (no. 2): pp. 98-100.

Jones, J.L. 1978 (December). *Mission Analysis for the Federal Fuels from Biomass Program. Volume 1: Summary and Conclusions.* Menlo Park, CA: Stanford Research Institute International. Report no. SAN-0115-T2. 76 p. Available from: National Technical Information Service, 5285 Port Royal Road, Springfield, VA 22161.

Hertzmark, D. 1980 (March). *Agricultural Sector Impacts of Making Ethanol from Grain.* Golden, CO: Solar Energy Research Institute. Report no. SERI/TR-352-554. 64p. Available from: National Technical Information Service, 5285 Port Royal Road Springfield, VA 22161

Lee, Linda K. 1978. *A Perspective on Cropland Availability.* Washington, DC: U.S. Department of Agriculture. Report no. 406. 23 p. Available from: Superintendent of Documents, U.S. Government Printing Office, Washington, DC 20402.

Lipinsky, E. S. et al. 1977 (July 29). *Fuels from Sugar Crops: First Quarterly Report.* Columbus, OH: Battelle. Report no. TID-28414. 140p. Available from: National Technical Information Service, 5285 Port Royal Road, Springfield, VA 22161.

Lipinsky, E. S. et al., 1977 (October 31). *Fuels from Sugar Crops—Second Quarterly Report.* Columbus, OH: Battelle. Report no. TID-27834. 160p. Available from: National Technical Information Service, 5285 Port Royal Road, Springfield, VA 22161

Lipinsky, E.S. et al. 1978. *Fuels from Sugar Crops—Third Quarterly Report.* Columbus, OH: Battelle. Report no. TID-28191. Available from: National Technical Information Service, 5285 Port Royal Road, Springfield, VA 22161.

Lipinsky, E. S. et al. 1977 (March). *Systems Study of Fuels from Sugarcane, Sweet Sorghum, and Sugar Beets. Volume 1: Comprehensive Evaluation.* Columbus, OH: Battelle Report no: BMI-1957 (Vol. 1). 167p. Available from: National Technical Information Service 5285 Port Royal Road, Springfield, VA 22161.

Lipinsky, E.S. et al. 1976 (December 31). *Systems Study of Fuels from Sugarcane, Sweet Sorghum, and Sugar Beets. Volume 2: Agriculture Considerations.* Columbus, OH: Battelle. Report no. BMI-1957 (Vol. 2) 245 p. Available from: National Technical Information Service, 5285 Port Royal Road, Springfield, VA 22161

Lipinsky, E. S. et al. 1976 (December 31). *Systems Study of Fuels from Sugarcane, Sweet Sorghum, and Sugar Beets. Volume 3: Conversion to Fuels and Chemical Feed Stocks.* Columbus, OH: Battelle. Report no. BMI-1957 (Vol. 3). 175 p. Available from: National Technical Information Service, 5285 Port Royal Road, Springfield, VA 22161

Lipinsky, E. S. et al. 1977 (March 31). *Systems Study of Fuels from Sugarcane, Sweet Sorghum, and Sugar Beets. Volume 4: Corn Agriculture.* Columbus, OH: Battelle. Report no. BMI-1957A (Vol. 4). 203 p. Available from: National Technical Information Service, 5285 Port Royal Road, Springfield, VA 22161.

Lipinsky, E. S. et al. 1977 (March 31) *Systems Study of Fuels from Sugarcane, Sweet Sorghum, Sugar Beets, and Corn. Volume 5: Comprehensive Evaluation of Corn.* Columbus, OH: Battelle Report no. BMI-1957A (Vol. 5). 168 p. Available from: National Technical Information Service, 5285 Port Royal Road, Springfield, VA 22161.

Nathan, R.A. 1978. *Fuels from Sugar Crops: Systems Study for Sugar Cane, Sweet Sorghum, and Sugar Beets.* Oak Ridge, TN: Technical Information Center, U.S. Department of Energy. Report no. TID-22781. 137p. Available from National Technical Information Service, 5285 Port Royal Road, Springfield, VA 22161.

U.S. Department of Agriculture 1979 *Agricultural Statistics, 1979.* Washington, DC: USDA. Stock no. 001-000-03775-7. Available from: Superintendent of Documents, U.S. Government Printing Office, Washington, DC 20402.

FERMENTATION

Engelbart, W. "Basic Data on Continuous Alcoholic Fermentation of Sugar Solutions and of Mashes from Starch Containing Raw Materials " *in: Proceedings of the International Symposium on Alcohol Fuel Technology: Methanol and Ethanol.* Wolfsburg. Federal Republic of Germany. November 21-23, 1977. Report no. CONF-771175, paper 5-3. Entire report available from: National Technical Information Service, 5285 Port Royal Road, Springfield, VA. 22161.

Lipinsky, E. S. et al. 1979 (June 4). *Systems Study of the Potential Integration of U.S. Corn Production and Cattle Feeding with Manufacture of Fuels via Fermentation.* Columbus, OH: Battelle. Report no. BMI-2033. 147 p. Available from National Technical Information Service, 5285 Port Royal Road, Springfield, VA 22161.

Miller, D.L. "Ethanol Fermentation and Potential." *in:* Wilke, C. R., ed. *Biotechnology and Bioengineering Symposium No. 5. Cellulose as a Chemical and Energy Resource Conference.* Berkeley, CA, June 25-27, 1974. New York NY: pp. 345-352. John Wiley and Sons.

INTERNATIONAL

Ribeiro, Filho F. A., "The Ethanol-Based Chemical Industry in Brazil" *in: Workshop on Fermentation Alcohol for Use as Fuel and Chemical Feedstock in Developing Countries.* Vienna, Austria. 26-30 March 1979. Paper No. ID/WG. 293/4 UNIDO. Available from: UN Publications, Room A 3315, New York, NY 10017.

Sharma, K.D. "Present Status of Alcohol and Alcohol Based Chemicals Industry in India." *in: Workshop on Fermentation Alcohol for Use as a Fuel and Chemical Feedstock in Developing Countries.* Vienna, Austria: 26-30 March 1979. Paper No. ID/WG.293/14 UNIDO Available from: UN Publications, Room A 3315, New York, NY 10017.

REGULATORY

Abeles, T. P. and King, Janna R. 1978 (February). *Paramenters for Legislature Consideration of Bioconversion Technologies.* St. Paul, MN: Minnesota Legislature Science and Technology Project. Report No. PB 284742/45T. 45 p. Available from: National Technical Information Project, 5285 Port Royal Road, Springfield, VA 22161.

Bureau of Alcohol, Tobacco and Firearms. Ethyl Alcohol for Fuel Use. Washington, DC: BATF. Brochure available from: BATF Distribution Center, 3800 S. Four Mile Run Drive, Arlington, VA 22206.

Bureau of Alcohol, Tobacco, and Firearms. 1979 (August). Alcohol Fuel and ATF. Washington, DC: BATF. Brochure No. ATF 5000.2. 4 p. Available from: ATF Distribution Center, 3800 S. Four Mile Run Drive, Arlington, VA 22206.

Denaturants for Ethanol/Gasoline Blends. 1978 (April). Baltimore, MD: Mueller Associates. Report No. HCP/M2098-01. 16 p. Available from: National Technical Information Service, 5285 Port Royal Road, Springfield, VA 22161.

"Fuel Use of Distilled Spirits—Implementing a Portion of the Crude Oil Windfall Profit Tax Act of 1980 (Pub. L96-223); Temporary and Proposed Rule." 1980 (20 June) *Federal Register.* Vol. 45 (No. 121): pp. 41837-41850.

TRANSPORTATION USE

Adt, R. R. Jr., et al. 1978. Effects of Blending Ethanol with Gasoline on Automotive Engines Steady State Performance and Regulated Emissions Chartersssssstics. Troy, MI: Report No. CONF-7805102. p. 68-76. Entire report available from: National Technical Information Center, 5285 Port Royal Road, Springfield, VA 22161.

Allsup, J.R. and Eccleston, D.B. 1979 (May). *Ethanol/Gasoline Blends as Automotive Fuel.* Asilomar, CA: International alcohol fuels technology. Report No. BETC/R1-79/2. 13 p. Available from: National Technical Information Service, 5285 Port Royal Road, Springfield, VA 22161.

Bernhardt, W. 1977. "Future Fuels and Mixture Preparation Methods for Spark Ignition Automobile Engines." *Progress in Energy and Combustion Science.* Vol. 3 (no. 3): pp. 139-150.

Bushnell, D., Simonsen, J. M. 1976 "Alcohol Assisted Hydrocarbon Facilities: A Comparison of Exhaust Emissions and Fuel Consumption Using Study State and Dynamic Engine Test Facilities." *Energy Communications.* Vol. 2 (no. 2): pp. 107-132.

Ecklund, E.E. 1978 (May). *Comparative Automotive Engine Operation When Fueled with Ethanol and Methanol.* Washington, DC: U.S. Department of Energy. Report No. HCP/W1737-01. 5 p. Available from: National Technical Information Service, 5285 Port Royal Road, Springfield, VA 22161.

Panchapakesan, N. R. and Gopalakrishnan, K. V. "Factors that Improve the Performance of an Ethanol-Diesel Oil Dual Fuel Engine." *in: Proceedings of the International Symposium on Alcohol Fuel Technology: Methanol and Ethanol.* Wolfsburg. Federal Republic of Germany November 21-23, 1977. Report No. CONF-771175, paper 2-2. Entire report available from: National Technical Information Service, 5285 Port Royal Road Springfield, VA 22161.

Rutan, Al. 1980 *Alcohol Car Conversion.* Minneapolis, MN: Rutan Publishing Co. 47 p.

Scott, W.M. "Alternative Fuels for Automotive Diesel Engines." *in:* Colucci, Joseph M. and Gallopoulos, Nicholas E., eds. 1977. *Future Automotive Fuels; Prospects, Performance, Perspective.* New York, NY: Plenum Press. pp. 263-290.

APPENDIX G.

Glossary

AAFCO: Association of American Feed Control Officials

ABSOLUTE ALCOHOL: completely dehydrated ethyl alcohol of the highest proof obtainable (200° proof); also "neat" alcohol. (See ANHYDROUS.)

ACID HYDROLYSIS: decomposition or alteration of a chemical substance by water in the presence of acid.

ACIDITY: the measure of how many hydrogen ions a solution contains per unit volume; may be expressed in terms of pH.

AFLATOXIN: the substance produced by some strains of the fungus Aspergillus flavus; the most potent carcinogen yet discovered; a persistent contaminant of corn that renders crops unsalable.

AFO: Office of Alcohol Fuels.

ALCOHOL: the family name of a group of organic chemical compounds composed of carbon, hydrogen, and oxygen; a series of molecules that vary in chain length and are composed of a hydrocarbon plus a hydroxyl group, CH_3-(CH_2)n-OH; includes methanol, ethanol, isopropyl alcohol, and others; see ETHANOL.

defined in the Crude Oil Windfall Profit Tax Act of 1980 (26 USC 44E P.L. 96-223) to include "methanol and ethanol but does not include alcohol produced from petroleum, natural gas, or coal or alcohol with a proof less than 150."

defined in the Energy Security Act (42 USC 8802, P.L. 96-294) as "alcohol (including methanol and ethanol) which is produced from biomass and which is suitable for use by itself or in combination with other substances as a fuel or as a substitute for petroleum or petrochemical feed stocks."

ALCOHOL FUEL PLANT: under BATF regulations, a distilled spirits plant established solely for producing, processing, and using or distributing distilled spirits to be used extensively for fuel use.

ALCOHOL FUEL PRODUCER'S PERMITS: the document issued by BATF pursuant to the Crude Oil Windfall Profit Tax Act (26 USC 5181, P.L. 96-223) authorizing the person named to engage in business as an alcohol fuel production facility.

ALDEHYDES: any of a class of highly reactive organic chemical compounds obtained by controlled oxidation of primary alcohols, characterized by the common group CHO, and used in the manufacture of resins, dyes, and organic acids.

ALKALI: soluble mineral salt of a low-density, low-melting point, highly reactive metal; characteristically "basic" in nature.

ALPHA-AMYLASE: enzyme which liquefies starch by conversion to dextrins.

AMBIENT: the prevalent surrounding conditions usually expressed as functions of temperature, pressure, and humidity.

AMINO ACIDS: the naturally occurring, nitrogen-containing building blocks of protein.

AMYLACEOUS FEEDSTOCKS: materials, such as cereal grains and potatoes, that are composed of saccharides in the form of starches.

AMYLASE: any of the enzymes that accelerate the hydrolysis of starch and glycogen.

AMYLODEXTRINS: see DEXTRINS.

ANAEROBIC DIGESTION: a type of bacterial degradation of organic matter that occurs in the absence of air (oxygen) and produces primarily carbon dioxide and methane.

ANHYDROUS: devoid of water; refers to a compound that does not contain water either absorbed on its surface or as water of crystallization.

ANHYDROUS ETHANOL: 100-percent alcohol, neat alcohol, 200°-proof alcohol.

APPARENT PROOF: the proof indicated by a hydrometer after correction for temperature but without correction of the obscuration caused by the presence of solids.

ATF: Should refer to BATF (Bureau of Alcohol Tobacco and Firearms) for proper usage.

ATMOSPHERIC PRESSURE: pressure of the air (and atmosphere surrounding us) which changes from day to day; it is equal to 14.7 psia.

AZEOTROPE: the chemical term for two or more liquids that, at a certain concentration, boil as though they are a single substance; alcohol and water cannot be separated further than 194.4° proof because at this concentration, alcohol and water form an azeotrope and vaporize together.

AZEOTROPIC DISTILLATION: distillation in which a substance is added to the mixture to be separated in order to form an azeotropic mixture with one or more of the components of the original mixture; the azeotrope formed will have a boiling point different from the boiling point of the original mixture and will allow separation to occur.

BACKSET (also called set back): the liquid portion of the stillage recycled as part of the process liquid in mash preparation.

BACTERIAL SPOILAGE: occurs when bacterial contaminants take over the fermentation process in competition with the yeast.

BAGASSE: the cellulosic residue left after sugar is extracted from sugar cane.

BALLING HYDROMETER OR BRIX HYDROMETER: a triple-scale wine hydrometer designed to record the specific gravity of a solution containing sugar.

BANKABLE DEBT: debt which is sufficiently collateralized to allow financing through normal commercial bank channels.

BARREL: a liquid measure equal to 42 American gallons or about 306 pounds of crude oil; one barrel equals 5.6 cubic feet or 0.159 cubic meters.

BASIC HYDROLYSIS: decomposition or alteration of a chemical substance by water in the presence of alkali.

BATCH FERMENTATION: fermentation of a specific quantity of material conducted from start to finish in a single vessel.

BATF: Bureau of Alcohol, Tobacco, and Firearms, under the U.S. Department of the Treasury; responsible for the issuance of permits, for both experimental and commercial facilities, for the production of alcohol.

BEER: the product of fermentation by microorganisms; the raw fermented mash, which contains about 7 to 12% alcohol; usually refers to the alcohol solution remaining after yeast fermentation of sugars.

BEER STILL: the stripping section of a distillation column for concentrating ethanol, or the first column of a two (or more) column system, in which the first separation from the mash takes place.

BEER WELL: the surge tank used for storing beer prior to distillation.

BETA - AMYLASE: enzyme which converts dextrins into glucose.

BIOMASS: Organic matter, such as trees, crops, manure, and aquatic plants, that is available on a renewable basis.

defined in the Energy Security Act (42 USC 8802, P. L. 96-294) as "any organic matter which is available on a renewable basis, including agricultural crops and agricultural wastes and residues, wood and wood wastes and residues, animal wastes, municipal wastes, and aquatic plants."

BIOMASS ENERGY PROJECT: defined in the Energy Security Act (42 USC 8802, P. L. 96-294) as "any facility (or portion of a facility) located in the United States which is primarily for (a) the production of biomass fuel (and byproducts); or (b) the combustion of biomass for the purpose of generating industrial process heat, mechanical power, or electricity (including cogeneration).

BIOMASS FUEL: defined in the Energy Security Act (42 USC 8802, P. L. 96-294) as "any gaseous, liquid, or solid fuel produced by conversion of biomass."

BOD: Biochemical Oxygen Demand; a measure of organic water pollution potential.

BOILING POINT: the temperature at which the transition from the liquid to the gaseous phase occurs in a pure substance at fixed pressure.

BOND: a type of insurance which gives the government security against possible loss of distilled spirits tax revenue; not required for alcohol fuel plants producing less than 10,000 proof gallons per year.

BRITISH THERMAL UNIT (Btu): the amount of heat required to raise the temperature of one pound of water one degree Fahrenheit under stated conditions of pressure and temperature (equal to 252 calories, 778 foot-pounds, 1055 joules, and 0.293 watt-hours); it is a standard unit for measuring quantity of heat energy.

BULK DENSITY: the mass (weight) of a material divided by the actual volume it displaces as a whole substance, expressed in lb/ft^3; kg/m^3; etc.

CALORIE: the amount of heat required to raise the temperature of one gram of water one degree Centigrade.

CARBOHYDRATE: a chemical term describing certain neutral compounds made up of carbon, hydrogen, and oxygen; includes all starches and sugars; a general formula is $C_x(H_2O)y$.

CARBON DIOXIDE: a gas produced as a byproduct of fermentation; chemical formula is CO_2.

CASSAVA: a starchy root crop used for tapioca; can be grown on marginal croplands along the southern coast of the United States.

CATALYSIS: the effect produced by a small quantity of a substance (catalyst) on a chemical reaction, after which the substance (catalyst) appears unchanged.

CELL RECYCLE: the process of separating yeast from fully fermented beer and returning it to ferment a new mash; can be done with clear worts in either batch or continuous operations.

CELLULASE: an enzyme capable of decomposing cellulose into simpler carbohydrates.

CELLULOSE: the main polysaccharide in living plants, forms the skeletal structure of the plant cell wall; can be hydrolyzed to glucose.

CELLULOSIC FEEDSTOCKS: materials, such as wood, crop stalks, and newsprint, containing sugar units linked by bonds that are not easily ruptured.

CELSIUS (Centigrade): a temperature scale commonly used in the sciences; at sea level, water freezes at 0° C and boils at 100° C. C° = 5/9 (F − 32).

CENTRIFUGE: a rotating device for separating liquids of different specific gravities or for separating suspended colloidal particles according to particle-size fractions by centrifugal force.

CETANE NUMBER (cetane rating): measure of a fuel's ease of self-ignition; the higher the number the better the fuel for a diesel engine.

CFR: Code of Federal Regulations

COD: Chemical Oxygen Demand; a measure of water pollution.

COLLATERAL VALUE: the resale value of alcohol equipment and/or plants; specifically, the ability of the equipment to derive from sale the amount necessary to pay off the debt borrowed against the unit.

COLUMN: vertical, cylindrical vessel containing a series of perforated plates or packed with materials through which vapors may pass, used to increase the degree of separation of liquid mixtures by distillation or extraction.

COMPLETELY DENATURED ALCOHOL (CDA): ethyl alcohol which is at least 160° proof blended, pursuant to formulas prescribed by BATF, with sufficient quantities of various denaturants to make it unfit for and not readily recoverable for beverage use; this may then be distributed through retail outlets without permits. (Compare to SPECIALLY DENATURED ALCOHOL).

COMPOUND: a chemical term denoting a specific combination of two or more distinct elements.

CONCENTRATION: ratio of mass or volume of solute present in a solution to the amount of solvent.

CONDENSER: a heat-transfer device that reduces a fluid substance from its vapor phase to its liquid phase by reducing its temperature as it contacts cooling surfaces in its path.

CONTINUOUS FERMENTATION: a steady-state fermentation system that operates without interruption; each stage of fermentation occurs in a separate section of the fermenter, and flow rates are set to correspond with required residence times.

COOKER: a tank or vessel designed to cook a liquid or extract or digest solids in suspension; the cooker usually contains a source of heat, and is fitted with an agitator; its purpose is to aid in breaking down starches into fermentable sugars.

COPRODUCTS: the resulting substances and materials that accompany the production of ethanol by fermentation processes.

CORPORATE BONDS: bonds issued and sold to the public which are backed by the corporation which issues them; instruments which provide debt financing to the private sector from institutions such as insurance companies, mutual funds, pension funds, etc.

DCF-IROR: Discounted Cash Flow - Interest Rate of Return

DDG: see Distillers Dried Grains.

DDGS: see Distillers Dried Grains with Solubles.

DDS: see Dried Grains with Solubles.

DEHYDRATION: the process of removing water from any substance by exposure to high temperature or by chemical means.

DENATURANT: a substance added to ethanol to make it unfit for human consumption so that it is not subject to alcohol beverage taxes.

DENATURE: the process of adding a substance to ethyl alcohol to make it unfit for human consumption; the denaturing agent may be gasoline or other substances specified by the Bureau of Alcohol, Tobacco, and Firearms.

DEPARTMENT OF ENERGY: In October 1977, the Department of Energy (DOE) was created to consolidate the multitude of energy-oriented government programs and agencies; the Department carries out its mission through a unified organization that coordinates and manages energy conservation, supply development, information collection and analysis, regulation, research, development, and demonstration.

DESICCANT: a substance having an affinity for water; used for drying purposes.

DEWATERING: removal of the free water from a solid substance.

DEXTRINS: a polymer of D-Glucose which is intermediate in complexity between starch and maltose and formed by partial hydrolysis of starches.

DEXTROSE: the same as glucose.

DISACCHARIDES: the class of compound sugars which yield two monosaccharide units upon hydrolysis; examples are sucrose, maltose, and lactose.

DISPERSION: the distribution of finely divided particles in a medium.

DISTILLATE: that portion of a liquid which is removed as a vapor and condensed during a distillation process.

DISTILLATION: the process of separating the components of a mixture by differences in boiling point; a vapor is formed from the liquid by heating the liquid in a vessel and the vapor is successively collected and condensed into liquids.

DISTILLERS DARK GRAINS: see DISTILLERS DRIED GRAINS WITH SOLUBLES (DDGS).

DISTILLER DRIED GRAINS (DDG): the water-insoluble, dried distillers grains coproduct of the grain fermentation process which may be used as a high-protein (28 percent) animal feed. (see DISTILLERS GRAINS).

DISTILLERS DRIED GRAINS WITH SOLUBLES (DDGS): a grain mixture obtained by mixing distillers dried grains and distillers dried solubles.

DISTILLERS DRIED SOLUBLES (DDS): a mixture of water-soluble oils and hydrocarbons obtained by condensing the thin stillage fraction of the solids obtained from fermentation and distillation processes.

DISTILLERS FEEDS: coproducts resulting from the fermentation of cereal grains by the yeast *Saccharomyces cerevisiae;* the nonfermentable portion of grain mash.

DISTILLERS GRAIN: the nonfermentable portion of a grain mash comprised of protein, unconverted carbohydrates and sugars, and mineral material.

DOE: Department of Energy

DRAWBACK: a refund of part of the tax given when tax-paid alcohol is used to produce approved products unfit for beverage purposes.

DRY MILLING: a process of separating various components of grains, such as germ, bran, and starch without using water.

DSP (DISTILLED SPIRITS PLANT): a plant, including fuel alcohol plants, authorized by the Bureau of Alcohol, Tobacco, and Firearms to produce, store, or process ethyl alcohol in any of its forms.

ECONOMIC REGULATORY ADMINISTRATION (ERA): a regulatory agency within the Department of Energy administering petroleum pricing and allocation programs, oil and gas fuel conversion programs, and other programs as assigned by the Secretary of Energy.

EDA: Economic Development Administration

ENERGY CROPS: includes such agricultural crops as corn and sugar cane; also non-food crops such as poplar trees. (See BIOMASS).

ENERGY SECURITY ACT: (42 USC 8701, et seq., P. L. 96-294) June 30, 1980 legislation authorizing, inter alia, a U.S. biomass and alcohol fuel program; established independent Office of Alcohol Fuels within DOE, and authorized program including loan guarantees, price guarantees, and purchase agreements with producers of fuel alcohol.

ENRICHMENT: the increase of the more volatile component in the condensate of each successive stage above the feed plate.

ENSILAGE: immature green forage crops and grains which are preserved by alcohol formed by an anaerobic fermentation process.

ENTITLEMENT PROGRAM: a DOE program administered by the Economic Regulatory Administration which pays producers of ethanol for gasohol an entitlement per gallon through September, 1981.

ENZYMES: the group of catalytic proteins that are produced by living microorganisms; enzymes mediate and promote the chemical processes of life without themselves being altered or destroyed.

EPA: Environmental Protection Agency

EQUITY CAPITAL: that portion of the total debt of a corporation in which stock is given in return for invested capital.

ETHANOL: chemical formula C_2H_5OH; the alcohol product of fermentation that is used in alcoholic beverages and for industrial purposes; blended with gasoline to make gasohol; also known as ethyl alcohol or grain alcohol.

ETHYL ALCOHOL: see ETHANOL.

EVAPORATION: conversion of a liquid to the vapor state by the addition of latent heat of vaporization; usually refers to vaporization into the atmosphere.

EXCISE TAX, GASOLINE: a tax collected at the pump to support the construction and maintenance of highways. Gasohol is exempt through 1992 from the $.04 Federal excise tax and in some States from the State excise tax.

FACULTATIVE (ANAEROBE): a microorganism that grows equally well under aerobic and anaerobic conditions.

FAHRENHEIT SCALE: a temperature scale in which the boiling point of water is 212 and its freezing point 32°; to convert °F to °C, subtract 32, multiply by 5, and divide the product by 9 (at sea level). $C° = (F° - 32) \times 5/9$.

FDA: Food and Drug Administration

FEED PLATE: the theoretical position in a distillation column above which enrichment occurs and below which stripping occurs.

FEEDSTOCK: the base raw material that is the source of sugar for fermentation.

FERMENTABLE SUGAR: sugar (usually glucose) derived from starch or cellulose that can be converted to ethanol (also known as reducing sugar or monosaccharide).

FERMENTATION: a microorganically mediated enzymatic transformation of organic substances, especially carbohydrates, generally accompanied by the evolution of a gas; the conversion of simple sugars to ethanol with the aid of enzymes.

FERMENTATION ETHANOL: ethyl alcohol produced from the enzymatic transformation of organic substances.

FLASH HEATING: very rapid heating of material by exposure of small fractions to high temperature and using high flow rates.

FLASH POINT: the temperature at which a combustible liquid will ignite when a flame is introduced; anhydrous ethanol will flash at 51° F, 90°-proof ethanol will flash at 78° F.

FLOCCULATION: the aggregation of fine suspended particles to form floating clusters or clumps.

FmHA: Farmers Home Administration

FOSSIL FUEL: any naturally occurring fuel of an organic nature which originated in a past geologic age (such as coal, crude oil, or natural gas).

FRACTIONAL DISTILLATION: a process of separating alcohol and water (or other mixtures) by boiling and drawing off vapors from different levels of the distilling column.

FRUCTOSE: a fermentable monosaccharide (simple) sugar of chemical formula $C_6H_{12}O_6$; fructose is a ketohexose.

FUEL GRADE ALCOHOL: usually refers to ethanol of 160° to 200° proof, although 100° to 200° proof can be used with diesel fuel. See DIESELHOL.

FUSEL OIL: a clear, colorless, poisonous liquid mixture of alcohols obtained as a byproduct of grain fermentation; major constituents are amyl, isoamyl, propyl, isopropyl, butyl, and isobutyl alcohols.

GASOHOL (GASAHOL): a registered trademark held by the State of Nebraska for a fuel mixture of agriculturally derived 10 percent anhydrous fermentation ethanol and 90 percent unleaded gasoline; it is often incorrectly used to mean any mixture of alcohol and gasoline to be used for motor fuel.

GASOLINE: a volatile, flammable liquid obtained from petroleum that has a boiling range of approximately 200° to 216° C and is used as fuel for spark-ignition internal combustion engines.

GELATINIZATION: the rupture of starch granules by heat to form a gel of soluble starch and dextrins.

GLUCOSE: a monosaccharide; occurs free or combined and is the most common sugar; chemical formula $C_6H_{12}O_6$; glucose is an aldohexose.

GLUCOSIDASE: an enzyme that hydrolyzes polymers of glucose monomers (glucoside); specific glucosidases must be used to hydrolyze specific glucosides; e.g., β-glucosidases are used to hydrolyze cellulose; α-glucosidases are used to hydrolyze starch.

GRAIN ALCOHOL: see ETHANOL.

GUARANTEED DEBT: 100% versus 90%: a distinction made to the structure of government-guaranteed debt; i.e., under the FmHA program the government guarantees 90% of the debt which the lender loans to the corporation; contrast to a 100% guarantee which eliminates risk from the bank altogether.

HEAT EXCHANGER: a device that transfers heat from one fluid (liquid or gas) to another, or to the environment.

HEAT OF CONDENSATION: the same as the heat of vaporization, except that the heat is given up as the vapor condenses to a liquid at its boiling point.

HEAT OF VAPORIZATION: the heat input required to change a liquid at its boiling point to a vapor at the same temperature (e.g., water at 212° F to steam at 212° F).

HEATING VALUE: the amount of heat obtainable from a fuel and expressed, for example, in Btu/lb.

HEXOSE: any of various simple sugars that have six carbon atoms per molecule.

HHV: higher heating value; the heat released during combustion of fuel if all products are cooled to room temperature and water is condensed to liquid.

HUD: Housing and Urban Development

HYDRATED: chemically combined with water.

HYDROCARBON: a chemical compound containing hydrogen and carbon.

HYDROLYSIS: the decomposition or alteration of a substance by chemically adding a water molecule to the unit at the point of bonding.

HYDROMETER: a long-stemmed glass tube with a weighted bottom; it floats at different levels depending on the relative weight (specific gravity) of the liquid; the specific gravity or other information is read where the calibrated stem emerges from the liquid.

INDOLENE: a standard mixture of chemicals used in comparative tests of automotive fuels.

INDUSTRIAL ALCOHOL: ethyl alcohol produced and sold for other than beverage purposes; depending on the use, may or may not be denatured.

INDUSTRIAL REVENUE BONDS (IDR BONDS): debt incurred through industrial revenue authorities in numerous states; since such authorities are permitted to issue bonds for sale to the public which are payable in tax-exempt interest rates, IDR or tax-exempt bonds normally reflect an interest rate substantially below the prime rate.

INEL: Idaho National Engineering Laboratory

INOCULUM: a small amount of bacteria produced from a pure culture which is used to start a new culture.

INULIN: a polymeric carbohydrate comprised of fructose monomers found in the roots of many plants, particularly Jerusalem artichokes.

INVENTORY: refers to the supplies of all factors of production which must be maintained in storage; i.e., grain, wood, enzymes, etc.

LACTIC ACID: $C_3H_6O_3$, the acid formed from milk sugar (lactose) and produced as a result of fermentation of carbohydrates by bacteria called Lactobacillus.

LACTOSE: a crystalline disaccharide made from whey and used in pharmaceuticals, infant foods, bakery products, and confections; also called "milk sugar", $C_{12}H_{22}O_{11}$.

LEADED GASOLINE: gasoline containing tetraethyllead to raise octane value.

LEAN FUEL MIXTURE: an excess of air in the air/fuel ratio; gasohol has a leaning effect over gasoline because the alcohol adds oxygen to the system.

LEASING: a form of financing whereby debt and ownership are retained by a third party who is normally not involved in the management of the operation; allows for the loan of both debt and equity capital to the market.

LHV: lower heating value; the heat released during combustion of fuel if all products are cooled to room temperature while water remains as steam; the lower heating value is more representative of typical fuel burning than HHV.

LIGNIFIED CELLULOSE: cellulose polymer wrapped in a polymeric sheath and extremely resistant to hydrolysis because of the strength of its linkages.

LIGNIN: a polymeric, noncarbohydrate constituent of wood that functions as a binder and support for cellulose fibers.

LIMITED PARTNERSHIP: legal mechanism in which investors can limit their liability by investing in a "managing partner" who operates the company; the managing partner is usually a corporation and limited partners are allocated tax credits and depreciation in a ratio which may exceed their pro-rata share; basically used for tax-shelter-type investments, since the limited partners can be allocated depreciation, tax credits/losses which exceed their original investments.

LINKAGE: the bond or chemical connection between constituents of a molecule.

LIQUEFACTION: the change in the phase of a substance to the liquid state; in the case of fermentation, the conversion of water-insoluble carbohydrate to water-soluble carbohydrate.

MALT: barley softened by steeping in water, allowed to germinate, and used especially in brewing and distilling as a source of amylase.

MALTOSE: a disaccharide of glucose.

MASH: a mixture of grain and other ingredients with water to prepare wort for brewing operations.

MEAL: a granular substance produced by grinding.

MEMBRANE: a sheet polymer which separates components of solutions by permitting passage of certain substances but preventing passage of others.

METHANOL: a light volatile, flammable, poisonous, liquid alcohol, CH_3OH, formed in the destructive distillation of wood or made synthetically and used especially as a fuel, a solvent, an antifreeze, or a denaturant for ethyl alcohol, and in the synthesis of other chemicals; methanol can be used as fuel for motor vehicles; also known as methyl alcohol or wood alcohol.

METHYL ALCOHOL: also known as methanol or wood alcohol; see METHANOL.

MOLECULAR SIEVE: a compound which separates molecules by selective penetration into the sieve space on the basis of size, charge, or both.

MOLECULE: the chemical term for the smallest particle of matter that is the same chemically as the whole mass.

MONOMER: a simple molecule which is capable of combining with a number of like or unlike molecules to form a polymer.

MONOSACCHARIDES: see FERMENTABLE SUGAR.

MULTIPLE-EFFECT EVAPORATOR: a series of evaporators in which the vapors removed from each unit are used to supply heat to the next unit in the series.

MUNICIPAL SOLID WASTE (MSW): combined residential and commercial wastes generated within a municipal area and consisting of any materials, including food wastes, that are discarded or rejected as spent, worthless, useless, or in excess and that are not wet enough to be free-flowing.

MUNICIPAL WASTE: defined in the Energy Security Act (42 USC 8802, P. L. 96-294) as "any organic matter, including sewage, sewage sludge, and industrial or commercial waste, and mixtures of such matter and inorganic refuse (i) from any publicly or privately operated municipal waste collection or similar disposal system, or (ii) from similar wastes flows (other than such flows which constitute agricultural wastes or residues, or wood wastes or residues from wood harvesting activities or production of forest products)."

MUNICIPAL WASTE ENERGY PROJECT: defined in the Energy Security Act (42 USC 8802, P. L. 96-294) as "any facility (or portion of a facility) located in the United States primarily for (i) the production of biomass fuel (and byproducts) from municipal waste; or (ii) the combustion of municipal waste for the purpose of generating steam or forms of useful energy, including industrial process heat, mechanical power, or electricity (including cogeneration)."

NAFI: National Alcohol Fuels Institute.

NET ENERGY BALANCE: the amount of energy available from fuel when it is burned, less the amount of energy it takes to produce the fuel.

OCTANE NUMBER: a rating which indicates the tendency to knock when a fuel is used in a standard internal combustion engine under standard conditions.

OFFERING STATEMENT: a prospectus of an investment in the form of an offering of stock or partnership in the investment in return for cash invested into the company.

OFFICE OF ALCOHOL FUELS: an independent Office within the Department of Energy responsible for administration of all alcohol fuels programs in the Department and coordination of related programs in other Federal agencies.

OSHA: Occupational Safety and Health Administration

OSMOTIC PRESSURE: applied pressure required to prevent passage of a solvent across a membrane which separates solutions of different concentrations.

OVERHEAD: the relatively low-boiling-point liquids removed from the top of a distillation unit.

OVER-THE-COUNTER (OTC) STOCK MARKET: market for new or speculative stocks which are traded in the public sector but not to the extent which New York Stock Exchange or American Stock Exchange require for active trading.

PACKED DISTILLATION COLUMN: a column or tube constructed with a packing of ceramics, steel, copper, or fiberglass-type material to increase surface area through which vapors or liquid may pass.

PERSONAL VERSUS CORPORATE DEBT: under any loan agreement which involves high-risk capital speculation or unknown collateral values, personal endorsements are normally required by the lender, which allows the lender to have lien against other assets of the owners; corporate debt restricts the liability of the debt to the corporation and the assets in the corporation.

pH: a term used to describe the free hydrogen ion concentration of a system; a solution of pH less than 7 is acid; pH of 7 is neutral; pH over 7 is alkaline.

PLATE DISTILLATION COLUMN (sieve tray column): a distillation column constructed with perforated plates or screens.

POLYMER: a substance made of molecules comprised of long chains or cross-linked simple molecules.

POUNDS PER SQUARE INCH ABSOLUTE (psia): the measurement of pressure referred to a complete vacuum or 0 pressure.

POUNDS PER SQUARE INCH GAUGE (psig): the measurement of pressure expressed as a quantity measured from above atmospheric pressure.

POUND OF STEAM: one pound (mass) of water in the vapor phase, not to be confused with the steam pressure which is expressed in pounds per square inch.

PRACTICAL YIELD: the amount of product that can actually be derived under normal operating conditions; i.e., the amount of sugar that normally can be obtained from a given amount of starch or the amount of alcohol that normally can be obtained is usually less than theoretical yield.

PRIVATE PLACEMENTS: a security offering which is limited to a small group of investors and hence is not in the public markets or under the full scrutiny of the SEC; such investments may be in Sub-chapter S corporations, standard corporations, limited partnerships, etc.

PROCESS GUARANTEE: refers to the financial ability of an engineering company to successfully make its process perform, within given tolerance levels, in the event that the process does not meet production levels originally agreed to in the contract.

PROOF: a measurement of the alcohol concentration in an alcohol-water mixture, equal to twice the percentage by volume of the alcohol; e.g., 80-percent alcohol equals 160° proof, 100-percent alcohol equals 200° proof.

PROOF GALLON: a U.S. gallon of liquid which is 50-percent ethyl alcohol by volume or the alcohol equivalent thereof; also one tax gallon.

PROTEIN: any of a class of high-molecular-weight polymer compounds comprised of a variety of amino acids joined by peptide linkages.

PURE ETHYL ALCOHOL: ethyl alcohol that has not been denatured and is usually sold as 190° proof and 200° proof (absolute).

QUAD: one quadrillion (10^{15} or 1,000,000,000,000,000) Btu's (British thermal units).

RECTIFICATION: with regard to distillation, the selective increase of the concentration of a component in a mixture by successive evaporation and condensation.

RECTIFYING COLUMN: the portion of a distillation column above the feed tray in which rising vapor is enriched by interaction with a countercurrent falling stream of condensed vapor.

REFLUX: the condensate returned to a rectifying column to maintain the liquid-vapor equilibrium.

RELATIVE DENSITY: see SPECIFIC GRAVITY.

RENEWABLE RESOURCES: renewable energy; resources that can be replaced after use through natural means; example: solar energy, wind energy, energy from growing plants.

ROAD OCTANE: a numerical value for automotive anti-knock properties of a gasoline; determined by operating a car over a stretch of level road.

S-18: a relatively new program of the Securities and Exchange Commission which is designed to allow small stock offerings to go to the public sector through private placement, or through brokers up to $5 million on interstate placements; formerly, this was limited to $750,000 within a single state.

SACCHARIDE: a simple sugar or a compound that can be hydrolyzed to simple sugar units.

SACCHARIFY: to hydrolyze a complex carbohydrate into simpler soluble fermentable sugars, such as glucose.

SACCHAROMYCES: a class of single-cell yeasts which selectively consume simple sugars.

SBA: Small Business Administration

SBIC: Small Business Investment Company. Venture capital companies which are licensed by the Small Business Administration (SBA) for the purpose of investing in small businesses; requires a minimum of $500,000 in capital for which the SBA will lend an additional $1.5 million in investment capital to help small businesses; SBIC's can either lend to a small business or buy up to 49% of the stock in a small business, or a combination thereof.

SCRUBBING EQUIPMENT: equipment for countercurrent liquid-vapor contact of flue gases to remove chemical contaminants and particulates.

SECONDARY MARKET: market of institutional investors who buy government guaranteed debt for their portfolio.

SEIDB: Solar Energy Information Data Bank

SERI: Solar Energy Research Institute. The Solar Energy Research Development and Demonstration Act of 1974 called for the establishment of SERI, whose general mission is to support DOE's solar energy program and foster the widespread use of all aspects of solar technology, including direct solar conversion (photovoltaics), solar heating and cooling, solar thermal power generation, wind conversion, ocean thermal conversion, and biomass conversion.

SET BACK: the liquid portion of the stillage that is recycled as a portion of the process liquid in the mash preparation.

SETTLING TIME: in a controlled system, the time required for entrained or colloidal material to separate from the liquid.

SIGHT GAUGE: a clear calibrated cylinder through which liquid level can be observed and measured.

SIMPLE SUGARS: see FERMENTABLE SUGARS.

SMALL-SCALE BIOMASS ENERGY PROJECT: defined in the Energy Security Act (42 USC 8802, P. L. 96-294) as "a biomass energy project with an anticipated annual production capacity of not more than 1,000,000 gallons of ethanol per year, or its energy equivalent of other forms of biomass energy."

SPECIAL FUEL: defined in the Crude Oil Windfall Profit Tax Act of 1980 (26 USC 44E, P. L. 96-223) as "any liquid fuel (other than gasoline) which is suitable for use in an internal combustion engine.

SPECIALLY DENATURED ALCOHOL (SDA): ethyl alcohol to which sufficient quantities of various denaturants have been added, pursuant to formulas prescribed by Federal regulations, to render it unfit for beverage purposes without impairing its usefulness for other purposes; specially denatured alcohol may be distributed only to persons holding BATF permits.

SPECIFIC GRAVITY: the ratio of the mass of a solid or liquid to the mass of an equal volume of distilled water at 4° C.

SPECIFIC PERFORMANCE BONDS: refers to the ability of a contractor, through a third-party insurer, to eliminate the risk of non-performance on a construction job.

SPENT GRAINS: the nonfermentable solids remaining after fermentation of a grain mash.

STANDARD CORPORATION: a corporation in which the investment tax credits, depreciation, or corporate losses stay within the corporate shell and cannot be deducted from personal income taxes.

STARCH: a carbohydrate polymer comprised of glucose monomers linked together by a glycosidic bond and organized in repeating units; starch is found in most plants and is a principal energy storage product of photo synthesis; starch hydrolyzes to several forms of dextrin and glucose.

STILL: an apparatus for distilling liquids, particularly alcohols; it consists of a vessel in which the liquid is vaporized by heat, and a cooling device in which the vapor is condensed.

STILLAGE: the nonfermentable residue from the fermentation of a mash to produce alcohol.

STOICHIOMETRIC RATIO: the ratio of chemical substances necessary for a reaction to occur completely.

STOVER: the dried stalks and leaves of a crop remaining after the grain has been harvested.

STRIPPING SECTION: the section of a distillation column below the feed in which the condensate is progressively decreased in the fraction of more volatile component by stripping.

SUB-CHAPTER S CORPORATION: a corporation which is limited to 15 or fewer investors who can use the tax credits, depreciation or losses which the corporation incurs in such a manner as to decrease personal income tax liability; the Sub-s election can be rescinded to convert the corporation at a later date to a standard 1244 corporation in which any loss investors incur can be deducted from personal income taxes.

SUCROSE: a crystalline disaccharide carbohydrate found in many plants, mainly sugar cane, sugar beets, and maple trees; $C_{12}H_{22}O_{11}$.

SURETY BOND: a type of insurance which satisfies the government's bonding requirements on distilled spirits production (see BOND); obtainable from U.S. Treasury-authorized insurance companies, surety bonds usually carry an annual premium of 1 to 2% of face value.

SURFACTANT: surface-active agent, a substance that alters the properties, especially the surface tension, at the point of contact between phases; e.g., detergents and wetting agents are typical surfactants.

TAX-FREE ALCOHOL: pure ethyl alcohol withdrawn free of tax for government, for hospital use, for science, or for humanitarian reasons; it cannot be used in foods or beverages; all purchasers must obtain BATF permits, post bonds, and exert controls upon storage and use of Tax-Free Alcohol.

TAX-PAID ALCOHOL: pure ethyl alcohol which has been released from Federal bond by payment of the Federal tax of $21.00 per gallon at 200° proof or $19.95 per gallon at 190° proof.

TETRAETHYLLEAD (TEL): an octane enhancer for gasoline now under environmental restriction.

THERMAL EFFICIENCY: energy heating value; the ratio of energy output to energy input.

THERMOPHILIC: capable of growing and surviving at high temperatures.

THIN STILLAGE: the water-soluble fraction of a fermented mash plus the mashing water.

TRAY: one of several types of horizontal pieces in a distillation column.

UDAG: Urban Development Action Grant

USDA: U.S. Department of Agriculture

VACUUM DISTILLATION: the separation of two or more liquids under reduced vapor pressure; reduces the boiling points of the liquids being separated.

VAPORIZE: to change from a liquid or a solid to a vapor, as in heating water to steam.

VAPOR PRESSURE: the pressure at any given temperature of a vapor in equilibrium with its liquid or solid form.

VOLUMETRIC FUEL ECONOMY: miles per gallon.

WET MILLING: a process similar to dry milling except that the various components of grain are separated in water.

WHEY: the watery part of milk separated from the curd in the process of making cheese; it is produced commercially in large quantities and can be used as a fertilizer, animal feed, or feedstock in the production of ethanol.

WHOLE STILLAGE: the undried "bottoms" from the beer well comprised of nonfermentable solids, distillers solubles, and the mashing water.

WINE GALLON: a United States gallon of liquid measure equivalent to the volume of 231 cubic inches.

WOOD ALCOHOL: see METHANOL

WORKING CAPITAL: capital which is used for initial training, inventory and labor.

WORT: the liquid remaining from a brewing mash preparation following the filtration of fermentable beer.

YEAST: single-cell microorganisms (fungi) that produce alcohol and CO_2 under anaerobic conditions and acetic acid and CO_2 under aerobic conditions; the microorganism that is capable of changing sugar to alcohol by fermentation.

ZYMOSIS: see FERMENTATION

www.ingramcontent.com/pod-product-compliance
Lightning Source LLC
Chambersburg PA
CBHW080514220326
41599CB00032B/6079